Lecture Notes in Mathematics

Volume 2387

This series reports on new developments in all areas of mathematics and their applications - quickly, informally and at a high level. Mathematical texts analysing new developments in modelling and numerical simulation are welcome. The type of material considered for publication includes:

1. Research monographs
2. Lectures on a new field or presentations of a new angle in a classical field
3. Summer schools and intensive courses on topics of current research.

Texts which are out of print but still in demand may also be considered if they fall within these categories. The timeliness of a manuscript is sometimes more important than its form, which may be preliminary or tentative. Please visit the LNM Editorial Policy (https://drive.google.com/file/d/1MOg4TbwOSokRnFJ3ZR3ciEeKs9hOnNX_/view?usp=sharing)

Titles from this series are indexed by Scopus, Web of Science, Mathematical Reviews, and zbMATH.

Jesús A. Álvarez López • Yuri A. Kordyukov •
Eric Leichtnam

A Trace Formula for Foliated Flows

Jesús A. Álvarez López
Dept of Math & Galician Centre for Math Research & Technology (CITMAga)
University of Santiago de Compostela (USC)
Santiago de Compostela, Spain

Yuri A. Kordyukov
Institute of Mathematics
Ufa Federal Research Centre of Russian Academy of Sciences
Ufa, Russia

Eric Leichtnam
Institut de Mathématiques de Jussieu-Paris Rive Gauche
Paris, France

ISSN 0075-8434 ISSN 1617-9692 (electronic)
Lecture Notes in Mathematics
ISBN 978-3-032-15412-5 ISBN 978-3-032-15413-2 (eBook)
https://doi.org/10.1007/978-3-032-15413-2

Mathematics Subject Classification: 58J32, 57R30, 58A14, 35K05, 57R30, 35S05

This work was supported by Agencia Estatal de Investigación (MTM2017-89686-P, PID2020-114474GB-I00). and Xunta de Galicia (ED431C 2019/10).

This Springer imprint is published by the registered company Springer Nature Switzerland AG
The registered company address is: Gewerbestrasse 11, 6330 Cham, Switzerland

Foreword

Classical index theory deals with complexes of elliptic operators on compact manifolds. In this case, the index is an integer. If the differential operator is not elliptic, but there is an action by a Lie group G whose orbits are transversal to the directions where the operator is elliptic, one can still define an index. In this case, it is no longer an integer but a Schwartz distribution on G. For compact Lie groups G, this was done in a lecture note by Atiyah, who also gave a method to calculate this index. For non-compact G, the index was first defined by Hörmander, but no general method is known to compute it. Several years ago, Álvarez López and Kordyukov proved an explicit index formula in the case where the non-compact Lie group $\mathbb{R}$ of real numbers acts transversally to a codimension one foliation, mapping leaves to leaves. The $\mathbb{R}$-action is then transversal to the de Rham complex of the foliation. The resulting transversal index formula was suggested by a distributional trace formula of Guillemin and Sternberg. There is also a striking analogy with the explicit formulas in number theory in the sense of André Weil, at least if one disregards the contributions from the infinite primes. The question arose as to what the analogue of the general Weil "explicit formula" should be in geometry. Many years ago, I had sent Jesus and Yuri a letter with a suggestion coming from the infinite primes in Weil's explicit formulas, on how the contributions from the fixed points in geometry might look. The resulting distributional equation corresponds to an index formula for an $\mathbb{R}$-action which is not everywhere transversal to the given codimension one foliation. This is a fundamentally more complicated situation than in transversal index theory. In particular, smoothing the $\mathbb{R}$-action operators on the spaces of harmonic forms along the leaves by integrating along a Schwartz test function on $\mathbb{R}$ no longer gives a trace class operator. Thus, the Hörmander definition of the transversal index breaks down. Using the Melrose b-calculus, ideas from Witten's deformation, and many more theories and new ideas, the authors have admirably succeeded in overcoming all analytical obstacles, arriving at an index theorem which formally now looks quite identical with the Weil explicit formulas. The work in this lecture note is an impressive extension of classical analysis to obtain for the first time non-transversal index theorems in a situation where both fixed points and also infinitely many periodic orbits are present. The main result may also be viewed as a great sharpening of the formal index formulas of Guillemin-Sternberg and Guillemin in a special situation. Speculatively, there is the fascinating possibility of a deeper relation between global analysis on certain spaces and number theory, where the explicit formulas would actually become a non-transversal index theorem in the sense of the authors. An extension of their work and methods would then lead to a much deeper understanding of the Riemann zeta function and more general Hasse-Weil zeta functions than presently possible.

Christopher Deninger

Abstract

Let $\mathcal{F}$ be a transversely oriented foliation of codimension one on a closed manifold M, and let $\phi = \{\phi^t\}$ be a foliated flow on $(M, \mathcal{F})$. Assume the closed orbits of ϕ are simple and its preserved leaves are transversely simple. In this case, there are finitely many preserved leaves, which are compact. Let M^0 denote their union, $M^1 = M \setminus M^0$ and $\mathcal{F}^1 = \mathcal{F}|_{M^1}$. We consider two topological vector spaces, $I(\mathcal{F})$ and $I'(\mathcal{F})$, consisting of the leafwise currents on M that are conormal and dual-conormal to M^0, respectively. They become topological complexes with the differential operator $d_{\mathcal{F}}$ induced by the de Rham derivative on the leaves, and they have an $\mathbb{R}$-action $\phi^* = \{\phi^{t\,*}\}$ induced by ϕ. Let $\bar{H}^\bullet I(\mathcal{F})$ and $\bar{H}^\bullet I'(\mathcal{F})$ denote the corresponding leafwise reduced cohomologies, with the induced $\mathbb{R}$-action $\phi^* = \{\phi^{t\,*}\}$. $\bar{H}^\bullet I(\mathcal{F})$ and $\bar{H}^\bullet I'(\mathcal{F})$ are shown to be the central terms of short exact sequences in the category of continuous linear maps between locally convex spaces, where the other terms are described using Witten's perturbations of the de Rham complex on M^0 and leafwise Witten's perturbations for $\mathcal{F}^1$. This is used to define some kind of Lefschetz distribution $L_{\mathrm{dis}}(\phi)$ of the actions ϕ^* on both $\bar{H}^\bullet I(\mathcal{F})$ and $\bar{H}^\bullet I'(\mathcal{F})$, whose value is a distribution on $\mathbb{R}$. Its definition involves several renormalization procedures, the main one is the b-trace of some smoothing b-pseudodifferential operator on the compact manifold with boundary obtained by cutting M along M^0. We also prove a trace formula describing $L_{\mathrm{dis}}(\phi)$ in terms of infinitesimal data from the closed orbits and preserved leaves. This solves a conjecture of C. Deninger involving two leafwise reduced cohomologies instead of a single one. This memoir is the conclusion of a program started about 10 years ago by the three authors.

Declarations

Competing Interests The authors have no competing interests to declare that are relevant to the content of this manuscript.

Contents

Chapter 1
Introduction

1.1 Deninger's Program

Let $(M, \mathcal{F})$ be a smooth foliated manifold. The leafwise cohomology, $H^\bullet(\mathcal{F})$, is defined with the complex of differential forms on the leaves that are smooth on M, $C^\infty(M; \Lambda\mathcal{F})$ ($\Lambda\mathcal{F} = \bigwedge T^*\mathcal{F} \otimes \mathbb{C}$), equipped with de Rham differential operator along the leaves, $d_\mathcal{F}$. This differential complex is not elliptic, it is only leafwise elliptic. Therefore $H^\bullet(\mathcal{F})$ may be of infinite dimension and non-Hausdorff with the topology induced by the C^∞ topology. Thus it makes sense to consider the reduced leafwise cohomology, $\bar{H}^\bullet(\mathcal{F}) = H^\bullet(\mathcal{F})/\bar{0}$. (The reduced cohomology is defined and denoted in a similar way for any complex with a compatible topology, called a topological complex.)

A flow $\phi = \{\phi^t\}$ on M is said to be foliated if it maps leaves to leaves; equivalently, its infinitesimal generator Z is an infinitesimal transformation of $(M, \mathcal{F})$, or the induced section $\overline{Z}$ of the normal bundle $N\mathcal{F} = TM/T\mathcal{F}$ is parallel with respect to the Bott partial connection. In this case, there is an induced $\mathbb{R}$-action $\phi^* = \{\phi^{t\,*}\}$ on $(C^\infty(M; \Lambda\mathcal{F}), d_\mathcal{F})$, which induces an $\mathbb{R}$-action $\phi^* = \{\phi^{t\,*}\}$ on $\bar{H}^\bullet(\mathcal{F})$. Moreover, ϕ induces a local flow $\bar{\phi}$ on local transversals of $\mathcal{F}$. Some leaves may be preserved by ϕ, which correspond to the fixed points of $\bar{\phi}$. If these fixed points of $\bar{\phi}$ are simple, then the leaves preserved by ϕ are called transversely simple (Sect. 4.1.2).

Assume M is closed, $\operatorname{codim}\mathcal{F} = 1$, the closed orbits are simple, the preserved leaves are transversely simple, and ϕ is transverse to the non-preserved leaves. With these conditions, C. Deninger has conjectured that the supertrace of ϕ^* on $\bar{H}^\bullet(\mathcal{F})$ makes sense as a distribution $L_{\text{dis}}(\phi)$ on $\mathbb{R}$ (its Lefschetz distribution), and it has an expression involving infinitesimal data from the preserved leaves and closed orbits (a dynamical Lefschetz trace formula).

This problem is a part of a program proposed by Deninger, whose goal is the study of arithmetic zeta functions by finding an interpretation of the explicit formulae as a dynamical Lefschetz trace formula for some $(M, \mathcal{F}, \phi)$ of this type

J. A. Álvarez López et al., *A Trace Formula for Foliated Flows*, Lecture Notes in Mathematics 2387, https://doi.org/10.1007/978-3-032-15413-2_1

[Den98, Den01, Den02, Den05, Den08]. The precise expression of the trace formula was previously suggested by Guillemin [Gui77]. Further developments of these ideas were made in [DS02, Müm06, Kop06, Lei08, Kop11, Lei14, KP15, Kim17, Den26, Den23].

It became clear that more generality is needed to draw arithmetic consequences (perhaps foliated flows on possibly singular foliated spaces of arithmetic nature). But, even for $(M, \mathcal{F}, \phi)$ as above, this problem is difficult and interesting; for instance, $\bar{H}^\bullet(\mathcal{F})$ is not appropriate in general [DS01]. Besides its own interest, a solution might provide techniques to deal with more general settings. Moreover, we believe that the techniques developed in this paper will be useful in arithmetic once the appropriate framework allowing to interpret the Weil's explicit formulae for arithmetic zeta functions as Lefschetz trace formulae will have been discovered.

1.2 Case with No Preserved Leaves

The first two authors proved such a trace formula when ϕ has no preserved leaves [ÁLK02], and extended it for transverse actions of Lie groups [ÁLK08]. In this case, $\mathcal{F}$ is Riemannian; i.e., it is locally described by Riemannian submersions for some Riemannian metric g on M (a bundle-like metric). Using g, we get the leafwise coderivative $\delta_{\mathcal{F}}$ and the leafwise Laplacian $\Delta_{\mathcal{F}}$. Then the leafwise heat operator defines a continuous map [ÁLK01]

$$C^\infty(M; \Lambda\mathcal{F}) \times [0, \infty] \to C^\infty(M; \Lambda\mathcal{F}) , \quad (\alpha, u) \mapsto e^{-u\Delta_{\mathcal{F}}}\alpha . \tag{1.2.1}$$

It follows that there is a leafwise Hodge decomposition

$$C^\infty(M; \Lambda\mathcal{F}) = \ker \Delta_{\mathcal{F}} \oplus \overline{\operatorname{im} d_{\mathcal{F}}} \oplus \overline{\operatorname{im} \delta_{\mathcal{F}}} , \tag{1.2.2}$$

and therefore the orthogonal projection $\Pi_{\mathcal{F}} = e^{-\infty\Delta_{\mathcal{F}}}$ to $\ker \Delta_{\mathcal{F}}$ induces a leafwise Hodge isomorphism

$$\bar{H}^\bullet(\mathcal{F}) \cong \ker \Delta_{\mathcal{F}} . \tag{1.2.3}$$

This is surprising because $\Delta_{\mathcal{F}}$ is only leafwise elliptic; somehow, the transverse rigidity of Riemannian foliations makes up for the lack of transverse ellipticity. These properties may fail for non-Riemannian foliations [DS01].

Furthermore, for all $f \in C^\infty_{\mathrm{c}}(\mathbb{R})$ and $0 < u \leq \infty$, the operator

$$P_{u,f} = \int_{\mathbb{R}} \phi^{t*} e^{-u\Delta_{\mathcal{F}}} f(t)\, dt \tag{1.2.4}$$

is smoothing, and therefore of trace class, its supertrace $\operatorname{Str} P_{u,f}$ depends continuously on f and is independent of u, and the limit of $\operatorname{Str} P_{u,f}$ as $u \downarrow 0$ gives the expected contribution of the closed orbits [ÁLK02, ÁLK08]. By (1.2.3) and (1.2.4), the mapping $f \mapsto \operatorname{Str} P_{\infty,f}$ can be considered as the Lefschetz distribution $L_{\mathrm{dis}}(\phi)$, solving the problem in this case.

1.3 General Case

This publication is a continuation of the works [ÁLK01, ÁLK02, ÁLK08], recalled in Sect. 1.2. Our main goal is to propose an extension of the trace formula to the case where there are (compact) leaves preserved by ϕ, which are very relevant in Deninger's program. Examples of foliations with such foliated flows can be easily constructed by using foliation surgeries.

1.3.1 Ingredients of the Trace Formula

Assume $\mathcal{F}$ is transversely oriented for the sake of simplicity. Thus, by Frobenius theorem, $\mathcal{F}$ is defined by a 1-form ω with $d\omega = \eta \wedge \omega$ ($T\mathcal{F} = \ker \omega$). Except in trivial cases, the existence of leaves preserved by ϕ prevents $\mathcal{F}$ from being Riemannian (it is impossible to choose $\eta = 0$), yet $\mathcal{F}$ has a precise description [ÁLKL22]. For instance, there is a finite number of preserved leaves, which are compact. Let M^0 denote the union of the leaves preserved by ϕ, $M^1 = M \setminus M^0$ and $\mathcal{F}^1 = \mathcal{F}|_{M^1}$.

All versions of leafwise reduced cohomologies we will consider have an action $\phi^* = \{\phi^{t\,*}\}$ induced by ϕ, which is invariant by leafwise homotopy equivalences. Thus, up to leafwise homotopies, we can assume $\phi^t = \mathrm{id}$ on M^0. Then, for every leaf $L \subset M^0$, there is some $\varkappa_L \in \mathbb{R}^\times$ such that, on the normal bundle $NL = T_L M / TL$, the normal tangent map is $\phi^t_* = e^{\varkappa_L t}$. The numbers $\varkappa_L$ will be ingredients of the trace formula. Moreover $\mathcal{F}^1$ becomes a transversely complete $\mathbb{R}$-Lie foliation with the restriction of $\overline{Z}$. So $\mathcal{F}$ is a particular case of foliation almost without holonomy [Hec72, Hec78].

Take a Riemannian metric g on M so that ω is the transverse volume form. The corresponding leafwise metric is denoted by $g_{\mathcal{F}}$. We can suppose η vanishes on $T\mathcal{F}^\perp$, and therefore it can be considered as a leafwise form, and we have $d_{\mathcal{F}}\eta = 0$. Furthermore, on some tubular neighborhood $T \equiv (-\epsilon, \epsilon) \times M^0$ ($\epsilon > 0$) of M^0 in M, we can suppose η and $g_{\mathcal{F}}$ are lifts of their restrictions to M^0, and the fibers of the projection $\varpi : T \to M^0$ are orthogonal to the leaves and agree with the orbits of ϕ. Thus there are no closed orbits of ϕ in T. The projection $\rho : T \to (-\epsilon, \epsilon)$ is a defining function of M^0 on T ($d\rho \neq 0$ on $M^0 = \rho^{-1}(0)$), which can be assumed to satisfy $d_{\mathcal{F}}\rho = \rho\eta$ on T and $\phi^{t*}\rho = e^{\varkappa_L t}\rho$ around every leaf $L \subset M^0$. We can choose any $\eta|_{M^0}$ in some fixed real cohomology class $\xi \in H^1(M^0)$ determined by $\mathcal{F}$, and there is no restriction on the choice of $g|_{M^0}$.

For every closed orbit c of ϕ, let $\ell(c)$ denote its smallest positive period. The condition on c to be simple means that $\mathrm{id} - \phi_*^{k\ell(c)} : T_p\mathcal{F} \to T_p\mathcal{F}$ is an isomorphism for any $p \in c$ and $k \in \mathbb{Z}^\times$, whose determinant is independent of p, and its sign is denoted by $\epsilon_c(k)$. The integers $\ell(c)$ and $\epsilon_c(k)$ will be also ingredients of the trace formula.

Let g^1 be the bundle-like metric of $\mathcal{F}^1$ such that it defines the same orthogonal complement $(T\mathcal{F}^1)^\perp$ as g, its restriction to $T\mathcal{F}^1$ is $g_\mathcal{F}$, and $\overline{Z}|_{M^1}$ is of norm one with the induced Euclidean structure on $N\mathcal{F}^1$. Then $\mathcal{F}^1$ has bounded geometry with g^1 in the sense of [San08, ÁLKL14]. Let ω^1 denote the transverse volume form of $\mathcal{F}^1$ defined by g^1 and the transverse orientation given by $Z|_{M^1}$. The transverse density $|\omega^1|$ can be considered as an invariant transverse measure of $\mathcal{F}^1$.

By cutting M along M^0, we get a compact manifold with boundary $\boldsymbol{M}$ with a foliation $\boldsymbol{\mathcal{F}}$ tangent to $\partial\boldsymbol{M}$. This allows us to apply tools from b-calculus [Mel93, Mel96]. For instance, g^1 and ω^1 are restrictions to $M^1 \equiv \mathring{\boldsymbol{M}}$ of a b-metric $\boldsymbol{g}_\mathrm{b}$ and a b-form $\boldsymbol{\omega}_\mathrm{b}$ on $\boldsymbol{M}$, and therefore $|\omega^1|$ is the restriction of the b-density $|\boldsymbol{\omega}_\mathrm{b}|$.

We can suppose there is some boundary-defining function $\boldsymbol{\rho}$ on $\boldsymbol{M}$ ($\boldsymbol{\rho} \geq 0$ and $d\boldsymbol{\rho} \neq 0$ on $\partial\boldsymbol{M} = \boldsymbol{\rho}^{-1}(0)$) such that the lift $\boldsymbol{\eta}$ of η to $\boldsymbol{M}$ satisfies $d_{\boldsymbol{\mathcal{F}}}\boldsymbol{\rho} = \boldsymbol{\rho}\boldsymbol{\eta}$ on $\mathring{\boldsymbol{M}}$, and $\boldsymbol{\rho}$ is the lift of $|\rho|$ on a collar neighborhood $\boldsymbol{T} \equiv [0, \epsilon) \times \partial\boldsymbol{M}$ of $\partial\boldsymbol{M}$. The lift of ϕ to $\boldsymbol{M}$ is a foliated flow $\boldsymbol{\phi} = \{\boldsymbol{\phi}^t\}$ of $(\boldsymbol{M}, \boldsymbol{\mathcal{F}})$.

We will use the b-integral ${}^{\boldsymbol{\nu}}\!\!\int_{\boldsymbol{M}}$, depending on the choice of a trivialization $\boldsymbol{\nu}$ of $N\partial\boldsymbol{M}$ satisfying $d\boldsymbol{\rho}(\boldsymbol{\nu}) = 1$. We can apply ${}^{\boldsymbol{\nu}}\!\!\int_{\boldsymbol{M}}$ to b-densities on $\boldsymbol{M}$; the usual integral of their restrictions to $\mathring{\boldsymbol{M}}$ may not be defined. Assume $\dim\mathcal{F}$ is even, which is the relevant case in Deninger's program. Then the product of the leafwise Euler density $e(\boldsymbol{\mathcal{F}})$ and $|\boldsymbol{\omega}_b|$ is the restriction of a b-density on $\boldsymbol{M}$, obtaining a b-calculus version of the Connes' $|\boldsymbol{\omega}_\mathrm{b}|$-Euler characteristic of $\boldsymbol{\mathcal{F}}$,

$$
{}^{\mathrm{b}}\chi_{|\boldsymbol{\omega}_\mathrm{b}|}(\boldsymbol{\mathcal{F}}) = {}^{\boldsymbol{\nu}}\!\!\int_{\boldsymbol{M}} e(\boldsymbol{\mathcal{F}})\,|\boldsymbol{\omega}_\mathrm{b}|\;,
$$

which will be called the *b-Connes-Euler characteristic* of $\boldsymbol{\mathcal{F}}$ defined by $|\boldsymbol{\omega}_\mathrm{b}|$ (or of $\mathcal{F}^1$ defined by $|\omega^1|$). This number will be another ingredient of the trace formula, also denoted by ${}^{\mathrm{b}}\chi_{|\omega^1|}(\mathcal{F}^1)$. The b-integral can be used to define the b-trace ${}^{\mathrm{b}}\mathrm{Tr}$ of smoothing b-pseudodifferential operators on $\boldsymbol{M}$; these operators may not be of trace class. The corresponding concept of b-supertrace will be used, denoted by ${}^{\mathrm{b}}\mathrm{Str}$.

With this generality, (1.2.1)–(1.2.3) are not true for $C^\infty(M; \Lambda\mathcal{F})$. Using the space $C^{-\infty}(M; \Lambda\mathcal{F})$ of leafwise currents does not work either. Instead, we will use the topological complexes of leafwise currents that are conormal and dual-conormal at M^0 [KN65, Hör71], [Hör85, Section 18.2], [Mel96, Chapters 4 and 6], [ÁLKL24].

1.3.2 Conormal and Dual-Conormal Leafwise Currents

We first recall the definitions and some properties of conormal and dual-conormal distributions at M^0. Let $\mathrm{Diff}(M, M^0)$ be the filtered algebra of differential operators on $C^\infty(M)$ generated by $C^\infty(M)$ and the vector fields on M tangent to M^0, and let $H^s(M)$ be the Sobolev space of order $s \in \mathbb{R}$. A distribution $u \in C^{-\infty}(M)$ is said to be conormal at M^0 of Sobolev order s if $\mathrm{Diff}(M, M^0)u \subset H^s(M)$. These distributions form a Fréchet space $I^{(s)} = I^{(s)}(M, M^0)$ endowed with the projective topology given by the maps $P : I^{(s)} \to H^s(M)$ $(P \in \mathrm{Diff}(M, M^0))$. The spaces $I^{(s)}$ form an inductive spectrum defining an LF-space $I = I(M, M^0) = \bigcup_s I^{(s)}$, with continuous inclusions $C^\infty(M) \subset I \subset C^{-\infty}(M)$. (All inclusions considered here are continuous.) See [ÁLKL24] for the properties of I and of other related spaces.

All spaces of distributions considered here, and their properties, have straightforward extensions for distributional sections of vector bundles. In particular, for the density bundle $\Omega = \Omega M$, we get the strong dual $I'(M, L) = I(M, L; \Omega)'$, simply denoted by I'. The elements of I' are called dual-conormal distributions; in fact, $C^\infty(M) \subset I' \subset C^{-\infty}(M)$ with $I \cap I' = C^\infty(M)$.

Let also $K = K(M, M^0) \subset I$ be the closed subspace consisting of elements supported in M^0. On the other hand, via the lift to $\boldsymbol{M}$, we get another space, $J = J(M, M^0)$, which is isomorphic to the space of extendable distributions on $\boldsymbol{M}$ conormal at the boundary [Mel96, Chapter 4]. There are canonical injections $C^\infty(M) \subset J \subset C^\infty(M^1)$. Let $K' = K'(M, L)$ and $J' = J'(M, L)$ be defined like I'. We get $J' \subset C^{-\infty}(M)$. Moreover there are short exact sequences in the category of continuous linear maps between locally convex spaces [Wen03, Chapter 2],

$$0 \to K \xrightarrow{\iota} I \xrightarrow{R} J \to 0\,, \tag{1.3.1}$$

$$0 \leftarrow K' \xleftarrow{R'} I' \xleftarrow{\iota'} J' \leftarrow 0\,, \tag{1.3.2}$$

where ι is the inclusion map and R is defined by restriction to M^1, and (1.3.2) is the transpose of the version of (1.3.1) with Ω ($R' = \iota^{\mathrm{t}}$ and $\iota' = R^{\mathrm{t}}$). These sequences are relevant because K, J, K' and J' have better descriptions than I and I'. So (1.3.1) and (1.3.2) will play an important role.

Using the vector bundle $\Lambda\mathcal{F}$, we get the spaces of conormal and dual-conormal leafwise currents at M^0, $I(\mathcal{F}) = I(M, M^0; \Lambda\mathcal{F})$ and $I'(\mathcal{F}) = I'(M, M^0; \Lambda\mathcal{F})$, as well as the spaces $K(\mathcal{F})$, $J(\mathcal{F})$, $K'(\mathcal{F})$ and $J'(\mathcal{F})$, with a similar notation. All of them are topological complexes with $d_{\mathcal{F}}$, and have $\mathbb{R}$-actions $\phi^* = \{\phi^{t\,*}\}$ induced by ϕ, compatible with $d_{\mathcal{F}}$. They give rise to the conormal and dual-conormal leafwise reduced cohomologies, $\bar{H}^\bullet I(\mathcal{F})$ and $\bar{H}^\bullet I'(\mathcal{F})$, as well as the reduced cohomologies $\bar{H}^\bullet K(\mathcal{F})$, $\bar{H}^\bullet J(\mathcal{F})$, $\bar{H}^\bullet K'(\mathcal{F})$ and $\bar{H}^\bullet J'(\mathcal{F})$. All of them with induced $\mathbb{R}$-actions $\phi^* = \{\phi^{t\,*}\}$. The bars are omitted from the notation if the cohomologies are not reduced. There are versions of (1.3.1) and (1.3.2) for the spaces $K(\mathcal{F})$, $I(\mathcal{F})$, $J(\mathcal{F})$, $K'(\mathcal{F})$, $I'(\mathcal{F})$ and $J'(\mathcal{F})$, where ι, R, ι' and R' are cochain maps. The induced maps in cohomology (resp., reduced cohomology) are denoted by ι_*, R_*, ι'_* and R'_* (resp., $\bar{\iota}_*$, $\bar{R}_*$, $\bar{\iota}'_*$ and $\bar{R}'_*$).

1.3.3 Witten's Perturbed Complexes

To describe the reduced cohomologies of Sect. 1.3.2 with the $\mathbb{R}$-actions ϕ^*, we will use the Witten's perturbation $d_\mu = d + \mu\eta\wedge$ on $C^{\pm\infty}(L; \Lambda)$ ($\Lambda = \Lambda L = \bigwedge T^*L \otimes \mathbb{C}$), for $\mu \in \mathbb{R}$ and every leaf $L \subset M^0$. Its cohomology is denoted by $H^\bullet_\mu(L)$. The corresponding perturbed codifferential and Laplace operators are denoted by δ_μ and Δ_μ.

1.3.4 Leafwise Witten's Perturbed Complexes

Recall that $d_{\mathcal{F}}\rho = \rho\eta$ on $\mathring{M}$ and $\partial M = \rho^{-1}(0)$. We will also use the leafwise Witten's perturbation

$$d_{\mathcal{F},\mu} = d_{\mathcal{F}} + \mu\eta\wedge = \rho^{-\mu} d_{\mathcal{F}} \rho^\mu$$

on the Sobolev spaces $H^{\pm\infty}(\mathring{M}; \Lambda\mathcal{F}) \equiv H^{\pm\infty}(M^1; \Lambda\mathcal{F}^1)$ defined with $g_{\mathrm{b}} \equiv g^1$. Their reduced cohomologies are denoted by $\bar{H}^\bullet_\mu H^{\pm\infty}(\mathring{\mathcal{F}})$ ($\mathring{\mathcal{F}} = \mathcal{F}|_{\mathring{M}} \equiv \mathcal{F}^1$). They satisfy obvious versions of (1.2.1)–(1.2.3). We have the isomorphisms

$$\rho^\mu : (H^{\pm\infty}(\mathring{M}; \Lambda\mathcal{F}), d_{\mathcal{F},\mu}) \xrightarrow{\cong} (\rho^\mu H^{\pm\infty}(\mathring{M}; \Lambda\mathcal{F}), d_{\mathcal{F}}) . \tag{1.3.3}$$

Let also $\boldsymbol{\phi}^{t\,*}_\mu = \rho^{-\mu}\boldsymbol{\phi}^{t\,*}\rho^\mu$ on $H^{\pm\infty}(\mathring{M}; \Lambda\mathcal{F})$, which induces an endomorphism $\boldsymbol{\phi}^{t\,*}_\mu$ of $\bar{H}^\bullet_\mu H^{\pm\infty}(\mathring{\mathcal{F}})$. For $\mu < \mu'$, the inclusions

$$\rho^{\mu'} H^{\pm\infty}(\mathring{M}; \Lambda\mathcal{F}) \subset \rho^\mu H^{\pm\infty}(\mathring{M}; \Lambda\mathcal{F})$$

correspond via (1.3.3) to the maps

$$\rho^{\mu'-\mu} : H^{\pm\infty}(\mathring{M}; \Lambda\mathcal{F}) \to H^{\pm\infty}(\mathring{M}; \Lambda\mathcal{F}) . \tag{1.3.4}$$

The corresponding perturbed leafwise codifferential and Laplace operators are denoted by $\delta_{\mathcal{F},\mu}$ and $\Delta_{\mathcal{F},\mu}$. Finally, for $f \in C^\infty_{\mathrm{c}}(\mathbb{R})$, $\mu \in \mathbb{R}$ and $0 < u \le \infty$, we will use the operator

$$\boldsymbol{P}_{\mu,u,f} = \int_{\mathbb{R}} \boldsymbol{\phi}^{t*}_\mu \, e^{-u\Delta_{\mathcal{F},\mu}} \, f(t) \, dt$$

on $H^{\pm\infty}(\mathring{M}; \Lambda\mathcal{F})$, which is a version of (1.2.4).

1.3.5 Main Results Leading to the Trace Formula

Concerning the above reduced cohomologies, the following are our main achievements.

Theorem 1.3.1 *We have*[1]

$$K(\mathcal{F}) \equiv \bigoplus_{L,k} C^\infty(L;\Lambda)\;, \quad d_{\mathcal{F}} \equiv \bigoplus_{L,k} d_{-k-1}\;, \quad \phi^{t\,*} \equiv \bigoplus_{L,k} e^{-(k+1)\varkappa_L t}\;,$$

$$H^\bullet K(\mathcal{F}) \equiv \bar{H}^\bullet K(\mathcal{F}) \equiv \bigoplus_{L,k} H^\bullet_{-k-1}(L)\;, \quad \phi^{t\,*} \equiv \bigoplus_{L,k} e^{-(k+1)\varkappa_L t}\;,$$

where L runs over the set of leaves contained in M^0 and k runs over $\mathbb{N}_0$.

The first identity of Theorem 1.3.1 follows by considering the partial derivatives ∂_ρ^k ($k \in \mathbb{N}_0$) of leafwise currents of $(M,\mathcal{F})$ that are of Dirac type at the leaves $L \subset M^0$. It is a consequence of the properties of ρ, η and ϕ^* on T.

Now consider $\boldsymbol{\rho}$, $\boldsymbol{\eta}$ and $\boldsymbol{\phi}$ on $(\boldsymbol{M},\boldsymbol{\mathcal{F}})$.

Theorem 1.3.2 *Using* (1.3.3) *with $H^\infty(\mathring{\boldsymbol{M}};\Lambda\boldsymbol{\mathcal{F}})$, we get*

$$J(\mathcal{F}) = \bigcup_\mu \boldsymbol{\rho}^\mu H^\infty(\mathring{\boldsymbol{M}};\Lambda\boldsymbol{\mathcal{F}}) \equiv \varinjlim H^\infty(\mathring{\boldsymbol{M}};\Lambda\boldsymbol{\mathcal{F}})\;,$$

$$d_{\mathcal{F}} \equiv \varinjlim d_{\boldsymbol{\mathcal{F}},\mu}\;, \quad \phi^{t\,*} \equiv \varinjlim \boldsymbol{\phi}_\mu^{t\,*}\;,$$

where the inductive limits are defined with the maps (1.3.4) *as $\mu \downarrow -\infty$. Moreover, there are linear identities,*

$$\bar{H}^\bullet J(\mathcal{F}) \equiv \varinjlim \bar{H}^\bullet_\mu H^\infty(\mathring{\boldsymbol{\mathcal{F}}})\;, \quad \phi^{t\,*} \equiv \varinjlim \boldsymbol{\phi}_\mu^{t\,*}\;.$$

Theorem 1.3.3 *We have a short exact sequence*

$$0 \to H^\bullet K(\mathcal{F}) \xrightarrow{\bar{\iota}_*} \bar{H}^\bullet I(\mathcal{F}) \xrightarrow{\bar{R}_*} \bar{H}^\bullet J(\mathcal{F}) \to 0\;.$$

Theorem 1.3.4 *Using L and k like in Theorem 1.3.1, we have*

$$K'(\mathcal{F}) \equiv \prod_{L,k} C^{-\infty}(L;\Lambda)\;, \quad d_{\mathcal{F}} \equiv \prod_{L,k} d_k\;, \quad \phi^{t\,*} \equiv \prod_{L,k} e^{k\varkappa_L t}\;,$$

$$H^\bullet K'(\mathcal{F}) \equiv \bar{H}^\bullet K'(\mathcal{F}) \equiv \prod_{L,k} H^\bullet_k(L)\;, \quad \phi^{t\,*} \equiv \prod_{L,k} e^{k\varkappa_L t}\;.$$

[1] With some abuse of notation, we write $\bigoplus_m A = \bigoplus_m A_m$ and $\prod_m A = \prod_m A_m$ if $A_m = A$ for all m.

The identity of Theorem 1.3.4 is a consequence of the version of Theorem 1.3.1 for $K(\mathcal{F}; \Omega M)$. The shift in the role played by k is due to the introduction of ΩM.

Theorem 1.3.5 *Using* (1.3.3) *with* $H^{-\infty}(\mathring{M}; \Lambda\mathcal{F})$, *we get*

$$J'(\mathcal{F}) = \bigcap_{\mu} \rho^{\mu} H^{-\infty}(\mathring{M}; \Lambda\mathcal{F}) \equiv \varprojlim H^{-\infty}(\mathring{M}; \Lambda\mathcal{F}) ,$$

$$d_{\mathcal{F}} \equiv \varprojlim d_{\mathcal{F},\mu} , \quad \phi^{t\,*} \equiv \varprojlim \boldsymbol{\phi}^{t\,*}_{\mu} .$$

where the projective limits are defined with the maps (1.3.4) *as* $\mu \uparrow +\infty$. *Moreover, there are linear identities,*

$$\bar{H}^{\bullet} J'(\mathcal{F}) \equiv \varprojlim \bar{H}^{\bullet}_{\mu} H^{-\infty}(\mathring{\mathcal{F}}) , \quad \phi^{t\,*} \equiv \varprojlim \boldsymbol{\phi}^{t\,*}_{\mu} .$$

There is no essential difference between $J(\mathcal{F})$ and $J(\mathcal{F}; \Omega M)$ because $\mathring{\mathcal{F}}$ has the invariant transverse density $|\omega_{\mathrm{b}}|$. Thus Theorem 1.3.5 follows from Theorem 1.3.2.

Theorem 1.3.6 *We have a short exact sequence*

$$0 \leftarrow H^{\bullet} K'(\mathcal{F}) \overset{\bar{R}'_*}{\longleftarrow} \bar{H}^{\bullet} I'(\mathcal{F}) \overset{\bar{\iota}'_*}{\leftarrow} \bar{H}^{\bullet} J'(\mathcal{F}) \leftarrow 0 .$$

Recall the definition of $\boldsymbol{P}_{\mu,u,f}$ given in Sect. 1.3.4.

Theorem 1.3.7 $\boldsymbol{P}_{\mu,u,f}$ *is a smoothing b-pseudodifferential operator, and the map* $f \mapsto {}^{\mathrm{b}}\mathrm{Str}\, \boldsymbol{P}_{\mu,u,f}$ *defines a distribution on* $\mathbb{R}$.

Now we will use the integers $\ell(c)$ and $\epsilon_c(k)$ associated to every closed orbit c, and the b-Connes-Euler characteristic ${}^{\mathrm{b}}\chi_{|\omega_{\mathrm{b}}|}(\mathcal{F}) = {}^{\mathrm{b}}\chi_{|\omega^1|}(\mathcal{F}^1)$.

Theorem 1.3.8 *We have*

$$\lim_{u \downarrow 0} {}^{\mathrm{b}}\mathrm{Str}\, \boldsymbol{P}_{\mu,u,f} = {}^{\mathrm{b}}\chi_{|\omega^1|}(\mathcal{F}^1)\, f(0) + \sum_{c} \ell(c) \sum_{k \in \mathbb{Z}^\times} \epsilon_c(k)\, f(k\ell(c)) ,$$

where c runs in the set of closed orbits of ϕ.

Recall that the definition of η was given in Sect. 1.3.1.

Theorem 1.3.9 *If* $\dim \mathcal{F}$ *is even, then we can choose η and g on M^0 so that*

$$f \mapsto \lim_{u_1 \uparrow +\infty,\ u_0 \downarrow 0} \left({}^{\mathrm{b}}\mathrm{Str}\, \boldsymbol{P}_{\mu,u_1,f} - {}^{\mathrm{b}}\mathrm{Str}\, \boldsymbol{P}_{\mu,u_0,f} \right)$$

defines a tempered distribution Z_μ on $\mathbb{R}$, and $Z_\mu \to 0$ as $\mu \to \pm\infty$.

In Theorem 1.3.9, for more general choices of η and g on M^0, the limits of Z_μ as $\mu \to \pm\infty$ are multiples of the Dirac mass δ_0. These limits may not be zero because the b-trace does not vanish on commutators (it is not a trace).

This additional contribution of the b-trace shows up like the eta-invariant of manifolds with boundary [Mel93]. When $\dim\mathcal{F}$ is even, we can prescribe any limit of Z_μ as $\mu \to \pm\infty$ with appropriate choices of η and g on M^0 [ÁLKL25] (see Theorem 2.9.7); in particular, we can prescribe the zero limit. This makes ${}^{\mathrm{b}}\mathrm{Str}\, \boldsymbol{P}_{\mu,u,f}$ behave like a supertrace as $\mu \to \pm\infty$.

1.3.6 The Lefschetz Distribution

It seems there is no reasonable definition of $L_{\mathrm{dis}}(\phi)$ with a single leafwise reduced cohomology. However, $\bar{H}^\bullet I(\mathcal{F})$ and $\bar{H}^\bullet I'(\mathcal{F})$ together will do the job. Though this may look strange, we hope this idea will be valid in further developments of Deninger's program.

To begin with, by Theorem 1.3.3 and Theorem 1.3.6, it is enough to consider the actions ϕ^* on $H^\bullet K(\mathcal{F})$, $\bar{H}^\bullet J(\mathcal{F})$, $H^\bullet K'(\mathcal{F})$ and $\bar{H}^\bullet J'(\mathcal{F})$.

Let us try to define Lefschetz distributions $L_{\mathrm{dis},K}(\phi)$ and $L_{\mathrm{dis},K'}(\phi)$ of ϕ on $\bar{H}^\bullet K(\mathcal{F})$ and $\bar{H}^\bullet K'(\mathcal{F})$. By Theorems 1.3.1 and 1.3.4, and since all twisted cohomologies $H^\bullet_\mu(L)$ have the same Euler characteristic $\chi(L)$, it makes some sense to define, on $\mathbb{R}^\times$,

$$L_{\mathrm{dis},K}(\phi) = \sum_{\varkappa_L t>0} \chi(L) \sum_{k=0}^{\infty} e^{-(k+1)\varkappa_L t} = \sum_{\varkappa_L t>0} \frac{\chi(L)}{e^{\varkappa_L t}-1}\,,$$

$$L_{\mathrm{dis},K'}(\phi) = \sum_{\varkappa_L t<0} \chi(L) \sum_{k=0}^{\infty} e^{k\varkappa_L t} = \sum_{\varkappa_L t<0} \frac{\chi(L)}{1-e^{\varkappa_L t}}\,.$$

In each of these distributions, the conditions on the leaves $L \subset M^0$ guarantee that their contribution to the trace is defined; the other leaves in M^0 are omitted as a way of renormalization. Every L has a contribution to just one of these distributions on $\mathbb{R}^\pm$. Taking into account all contributions from leaves $L \subset M^0$ in $L_{\mathrm{dis},K}(\phi)$ and $L_{\mathrm{dis},K'}(\phi)$, we get a combined Lefschetz distribution on $\mathbb{R}^\times$,

$$L_{\mathrm{dis},K,K'}(\phi) := L_{\mathrm{dis},K}(\phi) + L_{\mathrm{dis},K'}(\phi) = \sum_L \frac{\chi(L)}{|e^{\varkappa_L t}-1|}\,.$$

By changing variables and using L'Hôspital's rule, it follows that every function $|e^{\varkappa_L t}-1|^{-1}$ on $\mathbb{R}^\times$ can be extended to a distribution W_L on $\mathbb{R}$ given by [Bar81]

$$\langle W_L, f\rangle = \int_0^\infty \left(\frac{f(t)+f(-t)}{|e^{\varkappa_L t}-1|} - \frac{2f(0)}{|\varkappa_L|\,t} \right) dt\,.$$

Thus $L_{\mathrm{dis},K,K'}(\phi)$ can be extended to $\mathbb{R}$ as the distribution

$$L_{\mathrm{dis},K,K'}(\phi) := \sum_L \chi(L)\, W_L\ .$$

Next, by Theorem 1.3.7, like in the case of (1.2.4), we can consider the mapping

$$f \mapsto \lim_{u\uparrow+\infty} {}^{\mathrm{b}}\mathrm{Str}\, \boldsymbol{P}_{\mu,u,f}$$

as the distributional supertrace of the action ϕ_μ^* on $\bar{H}_\mu^\bullet H^{\pm\infty}(\mathring{\boldsymbol{\mathcal{F}}})$. Since $\boldsymbol{P}_{\mu,u,f}$ is not of trace class, its b-supertrace is used here instead of the supertrace as a way of renormalization. By Theorems 1.3.2 and 1.3.5, it makes sense to define the Lefschetz distributions of ϕ on $\bar{H}^\bullet J(\mathcal{F})$ and $\bar{H}^\bullet J'(\mathcal{F})$, denoted by $L_{\mathrm{dis},J}(\phi)$ and $L_{\mathrm{dis},J'}(\phi)$, by

$$\langle L_{\mathrm{dis},J}(\phi), f\rangle = \lim_{\mu\downarrow-\infty}\lim_{u\uparrow+\infty} {}^{\mathrm{b}}\mathrm{Str}\, \boldsymbol{P}_{\mu,u,f}\ ,$$

$$\langle L_{\mathrm{dis},J'}(\phi), f\rangle = \lim_{\mu\uparrow+\infty}\lim_{u\uparrow+\infty} {}^{\mathrm{b}}\mathrm{Str}\, \boldsymbol{P}_{\mu,u,f}\ .$$

From now on, assume $\dim \mathcal{F}$ is even (the relevant case in Deninger's program is $\dim \mathcal{F} = 2$). By Theorems 1.3.8 and 1.3.9, we can choose η and g on M^0 so that

$$L_{\mathrm{dis},J}(\phi) = L_{\mathrm{dis},J'}(\phi) = {}^{\mathrm{b}}\chi_{|\omega^1|}(\mathcal{F}^1)\,\delta_0 + \sum_c \ell(c) \sum_{k\in\mathbb{Z}^\times} \epsilon_c(k)\,\delta_{k\ell(c)}\ .$$

Then the distribution

$$L_{\mathrm{dis},J,J'}(\phi) := L_{\mathrm{dis},J}(\phi) = L_{\mathrm{dis},J'}(\phi)$$

is considered as a common feature of the actions ϕ^* on $\bar{H}^\bullet I(\mathcal{F})$ and $\bar{H}^\bullet I'(\mathcal{F})$, and therefore, its repetition is avoided. Another explanation for this definition is that $L_{\mathrm{dis},J,J'}(\phi)$ can also be achieved like in Theorems 1.3.2 and 1.3.5 by using any of the following subspaces of $C^{-\infty}(\mathring{\boldsymbol{M}}; \Lambda\boldsymbol{\mathcal{F}})$, whose definitions combine $J(\mathcal{F})$ and $J'(\mathcal{F})$,

$$J(\mathcal{F}) + J'(\mathcal{F}) = \bigcup_\mu \rho^\mu H^{-\infty}(\mathring{\boldsymbol{M}}; \Lambda\boldsymbol{\mathcal{F}})\ ,$$

$$J(\mathcal{F}) \cap J'(\mathcal{F}) = \bigcap_\mu \rho^\mu H^{\infty}(\mathring{\boldsymbol{M}}; \Lambda\boldsymbol{\mathcal{F}})\ .$$

The last equality shows that $J(\mathcal{F}) \cap J'(\mathcal{F})$ consists of the restrictions to M^1 of the smooth leafwise forms on M that vanish to all orders at M^0; this agrees with the equality $I(\mathcal{F}) \cap I'(\mathcal{F}) = C^\infty(M, \Lambda\mathcal{F})$ (Sect. 1.3.2). Note also that $L_{\mathrm{dis},K}(\phi)$

and $L_{\mathrm{dis},K'}(\phi)$ collect the contributions of complementary sets of preserved leaves, whereas both $L_{\mathrm{dis},J}(\phi)$ and $L_{\mathrm{dis},J'}(\phi)$ collect the contributions of all closed orbits.

Finally, by Theorems 1.3.3 and 1.3.6, it makes sense to define the combined Lefschetz distribution of the actions ϕ^* on $\bar{H}^\bullet I(\mathcal{F})$ and $\bar{H}^\bullet I'(\mathcal{F})$ as

$$L_{\mathrm{dis}}(\phi) = L_{\mathrm{dis},I,I'}(\phi) := L_{\mathrm{dis},K,K'}(\phi) + L_{\mathrm{dis},J,J'}(\phi)\ .$$

By Theorems 1.3.8 and 1.3.9, the trace formula conjectured by Deninger is satisfied:

Theorem 1.3.10 *Assume* $\dim\mathcal{F}$ *is even. We can choose* η *and* g *on* M^0 *such that, using the preserved leaves L and the closed orbits c, we have*

$$L_{\mathrm{dis}}(\phi) = \sum_L \chi(L)\, W_L + {}^{\mathrm{b}}\chi_{|\omega^1|}(\mathcal{F}^1)\,\delta_0 + \sum_c \ell(c) \sum_{k\in\mathbb{Z}^\times} \epsilon_c(k)\,\delta_{k\ell(c)}\ .$$

1.4 Short Guide

Our arguments involve tools from two different sources: Analysis and Foliations. Concerning Analysis, we mainly use conormal and dual-conormal distributions, analysis on manifolds of bounded geometry and small b-calculus. Concerning Foliations, we mainly use local Reeb's stability, suspension foliations, Riemannian foliations, and differential forms and currents on foliated manifolds. For the readers' convenience, the needed basic concepts and results from those areas are recalled in Chaps. 2 and 3. The specialists on any of them may skip the corresponding chapter, except perhaps the notation. A few short proofs are also recalled in Chap. 2 because their arguments will be used.

Chapter 4 contains a more specific description of foliations with simple foliated flows, explaining all topological and geometric objects that will be used in our analysis. We specially focus on the case of suspension foliations, which describe $\mathcal{F}$ on a tubular neighborhood T of M^0.

Chapters 5 and 6 are devoted to the study of the actions ϕ^* on $\bar{H}^\bullet I(\mathcal{F})$ and $\bar{H}^\bullet I'(\mathcal{F})$, respectively, showing Theorems 1.3.1 to 1.3.6.

Finally, Chap. 7 is devoted to the study of ${}^{\mathrm{b}}\mathrm{Str}\,\boldsymbol{P}_{\mu,u,f}$, showing Theorems 1.3.7 to 1.3.9.

Chapter 2
Analytic Tools

2.1 Section Spaces and Operators on Manifolds

The field of coefficients is $\mathbb{K}$, equal to $\mathbb{R}$ or $\mathbb{C}$. We typically consider $\mathbb{K} = \mathbb{C}$, and the few cases where $\mathbb{K} = \mathbb{R}$ will be indicated without changing the notation.

2.1.1 *Topological Vector Spaces*

Let us recall some concepts and fix some conventions concerning topological vector spaces (TVSs); see [Edw65, Hor66, Köt69, Sch71, NB11, Wen03] for other concepts we use. We always consider (possibly non-Hausdorff) locally convex spaces (LCSs); the abbreviation LCHS is used in the Hausdorff case. Local convexity is preserved by all operations we use. For instance, we will use the (locally convex) inductive/projective limit of any inductive/projective spectrum (or system) of continuous linear maps between LCSs. If the inductive/projective spectrum is a sequence of continuous inclusions, then the inductive/projective limit is the union/intersection, always endowed with the inductive/projective limit topology. This applies to the locally convex direct sum and the topological product of LCSs. LF-spaces are not assumed to be strict. The (continuous) dual X' of any LCS X is always endowed with the strong topology.

Now fix an inductive spectrum of LCSs of the form $(X_k) = (X_0 \subset X_1 \subset \cdots)$, and let $X = \bigcup_k X_k$. The condition on (X_k) to be *acyclic* means that, for all k, there is some $k' \geq k$ such that, for all $k'' \geq k'$, the topologies of $X_{k'}$ and $X_{k''}$ coincide on some 0-neighborhood of X_k [Wen03, Theorem 6.1]. In this case, X is Hausdorff if and only if all X_k are Hausdorff [Wen03, Proposition 6.3]. It is said that (X_k) is *regular* if any bounded $B \subset X$ is contained and bounded in some step X_k. If moreover the topologies of X and X_k coincide on B, then (X_k) is said to be

J. A. Álvarez López et al., *A Trace Formula for Foliated Flows*, Lecture Notes in Mathematics 2387, https://doi.org/10.1007/978-3-032-15413-2_2

boundedly retractive. The conditions of being *compactly retractive* or *sequentially retractive* are similarly defined, using compact sets or convergent sequences.

If the steps X_k are Fréchet spaces, the above properties of (X_k) only depend on the LF-space X [Wen03, Chapter 6, p. 111], and therefore they are considered as properties of X. In this case, X is acyclic if and only if it is boundedly/compactly/sequentially retractive [Wen03, Proposition 6.4]. As a consequence, acyclic LF-spaces are complete and regular [Wen03, Corollary 6.5]. A topological vector subspace $Y \subset X$ is called a *limit subspace* if $Y \equiv \bigcup_k (Y \cap X_k)$ as TVSs.

Assume the steps X_k are LCHSs. It is said that (X_k) is *compact* if the inclusion maps are compact operators. Then (X_k) is acyclic, and so X is Hausdorff. Moreover X is a complete bornological DF Montel space [Kom67, Theorem 6']

The above concepts and properties also apply to an inductive/projective spectrum of LCSs consisting of continuous inclusions $X_r \subset X_{r'}$ for $r < r'$ in $\mathbb{R}$ because $\bigcap_r X_r = \bigcap_k X_{r_k}$ and $\bigcup_r X_r = \bigcup_k X_{s_k}$ for sequences $r_k \downarrow -\infty$ and $s_k \uparrow +\infty$.

In the category of continuous linear maps between LCSs, the exactness of a sequence $0 \to X \to Y \to Z \to 0$ means that it is exact as a sequence of linear maps and consists of topological homomorphisms [Wen03, Sections 2.1 and 2.2].

Given LCSs X and Y, let $L(X, Y)$ denote the LCS of continuous linear maps $X \to Y$ with the topology of uniform convergence over bounded subsets. If X and Y are Banach spaces, then $L(X, Y)$ is also a Banach space whose norm may be denoted by $\|\cdot\|_{X,Y}$, with possible simplifications to avoid redundant notation. If $X = Y$, then the notation $\operatorname{End}(X)$ is used, as well as $\|\cdot\|_X$ if X is a Banach space.

The following construction will be often used. Given a linear subspace $\mathcal{A}$ of closed operators, densely defined in X and with values in Y, we get the LCS

$$Z = \bigcap_{A \in \mathcal{A}} \operatorname{dom} A \tag{2.1.1}$$

with the projective topology given by the maps $A : Z \to Y$ ($A \in \mathcal{A}$). If Y is a Fréchet space, $L(X, Y) \subset \mathcal{A}$ and $\mathcal{A}/L(X, Y)$ is countably generated, then Z is easily seen to be a Fréchet space. If moreover Y is a Hilbertian space, then Z is easily seen to be a totally reflexive Fréchet space using [Val89, Theorem 4].

A *Hilbertian space* is a TVS X endowed with a family of Hilbert-space scalar products, all of them with equivalent norms defining the topology of X, but none of them is distinguished.

2.1.2 Smooth Functions on Open Subsets of $\mathbb{R}^n$

For any open $U \subset \mathbb{R}^n$ ($n \in \mathbb{N}_0 = \mathbb{N} \cup \{0\}$), we use the Fréchet space $C^\infty(U)$ of smooth ($\mathbb{K}$-valued) functions on U, whose topology is described by the semi-norms

$$\|u\|_{K,C^k} = \sup_{x \in K,\ |I| \le k} |\partial^I u(x)| , \tag{2.1.2}$$

for any compact $K \subset U$, $k \in \mathbb{N}_0$ and $I \in \mathbb{N}_0^n$, with standard multi-index notation. For any $S \subset U$, let $C_S^\infty(U) \subset C^\infty(U)$ be the topological vector subspace of smooth functions supported in S. The strict LF-space of compactly supported functions is

$$C_c^\infty(U) = \bigcup_K C_K^\infty(U) , \tag{2.1.3}$$

for compact subsets $K \subset U$.

Straightforward generalizations to the case of functions with values in $\mathbb{K}^l$ ($l \in \mathbb{N}$) can be given by

$$C_{\cdot/c}^\infty(U, \mathbb{K}^l) \equiv C_{\cdot/c}^\infty(U) \otimes \mathbb{K}^l . \tag{2.1.4}$$

(The notation $C_{\cdot/c}^\infty$ or $C_{c/\cdot}^\infty$ refers to both C^∞ and C_c^∞.)

2.1.3 Vector Bundles

We fix a smooth n-manifold M and a ($\mathbb{K}$-) vector bundle E of rank l over M. Let $E_x \subset E$ ($x \in M$) denote the fibers of E, 0_x the zero element of E_x, and 0_M the zero section of E. Let $\Omega^a E$ ($a \in \mathbb{R}$) be the line bundle of a-densities of E, and $o(E)$ the flat line bundle of its orientations; as usual, we write $\Omega E = \Omega^1 E$. Recall that $\Omega^a E \otimes \Omega^b E \equiv \Omega^{a+b} E$. We use the notation $\Lambda E = \bigwedge E^*$ for the exterior bundle of the dual bundle. We may denote $\Lambda^{\mathrm{top}} E = \Lambda^l E$, and use similar notation with other gradings and bigradings. For any submanifold $L \subset M$, we also write $E_L = E|_L$. As particular cases, we have the tangent and cotangent $\mathbb{R}$-vector bundles, TM and T^*M, and the associated $\mathbb{K}$-vector bundles $o(M) = o(TM) \otimes \mathbb{K}$, $\Lambda M = \Lambda TM \otimes \mathbb{K}$, $\Omega^a M = \Omega^a TM \otimes \mathbb{K}$ and $\Omega M = \Omega TM \equiv \Lambda^n M \otimes o(M)$.

2.1.4 Smooth and Distributional Sections

Concerning spaces of distributional sections, we follow the notation of [Mel96, Hör83, Hör85], with some minor changes to fit our notation for foliations. The precise references of the properties recalled here are given in [ÁLKL24, Section 2.4].

Consider the Fréchet space $C^\infty(M; E)$ of smooth sections of E, whose topology is described by semi-norms $\|\cdot\|_{K,C^k}$ defined like in (2.1.2), using charts (U, x) of M and diffeomorphisms of triviality $E_U \equiv U \times \mathbb{K}^l$ with $K \subset U$. Redundant notation is simplified as usual. For instance, in the case of the trivial vector bundle of rank 1 (resp., l), we write $C^\infty(M)$ (resp., $C^\infty(M, \mathbb{K}^l)$). We also write $C^\infty(L, E) = C^\infty(L, E_L)$ and $C^\infty(M; \Omega^a) = C^\infty(M; \Omega^a M)$. If M is fixed, the notation $C^\infty(E) = C^\infty(M; E)$ can be used, but it may be confusing because the space of smooth functions on E is also used. In particular, $\mathfrak{X}(M) = C^\infty(M; TM)$ is the

Lie algebra of vector fields. The subspace $C^\infty_S(M;E)$ is defined like in Sect. 2.1.2, and the strict LF-space $C^\infty_{\mathrm{c}}(M;E)$ is defined like in (2.1.3), using compact subsets $K \subset M$. There is a continuous inclusion $C^\infty_{\mathrm{c}}(M;E) \subset C^\infty(M;E)$.

The notation $C^\infty(M;E)$, or $C^\infty(E)$, is also used with any smooth fiber bundle E, obtaining a completely metrizable topological space with the weak C^∞ topology.

The space of distributional sections with arbitrary/compact support is

$$C^{-\infty}_{\cdot/\mathrm{c}}(M;E) = C^\infty_{\mathrm{c}/\cdot}(M;E^*\otimes\Omega)' . \tag{2.1.5}$$

The canonical pairing of any $u \in C^{-\infty}_{\cdot/\mathrm{c}}(M;E)$ and $v \in C^\infty_{\mathrm{c}/\cdot}(M;E^*\otimes\Omega)$ is denoted by $\langle u,v\rangle$ (or (u,v) if the notation $\langle\cdot,\cdot\rangle$ is used for other purposes). Integration of smooth densities on M and the canonical pairing of E and E^* define a continuous dense inclusion $C^\infty_{\cdot/\mathrm{c}}(M;E) \subset C^{-\infty}_{\cdot/\mathrm{c}}(M;E)$. If $U \subset M$ is open, the extension by zero defines a TVS-embedding $C^{\pm\infty}_{\mathrm{c}}(U;E) \subset C^{\pm\infty}_{\mathrm{c}}(M;E)$.

The above spaces of distributional sections can also be described in terms of the corresponding spaces of distributions as the algebraic tensor product of $C^\infty(M)$-modules [ÁLKL24, Eq. (2.5)]

$$C^{-\infty}_{\cdot/\mathrm{c}}(M;E) \equiv C^{-\infty}_{\cdot/\mathrm{c}}(M) \otimes_{C^\infty(M)} C^\infty(M;E) . \tag{2.1.6}$$

This tensor product has an induced topology so that this is a TVS-identity. Expressions like (2.1.6) hold for most of the LCSs of distributional sections we will consider, which are also $C^\infty(M)$-modules. Thus, from now on, we will often define and study those spaces for the trivial line bundle or density bundles, and then the notation for arbitrary vector bundles will be used without further comment, and the properties have straightforward extensions.

Given a smooth submersion $\phi : M \to M'$, a smooth/distributional section of E has *compact support in the vertical direction* if its support has compact intersections with the fibers of ϕ. They form the LCHSs $C^{\pm\infty}_{\mathrm{cv}}(M;E)$. Here, $C^\infty_{\mathrm{cv}}(M;E)$ has the inductive topology defined like in the case of $C^\infty_{\mathrm{c}}(M;E)$, using (2.1.2) and (2.1.3) with closed subsets $K \subset M$ whose intersection with the fibers is compact. $C^{-\infty}_{\mathrm{cv}}(M;E)$ has the projective topology defined by the (product) maps $f : C^{-\infty}_{\mathrm{cv}}(M;E) \to C^{-\infty}_{\mathrm{c}}(M;E)$, for $f \in C^\infty_{\mathrm{c}}(M)$. A version of (2.1.6) is also true for $C^{-\infty}_{\mathrm{cv}}(M;E)$ in this case.

Consider also the Fréchet space $C^k(M)$ ($k \in \mathbb{N}_0$) of C^k functions, with the semi-norms $\|\cdot\|_{K,C^k}$ given like in (2.1.2), the LF-space $C^k_{\mathrm{c}}(M)$ of C^k functions with compact support, defined like in (2.1.3), and the space $C'^{-k}_{\cdot/\mathrm{c}}(M)$ of distributions of order k with arbitrary/compact support, defined like in (2.1.5). There are continuous dense inclusions

$$C^{k'}_{\cdot/\mathrm{c}}(M) \subset C^{k}_{\cdot/\mathrm{c}}(M) , \quad C'^{-k'}_{\mathrm{c}/\cdot}(M) \supset C'^{-k}_{\mathrm{c}/\cdot}(M) \quad (k < k') , \tag{2.1.7}$$

with

$$\bigcap_k C^k_{\cdot/\mathrm{c}}(M) = C^\infty_{\cdot/\mathrm{c}}(M)\,, \quad \bigcup_k C'^{-k}_{\mathrm{c}}(M) = C^{-\infty}_{\mathrm{c}}(M)\,. \tag{2.1.8}$$

The space $\bigcup_k C'^{-k}(M)$ consists of the distributions with some order. If M is compact, then every $C^k(M)$ is a Banach space and $\bigcup_k C'^{-k}(M) = C^{-\infty}(M)$.

$C^\infty_{\mathrm{c}}(M)$ and $C^k_{\mathrm{c}}(M)$ are complete and Hausdorff. $C^\infty_{\cdot/\mathrm{c}}(M)$ and $C^k_{\cdot/\mathrm{c}}(M)$ are ultrabornological and barreled. $C^{\pm\infty}_{\cdot/\mathrm{c}}(M)$ is a Montel space (in particular, barreled) and reflexive. $C^\infty_{\cdot/\mathrm{c}}(M)$ is a Schwartz space, and therefore $C^{-\infty}_{\cdot/\mathrm{c}}(M)$ is ultrabornological. $C^\infty(M)$ is distinguished. $C^{\pm\infty}_{\cdot/\mathrm{c}}(M)$ is webbed.

The type of notation introduced in this section will be used with any LCHS and $C^\infty(M)$-module continuously included in $C^{\pm\infty}(M; E)$.

2.1.5 *Linear Operators on Section Spaces*

Let E and F be vector bundles over M, and let $A : C^\infty_{\mathrm{c}}(M; E) \to C^\infty(M; F)$ be a continuous linear map. The *transpose* of A is the continuous linear map

$$A^{\mathrm{t}} : C^{-\infty}_{\mathrm{c}}(M; F^* \otimes \Omega) \to C^{-\infty}(M; E^* \otimes \Omega)\,,$$
$$\langle A^{\mathrm{t}} v, u\rangle = \langle v, Au\rangle\,, \quad u \in C^\infty_{\mathrm{c}}(M; E)\,, \quad v \in C^{-\infty}_{\mathrm{c}}(M; F^* \otimes \Omega)\,.$$

For instance, the transpose of $C^\infty_{\mathrm{c}}(M; F^* \otimes \Omega) \subset C^\infty(M; E^* \otimes \Omega)$ is a continuous dense injection $C^{-\infty}_{\mathrm{c}}(M; E) \subset C^{-\infty}(M; E)$. If A^{t} restricts to a continuous linear map $C^\infty_{\mathrm{c}}(M; F^* \otimes \Omega) \to C^\infty(M; E^* \otimes \Omega)$, then $A^{tt} : C^{-\infty}_{\mathrm{c}}(M; E) \to C^{-\infty}(M; F)$ is a continuous extension of A, also denoted by A. The *Schwartz kernel*, $K_A \in C^{-\infty}(M^2; F \boxtimes (E^* \otimes \Omega))$, is determined by the condition $\langle K_A, v \otimes u\rangle = \langle v, Au\rangle$ for $u \in C^\infty_{\mathrm{c}}(M; E)$ and $v \in C^\infty_{\mathrm{c}}(M; F^* \otimes \Omega)$. The Schwartz kernel theorem [Hör71, Theorem 5.2.1] states that we have a linear isomorphism

$$L(C^\infty_{\mathrm{c}}(M; E), C^{-\infty}(M; F)) \xrightarrow{\cong} C^{-\infty}(M^2; F \boxtimes (E^* \otimes \Omega M))\,, \quad A \mapsto K_A\,. \tag{2.1.9}$$

Using that $(F^* \otimes \Omega)^* \otimes \Omega \equiv F$, we get

$$K_{A^{\mathrm{t}}} = R^* K_A \in C^{-\infty}(M^2; (E^* \otimes \Omega) \boxtimes F)\,,$$

where $R : M^2 \to M^2$ is given by $R(x, y) = (y, x)$. If K_A is C^∞, we can write

$$Au(x) = \int_M K_A(x, y) u(y)\,, \quad A^{\mathrm{t}} v(x) = \int_M K_A(y, x) v(y)\,, \tag{2.1.10}$$

for $u \in C^\infty_{\mathrm{c}}(M; E)$ and $v \in C^\infty_{\mathrm{c}}(M; F^* \otimes \Omega)$.

There are versions of the construction of A^{t} and A^{tt} when both the domain and codomain of A have compact support, or no support restriction. For example, for any open $U \subset M$, the transpose of the extension by zero $C^\infty_{\mathrm{c}}(U; E^* \otimes \Omega) \subset C^\infty_{\mathrm{c}}(M; E^* \otimes \Omega)$ is the restriction map

$$C^{-\infty}(M; E) \to C^{-\infty}(U, E) , \quad u \mapsto u|_U , \tag{2.1.11}$$

and the transpose of the restriction map $C^\infty(M; E^* \otimes \Omega) \to C^\infty(U, E^* \otimes \Omega)$ is the extension by zero

$$C^{-\infty}_{\mathrm{c}}(U; E) \subset C^{-\infty}_{\mathrm{c}}(M; E) . \tag{2.1.12}$$

Inclusion maps may be denoted by ι and restriction maps by R, without further comment. The *singular support* of any $u \in C^{-\infty}(M; E)$, $\operatorname{sing\,supp} u$, is the complement of the maximal open subset $U \subset M$ with $u|_U \in C^\infty(U; E)$.

2.1.6 Pull-Back and Push-Forward of Distributional Sections

Any smooth map $\phi : M' \to M$ induces the continuous linear pull-back map

$$\phi^* : C^\infty(M; E) \to C^\infty(M'; \phi^* E) . \tag{2.1.13}$$

If ϕ is a submersion, then it also induces the continuous linear push-forward map

$$\phi_* : C^\infty_{\mathrm{c}}(M'; \phi^* E \otimes \Omega_{\mathrm{fiber}}) \to C^\infty_{\mathrm{c}}(M; E) , \tag{2.1.14}$$

where $\Omega_{\mathrm{fiber}} = \Omega_{\mathrm{fiber}} M' = \Omega \mathcal{V}$ for the vertical subbundle $\mathcal{V} = \ker \phi_* \subset TM'$. Moreover, the map (2.1.14) has a continuous extension

$$\phi_* : C^\infty_{\mathrm{cv}}(M'; \phi^* E \otimes \Omega_{\mathrm{fiber}}) \to C^\infty(M; E) , \tag{2.1.15}$$

also called push-forward map. Using (2.1.14) and any partition of unity $\{\lambda_j\}$ of M consisting of compactly supported smooth functions, the map (2.1.15) is given by

$$\phi_* u = \sum_j \phi_*(\phi^* \lambda_j \cdot u) . \tag{2.1.16}$$

Since $\phi^* \Omega M \equiv \Omega(TM/\mathcal{V}) \equiv \Omega^{-1}_{\mathrm{fiber}} \otimes \Omega M'$, transposing the versions of (2.1.13) and (2.1.14) with $E^* \otimes \Omega M$ and using (2.1.5), we obtain continuous extensions of (2.1.14) and (2.1.13) [Hör71, Theorem 6.1.2],

$$\phi_* : C^{-\infty}_{\mathrm{c}}(M'; \phi^* E \otimes \Omega_{\mathrm{fiber}}) \to C^{-\infty}_{\mathrm{c}}(M; E) , \tag{2.1.17}$$

$$\phi^* : C^{-\infty}(M; E) \to C^{-\infty}(M'; \phi^* E) , \tag{2.1.18}$$

also called push-forward and pull-back maps. Again, (2.1.17) has a continuous extension,

$$\phi_* : C^{-\infty}_{\mathrm{cv}}(M'; \phi^* E \otimes \Omega_{\mathrm{fiber}}) \to C^{-\infty}(M; E) , \tag{2.1.19}$$

also called push-forward map, defined like (2.1.15) with (2.1.17).

If $\phi : M' \to M$ is a local diffeomorphism, we can omit Ω_{fiber} in the push-forward maps. If moreover ϕ is proper, the compositions $\phi_* \phi^*$ and $\phi^* \phi_*$ are defined on smooth or distributional sections with compact support or no support condition.

The spaces $C^\infty(M'; \phi^* E)$ and $C^\infty(M'; \phi^* E \otimes \Omega_{\mathrm{fiber}})$ become $C^\infty(M)$-modules via the homomorphism of algebras, $\phi^* : C^\infty(M) \to C^\infty(M')$, and we have

$$C^{\pm\infty}_{\cdot/\mathrm{c}}(M'; \phi^* E) = C^{\pm\infty}_{\cdot/\mathrm{c}}(M') \otimes_{C^\infty(M)} C^\infty(M; E) , \tag{2.1.20}$$

$$C^{\pm\infty}_{\cdot/\mathrm{c}}(M'; \phi^* E \otimes \Omega_{\mathrm{fiber}}) = C^{\pm\infty}_{\cdot/\mathrm{c}}(M'; \Omega_{\mathrm{fiber}}) \otimes_{C^\infty(M)} C^\infty(M; E) . \tag{2.1.21}$$

Using (2.1.6), (2.1.20) and (2.1.21), we can describe (2.1.13)–(2.1.19) as the $C^\infty(M)$-tensor products of their trivial-line-bundle versions with the identity map on the space $C^\infty(M; E)$. This kind of description is valid with other spaces of distributional sections with the obvious extensions of (2.1.20) and (2.1.21). Thus, in this chapter, we will mainly consider the pull-back and push-forward between spaces of distributions. Only the special case of the pull-back and push-forward between spaces of currents will be briefly indicated a few times.

2.1.7 Differential Operators

Let $\mathrm{Diff}(M) \subset \mathrm{End}(C^\infty_{\cdot/\mathrm{c}}(M))$ be the subalgebra and $C^\infty(M)$-submodule of differential operators, filtered by the order. Every $\mathrm{Diff}^m(M)$ ($m \in \mathbb{N}_0$) is spanned as $C^\infty(M)$-module by all compositions of up to m elements of $\mathfrak{X}(M)$, considered as the Lie algebra of derivations of $C^\infty_{\cdot/\mathrm{c}}(M)$. In particular, $\mathrm{Diff}^0(M) \equiv C^\infty(M)$.

On the other hand, let

$$P(T^*M) = \bigoplus_{m=0}^{\infty} P^{(m)}(T^*M) \subset C^\infty(T^*M)$$

be the subalgebra and $C^\infty(M)$-submodule of functions whose restriction to the fibers are polynomials, equipped with the grading given by the degree; in particular,

$$P^{(0)}(T^*M) \equiv C^\infty(M) , \quad P^{(1)}(T^*M) \equiv \mathfrak{X}(M) \otimes \mathbb{C} .$$

For every order m, the principal symbol exact sequence

$$0 \to \mathrm{Diff}^{m-1}(M) \hookrightarrow \mathrm{Diff}^m(M) \xrightarrow{\sigma_m} P^{(m)}(T^*M) \to 0 \tag{2.1.22}$$

is defined so that the principal symbol of any $X \in \mathfrak{X}(M) \subset \mathrm{Diff}^1(M)$ is $\sigma_1(X) = iX \in P^{(1)}(T^*M)$, and $\bigoplus_m \sigma_m$ induces an isomorphism of graded algebras and $C^\infty(M)$-modules,

$$\bigoplus_{m=0}^{\infty} \mathrm{Diff}^m(M)/\mathrm{Diff}^{m-1}(M) \xrightarrow{\cong} P(T^*M) .$$

For vector bundles E and F over M, the above concepts can be extended by taking the $C^\infty(M)$-tensor product with $C^\infty(M; F \otimes E^*)$, obtaining

$$\mathrm{Diff}^m(M; E, F) \subset L(C^\infty_{\cdot/\mathrm{c}}(M; E), C^\infty_{\cdot/\mathrm{c}}(M; F)) ,$$

$$P^{(m)}(T^*M; F \otimes E^*) \subset C^\infty(T^*M; \pi^*(F \otimes E^*)) ,$$

where $\pi : T^*M \to M$ is the projection. So $\mathrm{Diff}^0(M; E, F) \equiv C^\infty(M; F \otimes E^*)$. If $E = F$, we write $\mathrm{Diff}(M; E)$, which is a filtered algebra. The principal symbol σ_m on $\mathrm{Diff}^m(M; E, F)$ is given by the $C^\infty(M)$-tensor product of (2.1.22) with the identity map on $C^\infty(M; F \otimes E^*)$. Redundant notation is simplified like in Sect. 2.1.4. Recall that $A \in \mathrm{Diff}^m(M; E, F)$ is *elliptic* if $\sigma_m(A)(p, \xi)$ is an isomorphism for all $p \in M$ and $0 \neq \xi \in T^*_pM$. If E is a line bundle, then [ÁLKL24, Eq. (2.13)]

$$\mathrm{Diff}^m(M; E) \equiv \mathrm{Diff}^m(M) . \tag{2.1.23}$$

For $m = 0$, we get $C^\infty(M; E \otimes E^*) \equiv C^\infty(M)$.

For all $A \in \mathrm{Diff}^m(M; E)$, we have $A^{\mathrm{t}} \in \mathrm{Diff}^m(M; E^* \otimes \Omega)$, and therefore A has continuous extensions to an endomorphism A of $C^{-\infty}_{\cdot/\mathrm{c}}(M; E)$ (Sect. 2.1.5). A similar map is defined when $A \in \mathrm{Diff}^m(M; E, F)$.

The canonical coordinates of $\mathbb{R}^n \times \mathbb{R}^n \equiv \mathbb{R}^n \times \mathbb{R}^{n*} \equiv T^*\mathbb{R}^n$ are denoted by $(x, \xi) = (x^1, \dots, x^n, \xi^1, \dots, \xi^n)$. Let $dx = dx^1 \wedge \cdots \wedge dx^n$, $d\xi = d\xi^1 \wedge \cdots \wedge d\xi^n$, $D^I = D^I_x = (-i)^{|I|}\partial_I = (-i)^{|I|}\partial_{x,I}$ ($i = \sqrt{-1}$) and $\xi^I = \xi^{i_1} \cdots \xi^{i_n}$ ($I = (i_1, \dots, i_n) \in \mathbb{N}_0^n$). For any open $U \subset \mathbb{R}^n$ and $A = \sum_{|I| \le m} a_I(x)\, D^I \in \mathrm{Diff}^m(U)$, write $A = a(x, D)$ for $a(x, \xi) = \sum_{|I| \le m} a_I(x)\, \xi^I$, and then $\sigma_m(A) = \sum_{|I| = m} a_I(x)\, \xi^I$. We have

$$Au(x) = (2\pi)^{-n} \int_{\mathbb{R}^n} e^{i\langle x, \xi \rangle} a(x, \xi) \hat{u}(\xi)\, d\xi , \tag{2.1.24}$$

for all $u \in C^\infty_{\mathrm{c}}(U)$, where $\hat{u}$ is the Fourier transform of u. The local extension of this expression to the case where $A \in \mathrm{Diff}^m(M; E, F)$ is straightforward, using charts of M and local trivializations of E and F, and taking local coefficients a_I with values in $\mathbb{C}^{l'} \otimes \mathbb{C}^{l*} \equiv \mathbb{C}^{l \times l'}$ (l and l' are the ranks of E and F).

2.1.8 Symbols

For any open $U \subset \mathbb{R}^n$ and $l \in \mathbb{N}_0$, a *symbol* of *order* at most $m \in \mathbb{R}$ on $U \times \mathbb{R}^l$, or simply on U, is a function $a \in C^\infty(U \times \mathbb{R}^l)$ such that, for any compact $K \subset U$, $I \in \mathbb{N}_0^n$ and $J \in \mathbb{N}_0^l$,

$$\|a\|_{K,I,J,m} := \sup_{x \in K,\ \xi \in \mathbb{R}^l} \frac{|D_x^I D_\xi^J a(x,\xi)|}{(1+|\xi|)^{m-|J|}} < \infty\,. \tag{2.1.25}$$

They form a Fréchet space $S^m(U \times \mathbb{R}^l)$ with the semi-norms (2.1.25). There are continuous inclusions

$$S^m(U \times \mathbb{R}^l) \subset S^{m'}(U \times \mathbb{R}^l) \quad (m < m')\,, \tag{2.1.26}$$

giving rise to the LCSs

$$S^\infty(U \times \mathbb{R}^l) = \bigcup_m S^m(U \times \mathbb{R}^l)\,, \quad S^{-\infty}(U \times \mathbb{R}^l) = \bigcap_m S^m(U \times \mathbb{R}^l)\,.$$

$S^\infty(U \times \mathbb{R}^l)$ is an LF-space, and therefore barreled, ultrabornological and webbed [ÁLKL24, Proposition 3.1]. It is also a filtered algebra and $C^\infty(U)$-module with the pointwise multiplication. The homogeneous components of the corresponding graded algebra are denoted by $S^{(m)}(U \times \mathbb{R}^l)$. The Fréchet space $S^{-\infty}(U \times \mathbb{R}^l)$ is a filtered ideal and $C^\infty(U)$-submodule of $S^\infty(U \times \mathbb{R}^l)$. The notation $S^m(\mathbb{R}^l)$, $S^{\pm\infty}(\mathbb{R}^l)$ and $S^{(m)}(\mathbb{R}^l)$ is used when $U = \mathbb{R}^0 = \{0\}$.

Consider the first-factor projection $U \times \mathbb{R}^l \to U$ to define $C^\infty_{\text{cv}}(U \times \mathbb{R}^l)$. There are continuous inclusions

$$C^\infty_{\text{cv}}(U \times \mathbb{R}^l) \subset S^{-\infty}(U \times \mathbb{R}^l)\,, \quad S^\infty(U \times \mathbb{R}^l) \subset C^\infty(U \times \mathbb{R}^l)\,; \tag{2.1.27}$$

in particular, $S^\infty(U \times \mathbb{R}^l)$ is Hausdorff. The following properties hold [ÁLKL24, Corollaries 3.4–3.6 and Remark 3.8]: The topologies of $S^\infty(U \times \mathbb{R}^l)$ and $C^\infty(U \times \mathbb{R}^l)$ coincide on $S^m(U \times \mathbb{R}^l)$, however the second inclusion of (2.1.27) is not a TVS-embedding; $C^\infty_{\text{c}}(U \times \mathbb{R}^l)$ is dense in $S^\infty(U \times \mathbb{R}^l)$; and $S^\infty(U \times \mathbb{R}^l)$ is an acyclic Montel space, and therefore complete, boundedly/compactly/sequentially retractive and reflexive.

With more generality, a symbol of order m on a vector bundle E over M is a smooth function on E satisfying (2.1.25) via charts of M and local trivializations of E, with K contained in the domains of charts where E is trivial. As above, they form a Fréchet space $S^m(E)$ with the topology described by the semi-norms given by this version of (2.1.25). The version of (2.1.26) in this setting is true, obtaining the corresponding spaces $S^{\pm\infty}(E)$ and $S^{(m)}(E)$. The above properties have obvious extensions to this setting.

Given another vector bundle F over M, the $C^\infty(M)$-tensor product of the above spaces with $C^\infty(M;F)$ gives spaces $S^m(E;F)$, $S^{\pm\infty}(E;F)$ and $S^{(m)}(E;F)$, satisfying analogous properties. Now (2.1.27) becomes $C^\infty_{\mathrm{cv}}(E;\pi^*F) \subset S^{-\infty}(E;F)$ and $S^\infty(E;F) \subset C^\infty(E;\pi^*F)$, where $\pi : E \to M$ is the projection.

2.1.9 Pseudodifferential Operators

The notation of Sect. 2.1.8 is used here. For any $a \in S^m(U \times \mathbb{R}^n)$, the expression (2.1.24) defines a continuous linear map $A = a(x,D) : C^\infty_{\mathrm{c}}(U) \to C^\infty(U)$, with Schwartz kernel

$$K_A(x,y) = (2\pi)^{-n} \left(\int_{\mathbb{R}^n} e^{i\langle x-y,\xi\rangle} a(x,\xi)\, d\xi \right) |dy| ,$$

using an oscillatory integral, which is defined as a tempered distribution [Mel81, Eq. (4.2)], [Hör83, Section 7.8].

Take an atlas $\{U_k, x_k\}$ of M and an associated C^∞ partition of unity $\{f_k\}$. Via every chart (U_k, x_k), for all $a \in S^m(T^*U_k)$, the above procedure defines a continuous linear map $a(x_k, D_{x_k}) : C^\infty_{\mathrm{c}}(U_k) \to C^\infty(U_k)$.

Let $\Delta \subset M^2$ be the diagonal. A *pseudodifferential operator* of *order* at most m on M is a continuous linear map $A : C^\infty_{\mathrm{c}}(M) \to C^\infty(M)$ such that K_A is C^∞ on $M^2 \setminus \Delta$, and, for all k, the operator $f_k A : C^\infty_{\mathrm{c}}(U_k) \to C^\infty_{\mathrm{c}}(U_k)$ is of the form $a_k(x_k, D_{x_k})$ for some $a_k \in S^m(T^*U_k)$, which is supported in $\pi^{-1}(\operatorname{supp} f_k)$, where $\pi : T^*M \to M$ is the projection. They form a $C^\infty(M^2)$-module $\Psi^m(M)$ with the pointwise multiplication of their Schwartz kernels by smooth functions on M^2. Moreover $\sum_k a_k \in S^m(T^*M)$ defines a class $\sigma_m(A) \in S^{(m)}(T^*M)$, called the *principal symbol*, which is independent of the choices involved, obtaining an exact sequence of $C^\infty(M^2)$-modules,

$$0 \to \Psi^{m-1}(M) \hookrightarrow \Psi^m(M) \xrightarrow{\sigma_m} S^{(m)}(T^*M) \to 0 ,$$

where $S^{(m)}(T^*M)$ is a $C^\infty(M^2)$-module via the restriction linear map $C^\infty(M^2) \to C^\infty(\Delta) \equiv C^\infty(M)$. Then $\Psi(M) := \bigcup_m \Psi^m(M)$ is a filtered $C^\infty(M^2)$-module, and $\Psi^{-\infty}(M) := \bigcap_m \Psi^m(M)$ is the submodule of the operators with C^∞ Schwartz kernel (the *smoothing* operators). All of these concepts are independent of the choices involved. If $m \in \mathbb{N}_0$, then

$$\operatorname{Diff}^m(M) = \{ A \in \Psi^m(M) \mid \operatorname{supp} K_A \subset \Delta \} .$$

These concepts and properties can be extended to vector bundles by taking the $C^\infty(M^2)$-tensor product with $C^\infty(M^2; F \boxtimes E^*)$, like in the case of differential operators (Sect. 2.1.7). In this case, we use the notation $\Psi^m(M; E, F)$ (or $\Psi^m(M; E)$ if $E = F$), $S^{(m)}(T^*M; F \otimes E^*)$, etc. Recall that an operator $A \in \Psi^m(M; E, F)$ is called *elliptic* if $\sigma_m(A)$ has an inverse in $S^{(-m)}(T^*M; F, E)$; i.e., any representative of $\sigma_m(A)$ is an isomorphism at $(p, \xi) \in T^*M$ if ξ is far enough from 0_p in T_p^*M. The space $\Psi^m(M; E, F)$ is preserved by taking transposes. Thus any $A \in \Psi^m(M; E, F)$ has a continuous extension (Sect. 2.1.5)

$$A : C_c^{-\infty}(M; E) \to C^{-\infty}(M; F) ,$$

and $\operatorname{sing\,supp} Au \subset \operatorname{sing\,supp} u$ for all $u \in C_c^{-\infty}(M; E)$ (*pseudolocality*). Moreover $A \in \Psi^{-\infty}(M; E, F)$ just when it defines a continuous map

$$A : C_c^{-\infty}(M; E) \to C^\infty(M; F) .$$

It is said that A is *properly supported* if both factor projections $M^2 \to M$ have proper restrictions to $\operatorname{supp} K_A$. In this case, A defines continuous linear maps (Sect. 2.1.5)

$$A : C_c^\infty(M; E) \to C_c^\infty(M; F) , \quad A : C^{-\infty}(M; E) \to C^{-\infty}(M; F) ,$$

which gives sense to the composition of properly supported pseudodifferential operators. Any pseudodifferential operator is properly supported modulo smoothing operators, and the symbol map is multiplicative.

If $A \in \Psi^{-\infty}(M; E)$ and $P, Q \in \operatorname{Diff}(M; E)$, then

$$K_{PAQ}(x, y) = P_x\, Q^{\mathrm{t}}_y K_A(x, y) . \tag{2.1.28}$$

2.1.10 L^2 and L^∞ Sections

The Hilbert space $L^2(M; \Omega^{1/2})$ of square-integrable half-densities is the completion of $C_c^\infty(M; \Omega^{1/2})$ with the scalar product $\langle u, v\rangle = \int_M u\bar{v}$. The induced norm is denoted by $\|\cdot\|$.

If M is compact, $L^2(M; E)$ can be described as the $C^\infty(M)$-tensor product of $L^2(M; \Omega^{1/2})$ and $C^\infty(M; \Omega^{-1/2} \otimes E)$. It is a Hilbertian space with the scalar products $\langle u, v\rangle = \int_M (u, v)\,\omega$, determined by the choice of a Euclidean/Hermitian structure $(\cdot, \cdot)$ on E and a non-vanishing $\omega \in C^\infty(M; \Omega)$.

When M is not assumed to be compact, any choice of $(\cdot, \cdot)$ and ω can be used to define $L^2(M; E)$ and $\langle \cdot, \cdot\rangle$. Now $L^2(M; E)$ and the equivalence class of $\|\cdot\|$ depends on the choices involved. The independence still holds for sections supported in any compact $K \subset M$, obtaining the Hilbertian space $L^2_K(M; E)$. These spaces give rise to the strict LF-space $L^2_c(M; E)$ like in (2.1.3). We also get the Fréchet space

$$L^2_{\mathrm{loc}}(M; E) = \{ u \in C^{-\infty}(M; E) \mid C^\infty_c(M)\, u \subset L^2_c(M; E) \} ,$$

defining the topology like in (2.1.1). If M is compact, then $L^2_{\mathrm{loc/c}}(M; E) \equiv L^2(M; E)$ as TVSs. The spaces $L^2_{\mathrm{loc/c}}(M; E)$ satisfy the obvious version of (2.1.5).

Any $A \in \mathrm{Diff}^m(M; E)$ can be considered as a densely defined operator in $L^2(M; E)$. Its adjoint A^* is the closure of the *formal adjoint* $A^* \in \mathrm{Diff}^m(M; E)$, determined by the condition $\langle u, A^* v\rangle = \langle Au, v\rangle$ for all $u, v \in C^\infty_c(M; E)$.

We can also use $(\cdot, \cdot)$ to define the Banach space $L^\infty(M; E)$ of essentially bounded sections, with the norm $\|u\|_{L^\infty} = \operatorname{ess\,sup}_{x\in M} |u(x)|$. There is a continuous injection $L^\infty(M; E) \subset L^2_{\mathrm{loc}}(M; E)$. If M is compact, the equivalence class of $\|\cdot\|_{L^\infty}$ is independent of $(\cdot, \cdot)$.

2.1.11 Sobolev Spaces

Suppose first that M is compact. The *Sobolev space* of *order* $s \in \mathbb{R}$ is the Hilbertian space

$$H^s(M; E) = \{ u \in C^{-\infty}(M; E) \mid \Psi^s(M; E)\, u \subset L^2(M; E) \} , \tag{2.1.29}$$

with the topology like in (2.1.1). It can be equipped with any scalar product $\langle u, v\rangle_s = \langle (1 + P)^s u, v\rangle$, for any nonnegative symmetric elliptic $P \in \mathrm{Diff}^2(M; E)$ (by the elliptic estimate), where $\langle \cdot, \cdot\rangle$ is defined like in Sect. 2.1.10 and $(1 + P)^s$ is given by the spectral theorem. Let $\|\cdot\|_s$ denote the corresponding norm. We have

$$\Psi^s(M; E)\, L^2(M; E) = H^{-s}(M; E) = H^s(M; E^* \otimes \Omega)' . \tag{2.1.30}$$

If $s \in \mathbb{N}$, we can use $\mathrm{Diff}^s(M; E)$ instead of $\Psi^s(M; E)$ in (2.1.29) and the first equality of (2.1.30). There are dense compact inclusions (Rellich theorem)

$$H^s(M; E) \subset H^{s'}(M; E) \quad (s' < s) . \tag{2.1.31}$$

So the spaces $H^s(M, E)$ form a compact spectrum. Moreover, there are continuous dense inclusions, for $s > k + n/2$,

$$H^s(M; E) \subset C^k(M; E) \subset H^k(M; E) , \tag{2.1.32}$$

$$H^{-s}(M; E) \supset C'^{-k}(M; E) \supset H^{-k}(M; E) . \tag{2.1.33}$$

The first inclusion of (2.1.32) is the Sobolev embedding theorem, and (2.1.33) is the transpose of the version of (2.1.32) with $E^* \otimes \Omega M$. So

$$C^{\infty}(M;E)=\bigcap_{s}H^{s}(M;E)\quad C^{-\infty}(M;E)=\bigcup_{s}H^{s}(M;E)\,. \tag{2.1.34}$$

Any $A \in \Psi^m(M;E)$ defines a bounded operator $A : H^{s+m}(M;E) \to H^s(M;E)$. This can be considered as a densely defined operator in $H^s(M;E)$, which is closable because, after fixing a scalar product in $H^s(M;E)$, the adjoint of A in $H^s(M;E)$ is densely defined since it is induced by $\bar{A}^{\mathrm{t}} \in \Psi^m(M;\bar{E}^*\otimes\Omega)$ via the identity of real Hilbert spaces, $H^s(M;E)\equiv H^s(M;\bar{E})'=H^{-s}(M;\bar{E}^*\otimes\Omega)$, where the bar stands for the complex conjugate. In the case $s=0$, the adjoint of A is induced by the formal adjoint $A^*\in\Psi^m(M;E)$; if $A\in\mathrm{Diff}^m(M;E)$, then $A^*\in\mathrm{Diff}^m(M;E)$.

If M is not assumed to be compact, then $H^s(M;E)$ can be defined as the completion of $C^\infty_{\mathrm{c}}(M;E)$ with respect to the scalar product $\langle\cdot,\cdot\rangle_s$ defined by the above choices of $(\cdot,\cdot)$, ω and P; in this case, $H^s(M;E)$ and the equivalence class of $\|\cdot\|_s$ depend on the choices involved. With this generality, (2.1.29) and the first equality of (2.1.30) are wrong, but the second equality of (2.1.30) is true.

Like $L^2_{\mathrm{loc/c}}(M;E)$ (Sect. 2.1.10), we can define the Fréchet space $H^s_{\mathrm{loc}}(M;E)$ and the strict LF-space $H^s_{\mathrm{c}}(M;E)$, which satisfy the versions of the second equality of (2.1.30) (switching the support condition like in (2.1.5)) and (2.1.31)–(2.1.33). These spaces agree with $H^s(M;E)$ if M is compact. For any open $U\subset M$, the restriction map (2.1.11) defines a continuous linear map $H^s_{\mathrm{loc}}(M;E)\to H^s_{\mathrm{loc}}(U;E)$, and the extension by zero (2.1.12) defines a TVS-embedding $H^s_{\mathrm{c}}(U;E)\subset H^s_{\mathrm{c}}(M;E)$. In this case, any $A\in\Psi^m(M;E)$ defines continuous linear maps $A: H^s_{\mathrm{c}}(M;E)\to H^{s-m}_{\mathrm{loc}}(M;E)$. If $A\in\mathrm{Diff}^m(M;E)$, then it defines continuous linear maps $A: H^s_{\mathrm{c/loc}}(M;E)\to H^{s-m}_{\mathrm{c/loc}}(M;F)$.

For example, $H^s(\mathbb{R}^n)$ can be defined with $\langle u,v\rangle_s=\langle(1+\Delta)^s u,v\rangle$, involving the Laplacian $\Delta=-\sum_k\partial_k^2$ and the standard scalar product on $L^2(\mathbb{R}^n)$. Recall that the Fourier transform, $f\mapsto\hat{f}$, defines an automorphism of the Schwartz space $\mathcal{S}(\mathbb{R}^n)$, which extends to an automorphism of the space $\mathcal{S}(\mathbb{R}^n)'$ of tempered distributions [Hör83, Section 7.1], which in turn restricts to a TVS-isomorphism

$$H^s(\mathbb{R}^n)\xrightarrow{\cong}L^2(\mathbb{R}^n,(1+|\xi|^2)^s\,d\xi)\,,\quad f\mapsto\hat{f}\,. \tag{2.1.35}$$

We can use (2.1.35) to give an alternative description of $H^s_{\mathrm{c/loc}}(M;E)$ for arbitrary M and E. First, $H^s_K(\mathbb{R}^n)\subset H^s(\mathbb{R}^n)$ has the subspace topology for any compact $K\subset\mathbb{R}^n$. Next, for any open $U\subset\mathbb{R}^n$, we can describe $H^s_{\mathrm{c}}(U)$ by using $H^s_K(U)\equiv H^s_K(\mathbb{R}^n)$ for all compact $K\subset U$, and we can describe $H^s_{\mathrm{loc}}(U)$ by using $H^s_{\mathrm{c}}(U)$, as explained before. Then a locally finite atlas and a subordinated C^∞ partition of unity can be used in a standard way to describe $H^m_{\mathrm{c/loc}}(M)$. Finally, $H^s_{\mathrm{c/loc}}(M;E)$ can be described as the $C^\infty(M)$-tensor product of $H^s_{\mathrm{c/loc}}(M)$ with $C^\infty(M;E)$, or, equivalently, using local diffeomorphisms of triviality of E.

The norm on $L(H^m(M;E),H^{m'}(M;F))$ (resp., $\mathrm{End}(H^m(M;E))$) will be simply denoted by $\|\cdot\|_{m,m'}$ (resp., $\|\cdot\|_m$).

2.1.12 Weighted Spaces

Assume first that M is compact. Take any $h \in C^\infty(M)$ which is positive almost everywhere. Then the *weighted Sobolev space* $hH^s(M; E)$ is a Hilbertian space; a scalar product $\langle \cdot, \cdot \rangle_{hH^s}$ is given by $\langle u, v \rangle_{hH^s} = \langle h^{-1}u, h^{-1}v \rangle_s$, depending on the choice of a scalar product $\langle \cdot, \cdot \rangle_s$ on $H^s(M; E)$ (Sect. 2.1.11). The corresponding norm is denoted by $\|\cdot\|_{hH^s}$. In particular, we get the *weighted L^2 space* $hL^2(M; E)$. We have $h > 0$ just when $hH^s(M; E) = H^s(M; E)$; in this case, $\langle \cdot, \cdot \rangle_{hH^s}$ can be described like $\langle \cdot, \cdot \rangle_s$ using $h^{-2}\omega$ and hPh^{-1} instead of ω and P. Thus the notation $hH^s(M; E)$ for $h > 0$ is used when changing the density; e.g., if it is different from a distinguished choice, say a Riemannian volume.

If M is not compact, $hH^s(M; E)$ and $\langle u, v \rangle_{hH^s}$ depend on h and the chosen definitions of $H^s(M; E)$ and $\langle u, v \rangle_s$ (Sect. 2.1.11). We also get the weighted spaces $hH^s_{\text{c/loc}}(M; E)$, and the weighted Banach space $hL^\infty(M; E)$ with the norm $\|u\|_{hL^\infty} = \|h^{-1}u\|_{L^\infty}$. There is a continuous injection $hL^\infty(M; E) \subset hL^2_{\text{loc}}(M; E)$.

2.1.13 Topological Complexes

Recall that a complex (C, d) (over $\mathbb{C}$) consists of a ($\mathbb{Z}$-) graded vector space $C = C^\bullet$ and a linear map $d : C \to C$ which is homogeneous of degree 1 and satisfies $d^2 = 0$. If moreover C is a TVS and d is continuous, then (C, d) is called a *topological complex*. Then $\ker d$ and $\operatorname{im} d$ are topological graded subspaces, and the cohomology $H^\bullet(C, d) = \ker d / \operatorname{im} d$ becomes a graded TVS. Its maximal Hausdorff quotient, $\bar{H}^\bullet(C, d) := H^\bullet(C, d)/\overline{0} \equiv \ker d / \overline{\operatorname{im} d}$, is called the *reduced cohomology*. Let $[u] \in H^\bullet(C, d)$ and $\overline{[u]} \in \bar{H}^\bullet(C, d)$ denote the elements defined by any $u \in \ker d$. If C is a LCS, then $H^\bullet(C, d)$ and $\bar{H}^\bullet(C, d)$ are also LCSs because this property is inherited by subspaces and quotients [Sch71, Section II.4]. We may use the notation $Z = ZC = \ker d$, $B = BC = \operatorname{im} d$ and $\bar{B} = \bar{B}C = \overline{\operatorname{im} d}$.

We always assume C has finitely many nonzero homogeneous components, say $C = C^0 \oplus \cdots \oplus C^N$. So d is given by a finite sequence of *length* N,

$$C^0 \xrightarrow{d_0} C^1 \xrightarrow{d_1} \cdots \xrightarrow{d_{N-1}} C^N .$$

Negative or decreasing degrees may be also considered without any essential change. Continuous homomorphisms between topological complexes induce continuous linear maps between the corresponding cohomologies and reduced cohomologies. (Usually, the term *chain/cochain complex* is used for decreasing/increasing degrees, and *chain/cochain maps* for the corresponding homomorphisms, but we will ignore that difference.)

The *transpose* of (C, d) is the topological complex (C', d^{t}), graded by $(C')^r = (C^r)'$ $(r = 0, \ldots, N)$. For any $[f] \in H^\bullet(C', d^{\mathrm{t}})$, we have $fd = d^{\mathrm{t}}(f) = 0$, and therefore f induces an element of $H^\bullet(C, d)'$. This defines a canonical continuous linear map $H^\bullet(C', d^t) \to H^\bullet(C, d)'$.

Proposition 2.1.1 *The canonical map $H^\bullet(C', d^t) \to H^\bullet(C, d)'$ is:*

(i) *surjective if C is a LCHS; and*
(ii) *injective if C is a Fréchet space and* $\operatorname{im} d$ *is closed.*

Proof Property (i) is an easy consequence of the Hahn-Banach theorem [Sch71, Theorem II.4.2].

Property (ii) follows easily from the open mapping theorem [Sch71, Theorem III.2.1] and the Hahn-Banach theorem. □

Remark 2.1.2 Extensions of (ii) can be given by more general versions of the open mapping theorem (see e.g. [Bou14]).

2.1.14 *Differential Complexes*

Recall that a *differential complex* of *order* at most m is a topological complex of the form $(C^\infty(M; E), d)$, where E is a ($\mathbb{Z}$-) graded vector bundle and $d \in \operatorname{Diff}^m(M; E)$; it will be simply denoted by (E, d). Necessarily, it is of finite length, say $E = E^0 \oplus \cdots \oplus E^N$ and d is given by the sequence

$$C^\infty(M; E^0) \xrightarrow{d_0} C^\infty(M; E^1) \xrightarrow{d_1} \cdots \xrightarrow{d_{N-1}} C^\infty(M; E^N) .$$

The compactly supported version $(C^\infty_c(M; E), d)$ may be also considered, as well as the distributional versions $(C^{-\infty}_{\cdot/c}(M; E), d)$. Recall that (E, d) is called an *elliptic complex* of *order* m if moreover the symbol sequence,

$$0 \to E^0_p \xrightarrow{\sigma_m(d_0)(p,\xi)} E^1_p \xrightarrow{\sigma_m(d_1)(p,\xi)} \cdots \xrightarrow{\sigma_m(d_{N-1})(p,\xi)} E^N_p \to 0 , \tag{2.1.36}$$

is exact for all $p \in M$ and $0 \neq \xi \in T^*_pM$. If $N = 1$, this agrees with the ellipticity of $d_0 \in \operatorname{Diff}^m(M; E^0, E^1)$.

Equip E with a Hermitian structure so that its homogeneous components are orthogonal, and equip M with a Riemannian metric g, inducing a volume density on M. Consider the corresponding scalar product on $L^2(M; E)$. Then the formal adjoint $\delta = d^*$ also defines a differential complex, giving rise to symmetric differential operators $D = d + \delta$ and $\Delta = D^2 = d\delta + \delta d$. The ellipticity of the differential complex d is equivalent to the ellipticity of the differential complex δ, and it is also equivalent to the ellipticity of the differential operator D (or Δ).

In the rest of Sect. 2.1.14, suppose M is closed and d is elliptic. Then D and Δ have a discrete spectrum. Moreover, we have the following Hodge-type decomposition, and associated equalities and isomorphism:

$$\begin{cases} C^\infty(M;E) = \ker\Delta \oplus \operatorname{im}\delta \oplus \operatorname{im} d\,, \\ \operatorname{im}\delta \oplus \operatorname{im} d = \operatorname{im} D = \operatorname{im}\Delta\,, \\ \ker d \cap \ker\delta = \ker D = \ker\Delta \cong H^\bullet(C^\infty(M;E),d)\,. \end{cases} \tag{2.1.37}$$

Writing $C = C^\infty(M;E)$, it follows from (2.1.37) that $d : \operatorname{im}\delta \to \operatorname{im} d$ and $\delta : \operatorname{im} d \to \operatorname{im}\delta$ are TVS-isomorphisms.

Consider also the operators d, δ, D and Δ on $C^{-\infty}(M;E)$ (Sect. 2.1.7). Then $(C^{-\infty}(M;E),d)$ is another topological complex, and the analogue of (2.1.37) is satisfied with $C^{-\infty}(M;E)$. By ellipticity and since M is compact, Δ has the same kernel in $C^\infty(M;E)$ and in $C^{-\infty}(M;E)$, obtaining a canonical isomorphism $H^\bullet(C^\infty(M;E),d) \cong H^\bullet(C^{-\infty}(M;E),d)$.

2.2 Conormal Distributions

The space of conormal distributions plays a very important role in our work. We mainly follow [KN65, Hör71], [Hör85, Section 18.2], [Sim90, Chapters 3–5], [Mel96, Chapters 4 and 6], [MU08, Chapters 3 and 9], which are oriented to the role they play in pseudodifferential operators and generalizations of those operators. The study of its natural topology was begun in [Mel96, Chapters 4 and 6] and continued in [ÁLKL24].

For the sake of simplicity, we consider the case of the trivial line bundle first. But all definitions, properties and notation have obvious extensions for arbitrary vector bundles, like in Sects. 2.1.7 and 2.1.9, either by using local trivializations, or by taking $C^\infty(M)$-tensor products with spaces of smooth sections. When needed, the case of arbitrary vector bundles will be used without further comment.

2.2.1 *Differential Operators Tangent to a Submanifold*

Let L be a regular submanifold of M of codimension n' and dimension n'', which is a closed subset. Let $\mathfrak{X}(M,L) \subset \mathfrak{X}(M)$ be the Lie subalgebra and $C^\infty(M)$-submodule of vector fields tangent to L. Using $\mathfrak{X}(M,L)$ instead of $\mathfrak{X}(M)$, we can define the filtered subalgebra and $C^\infty(M)$-submodule $\operatorname{Diff}(M,L) \subset \operatorname{Diff}(M)$ like in Sect. 2.1.7. We have

$$A \in \operatorname{Diff}(M,L) \Rightarrow A^{\mathrm{t}} \in \operatorname{Diff}(M,L;\Omega)\,. \tag{2.2.1}$$

By the conditions on L, every $\operatorname{Diff}^m(M,L)$ ($m \in \mathbb{N}_0$) is locally finitely $C^\infty(M)$-generated, and therefore $\operatorname{Diff}(M,L)$ is countably $C^\infty(M)$-generated. The surjective restriction map $\mathfrak{X}(M,L) \to \mathfrak{X}(L)$, $X \mapsto X|_L$, induces a surjective linear restriction map of filtered algebras and $C^\infty(M)$-modules,

$$\mathrm{Diff}(M,L) \to \mathrm{Diff}(L)\,, \quad A \mapsto A|_L\,. \tag{2.2.2}$$

Let (U,x) be a chart of M adapted to L; i.e., it is a diffeomorphism

$$x = (x^1,\dots,x^n) \equiv (x',x'') : U \to U' \times U''\,,$$
$$x' = (x'^1,\dots,x'^{n'})\,, \quad x'' = (x''^1,\dots,x''^{n''})\,, \quad L_0 := L \cap U = \{x'=0\}\,,$$

for some open $U' \subset \mathbb{R}^{n'}$ and $U'' \subset \mathbb{R}^{n''}$. If L is of codimension one, then we will use the notation (x,y) instead of (x',x''). For every $m \in \mathbb{N}_0$, $\mathrm{Diff}^m(U,L_0)$ is $C^\infty(U)$-spanned by the operators $x'^I\partial_{x'}^J\partial_{x''}^K$ with $|J|+|K| \le m$ and $|I| = |J|$; we may use the generators $\partial_{x'}^J\partial_{x''}^K x'^I$ as well, with the same conditions on the multi-indices.

2.2.2 Conormal Distributions when M Is Compact

Suppose M is compact. Then the space of *conormal distributions* at L of *Sobolev order* at most $s \in \mathbb{R}$ is the LCS and $C^\infty(M)$-module

$$I^{(s)}(M,L) = \{\, u \in C^{-\infty}(M) \mid \mathrm{Diff}(M,L)\,u \subset H^s(M)\,\}\,, \tag{2.2.3}$$

with the topology like in (2.1.1). This is a totally reflexive Fréchet space [ÁLKL24, Proposition 4.1]. We have continuous inclusions

$$I^{(s)}(M,L) \subset I^{(s')}(M,L) \quad (s' < s)\,, \tag{2.2.4}$$

and consider the LCSs and $C^\infty(M)$-modules

$$I(M,L) = \bigcup_s I^{(s)}(M,L)\,, \quad I^{(\infty)}(M,L) = \bigcap_s I^{(s)}(M,L)\,.$$

Thus $I(M,L)$ is a Hausdorff LF-space (Sect. 2.1.1), and $I^{(\infty)}(M,L)$ is a Fréchet space and submodule of $I(M,L)$. The elements of $I(M,L)$ are called *conormal distributions* of M at L (or of (M,L)). The spaces $I^{(s)}(M,L)$ form what is called the *Sobolev-order filtration* of $I(M,L)$, or the *Sobolev-order inductive spectrum* defining $I(M,L)$. From (2.2.3), it follows that there are canonical continuous inclusions,

$$C^\infty(M) \subset I^{(\infty)}(M,L)\,, \quad I(M,L) \subset C^{-\infty}(M)\,. \tag{2.2.5}$$

Indeed, $C^\infty(M)$ is dense in $I(M,L)$ [Mel96, Eq. (6.2.12)], [ÁLKL24, Corollary 4.6].

$I(M, L)$ is barreled, ultrabornological, webbed, acyclic and a Montel space, and therefore complete, boundedly/compactly/sequently retractive and reflexive [ÁLKL24, Corollaries 4.2 and 4.7] (Sect. 2.1.1).

2.2.3 Filtration of $I(M, L)$ by the Symbol Order when M Is Compact

Take a chart of M adapted to L, $(U, x = (x', x''))$, like in Sect. 2.2.1. We use the identity $U'' \times \mathbb{R}^{n'} \equiv N^*U''$, and the symbol spaces $S^m(U'' \times \mathbb{R}^{n'}) \equiv S^m(N^*U'')$ (Sect. 2.1.8). The following holds true for $s, \bar{m} \in \mathbb{R}$ [Hör85, Theorem 18.2.8], [Mel96, Proposition 6.1.1], [MU08, Lemma 9.33], [ÁLKL24, Remark 4.4]:

- If $s < -\bar{m} - n'/2$, then the map $C^\infty_{\mathrm{cv}}(N^*U'') \to C^\infty(U)$, $a \mapsto u$, given by
$$u(x) = (2\pi)^{-n'} \int_{\mathbb{R}^{n'}} e^{i\langle x', \xi\rangle} a(x'', \xi)\, d\xi \;,$$
has a continuous extension $S^{\bar{m}}(N^*U'') \to I^{(s)}(U, L_0)$.
- If $\bar{m} > -s - n'/2$, then the map $C^\infty_{\mathrm{c}}(U) \to C^\infty(N^*U'')$, $u \mapsto a$, given by
$$a(x'', \xi) = \int_{\mathbb{R}^{n'}} e^{-i\langle x', \xi\rangle} u(x', x'')\, dx' \;,$$
induces a continuous linear map $I^{(s)}_{\mathrm{c}}(U, L_0) \to S^{\bar{m}}(N^*U'')$.

In what follows, it is convenient to use
$$a\,|d\xi| \in S^{\bar{m}}(N^*U''; \Omega N^*U'') \equiv S^{\bar{m}}(N^*L_0; \Omega N^*L_0) \;.$$

Assume M is compact. Take a finite cover of L by relatively compact charts (U_j, x_j) of M adapted to L, and write $L_j = L \cap U_j$. Let $\{h, f_j\}$ be a C^∞ partition of unity of M subordinated to the open covering $\{M \setminus L, U_j\}$. Then $I(M, L)$ consists of the distributions $u \in C^{-\infty}(M)$ such that $hu \in C^\infty(M \setminus L)$ and $f_j u \in I_{\mathrm{c}}(U_j, L_j)$ for all j. Every $f_j u$ is given by some $a_j \in S^\infty(N^*L_j; \Omega N^*L_j)$ as above. For

$$\bar{m} = m + n/4 - n'/2 \;, \tag{2.2.6}$$

the condition $a_j \in S^{\bar{m}}(N^*L_j; \Omega N^*L_j)$ describes the elements u of a $C^\infty(M)$-submodule $I^m(M, L) \subset I(M, L)$, which is independent of the choices involved [MU08, Proposition 9.33] (see also [Mel96, Definition 6.2.19] and [Sim90, Definition 4.3.9]). Moreover, applying the versions of semi-norms (2.1.2) on $C^\infty(M \setminus L)$ to hu and the versions of semi-norms (2.1.25) on $S^{\bar{m}}(N^*L_j; \Omega N^*L_j)$ to every a_j, we get semi-norms on $I^m(M, L)$, which becomes a Fréchet space [Mel96, Sections 6.2 and 6.10].

The version of (2.1.26) for the spaces $S^{\bar{m}}(N^*L_j; \Omega N^*L_j)$ gives continuous inclusions

$$I^m(M, L) \subset I^{m'}(M, L) \quad (m < m') . \tag{2.2.7}$$

The element $\sigma_m(u) \in S^{(\bar{m})}(N^*L; \Omega N^*L)$ represented by $\sum_j a_j \in S^{\bar{m}}(N^*L; \Omega N^*L)$ is called the *principal symbol* of u. This defines the exact sequence

$$0 \to I^{m-1}(M, L) \hookrightarrow I^m(M, L) \xrightarrow{\sigma_m} S^{(\bar{m})}(N^*L; \Omega N^*L) \to 0 .$$

We also get continuous inclusions

$$I^{(-m-n/4+\epsilon)}(M, L) \subset I^m(M, L) \subset I^{(-m-n/4-\epsilon)}(M, L) , \tag{2.2.8}$$

for all $m \in \mathbb{R}$ and $\epsilon > 0$ (cf. [Mel96, Eq. (6.2.5)], [MU08, Eq. (9.35)]). So

$$I(M, L) = \bigcup_m I^m(M, L) , \quad I^{(\infty)}(M, L) = I^{-\infty}(M, L) := \bigcap_m I^m(M, L) .$$

The spaces $I^m(M, L)$ form what is called the *symbol-order filtration* of $I(M, L)$, or the *symbol-order inductive spectrum* defining $I(M, L)$.

2.2.4 $I(M, L)$ for Non-compact M

If M is not assumed to be compact, the spaces and properties of Sects. 2.2.2 and 2.2.3 can be extended as follows [ÁLKL24, Sections 4.2.2 and 4.3.3].

We can similarly define the LCHS $I^{(s)}_{\cdot/\mathrm{c}}(M, L)$ by using $C^{-\infty}_{\cdot/\mathrm{c}}(M)$ and $H^s_{\mathrm{loc/c}}(M)$. Every $I^{(s)}(M, L)$ is a Fréchet space. We can describe $I^{(s)}_{\mathrm{c}}(M, L) = \bigcup_K I^{(s)}_K(M, L)$ like in (2.1.3), which is a strict LF-space, and therefore $I_{\mathrm{c}}(M, L) = \bigcup_s I^{(s)}_{\mathrm{c}}(M, L)$ is an LF-space; moreover $I_{\mathrm{c}}(M, L) = \bigcup_K I_K(M, L)$. We also have the LCHS $I^{(\infty)}_{\mathrm{c}}(M, L) = \bigcap_s I^{(s)}_{\mathrm{c}}(M, L)$. All of these spaces are modules over $C^\infty(M)$; $I_{\mathrm{c}}(M, L)$ is a filtered module and $I^{(\infty)}_{\mathrm{c}}(M, L)$ a submodule. The extension by zero defines a continuous inclusion $I_{\mathrm{c}}(U, L \cap U) \subset I_{\mathrm{c}}(M, L)$ for any open $U \subset M$. We also define the space $I^{(\infty)}(M, L)$ like in the compact case, as well as the space $\bigcup_s I^{(s)}(M, L)$, which consists of the conormal distributions with a Sobolev order. But now let (cf. [Hör85, Definition 18.2.6])

$$I(M, L) = \{ u \in C^{-\infty}(M) \mid C^\infty_{\mathrm{c}}(M)\, u \subset I_{\mathrm{c}}(M, L) \} , \tag{2.2.9}$$

which is a LCS with the topology like in (2.1.1). We have $I(M, L) = \bigcup_s I^{(s)}(M, L)$ if and only if L is compact; thus the spaces $I^{(s)}(M, L)$ form a filtration of $I(M, L)$ just when L is compact. There is an extension of (2.2.5) for non-compact M, taking arbitrary/compact support; in particular, $I_{\cdot/c}(M, L)$ is Hausdorff. The density of the smooth functions with arbitrary/compact support is also true.

The definition of $I^m(M, L)$ can be immediately extended assuming $\{U_j\}$ is locally finite. We can similarly define $I^m_K(M, L)$ for all compact $K \subset M$, and then define $I^m_c(M, L)$ like in (2.1.3). The space of conormal distributions with a symbol order is $\bigcup_m I^m(M, L)$, and let $I^{-\infty}_{\cdot/c}(M, L) = \bigcap_m I^m_{\cdot/c}(M, L)$. There are extensions of (2.2.7) and (2.2.8). So $\bigcup_m I^m(M, L) = \bigcup_s I^{(s)}(M, L)$, $I_c(M, L) = \bigcup_m I^m_c(M, L)$ and $I^{(\infty)}_{\cdot/c}(M, L) = I^{-\infty}_{\cdot/c}(M, L)$. $\bigcup_m I^m(M, L)$ and $I_{\cdot/c}(M, L)$ are acyclic Montel spaces, and $I(M, L)$ is a Montel space.

If M is the domain of a given smooth submersion, the LCHS $I_{cv}(M; E)$ can be defined like $C^{-\infty}_{cv}(M; E)$, using $I_c(M; E)$ instead of $C^{-\infty}_c(M; E)$.

2.2.5 *Pseudodifferential Operators vs Conormal Distributions*

Using the diagonal $\Delta \subset M^2$, the Schwartz kernel isomorphism (2.1.9) restricts to linear isomorphisms

$$\Psi^m(M; E, F) \xrightarrow{\cong} I^m(M^2, \Delta; F \boxtimes (E^* \otimes \Omega M)) , \quad A \mapsto K_A ,$$

and a similar one for the whole of $\Psi(M; E, F)$. Via them, $\Psi^m(M; E, F)$ and $\Psi(M; E, F)$ become LCHSs satisfying the properties of the corresponding spaces of conormal distributional sections. In this case, we have $\bar{m} = m$ in (2.2.6) and $\sigma_m(A) \equiv \sigma_m(K_A)$ for any $A \in \Psi^m(M; E, F)$ [Hör65, KN65], [Hör85, Chapter XVIII], [Sim90, Chapter 6].

2.2.6 *Dirac Sections at Submanifolds*

We have $\Omega NL \otimes \Omega L \equiv \Omega_L M$. The transpose of the restriction map $C^\infty_{c/\cdot}(M; E \otimes \Omega M) \to C^\infty_{c/\cdot}(L; E \otimes \Omega_L M)$ is a continuous inclusion

$$C^{-\infty}_{\cdot/c}(L; E \otimes \Omega^{-1} NL) \subset C^{-\infty}_{\cdot/c}(M; E) , \tag{2.2.10}$$

$$u \mapsto \delta^u_L , \quad \langle \delta^u_L, v \rangle = \langle u, v|_L \rangle , \quad v \in C^\infty_{c/\cdot}(M; E^* \otimes \Omega) .$$

By restriction of (2.2.10), we get a continuous inclusion [GS77, p. 310],

$$C^\infty_{\cdot/c}(L; E \otimes \Omega^{-1} NL) \subset C^{-\infty}_{\cdot/c}(M; E) ; \tag{2.2.11}$$

in this case, we can write $\langle \delta^u_L, v \rangle = \int_L u\, v|_L$. This is the subspace of *δ-sections* or *Dirac sections* at L. Actually, the inclusion (2.2.11) induces a continuous injection

[ÁLKL24, Corollary 4.9]

$$C^{\infty}_{\cdot/\mathrm{c}}(L; E \otimes \Omega^{-1}NL) \subset H^{s}_{\mathrm{loc/c}}(M; E) \quad (s < -n'/2)\,, \tag{2.2.12}$$

with

$$C^{\infty}_{\cdot/\mathrm{c}}(L; E \otimes \Omega^{-1}NL) \cap H^{-n'/2}_{\mathrm{loc/c}}(M; E) = 0\,.$$

For instance, for any $p \in M$ and $u \in E_p \otimes \Omega_p^{-1}M$, we get $\delta_p^u \in H^s_{\mathrm{c}}(M; E)$ if $s < -n/2$, with $\langle \delta_p^u, v\rangle = u \cdot v(p)$ for $v \in C^\infty(M; E^* \otimes \Omega)$, obtaining a continuous map

$$M \times C^\infty(M; E \otimes \Omega^{-1}) \to H^s_{\mathrm{c}}(M; E)\,, \quad (p, u) \mapsto \delta_p^{u(p)}\,. \tag{2.2.13}$$

As a particular case, the Dirac mass at any $p \in \mathbb{R}^n$ is $\delta_p = \delta_p^{1\otimes|dx|^{-1}} \in H^s_{\mathrm{c}}(\mathbb{R}^n)$.

The Schwartz kernel of any $A \in L(C^{-\infty}(M; E), C^\infty(M; F))$ has the following description: for all $q \in M$ and $u \in E_q \otimes \Omega_q^{-1}$,

$$K_A(\cdot, q)(u) = A\delta_q^u\,. \tag{2.2.14}$$

2.2.7 Differential Operators on Conormal Distributional Sections

Any $A \in \mathrm{Diff}^k(M; E)$ induces continuous linear maps [Mel96, Lemma 6.1.1]

$$A : I^{(s)}_{\cdot/\mathrm{c}}(M, L; E) \to I^{(s-k)}_{\cdot/\mathrm{c}}(M, L; E)\,, \tag{2.2.15}$$

which induce a continuous endomorphism A of $I_{\cdot/\mathrm{c}}(M, L; E)$. If $A \in \mathrm{Diff}(M, L; E)$, then it clearly induces a continuous endomorphism A of every $I^{(s)}_{\cdot/\mathrm{c}}(M, L; E)$.

By (2.2.10), for $A \in \mathrm{Diff}(M, L; E)$ and $u \in C^{\infty}_{\cdot/\mathrm{c}}(L; E \otimes \Omega^{-1}NL)$, we have [ÁLKL24, Eq. (4.17)]

$$A\delta_L^u = \delta_L^{A'u}\,, \quad A' = ((A^{\mathrm{t}})|_L)^{\mathrm{t}} \in \mathrm{Diff}(L; E \otimes \Omega^{-1}NL)\,, \tag{2.2.16}$$

where $A^{\mathrm{t}} \in \mathrm{Diff}(M, L; E^* \otimes \Omega)$ and $(A^{\mathrm{t}})|_L \in \mathrm{Diff}(L, E^* \otimes \Omega_L M)$ using the vector bundle versions of (2.2.1) and (2.2.2). By (2.2.16), $\mathrm{Diff}(M, L; E)$ preserves the subspace of Dirac sections given by (2.2.11). Thus (2.2.12) induces a continuous inclusion

$$C^{\infty}_{\cdot/\mathrm{c}}(L; E \otimes \Omega^{-1}NL) \subset I^{(s)}_{\cdot/\mathrm{c}}(M, L; E) \quad (s < -n'/2)\,. \tag{2.2.17}$$

2.2.8 Pull-Back of Conormal Distributions

If a smooth map $\phi : M' \to M$ is transverse to a regular submanifold $L \subset M$, which is a closed subset, then $L' := \phi^{-1}(L) \subset M'$ is a regular submanifold, which is a closed subset. The trivial-line-bundle version of (2.1.13) has continuous extensions

$$\phi^* : I^m(M, L) \to I^{m+k/4}(M', L') \quad (m \in \mathbb{R}) , \tag{2.2.18}$$

where $k = \dim M - \dim M'$ [Sim90, Theorem 5.3.8], [Mel96, Proposition 6.6.1]. Taking inductive limits and using (2.2.8), we get a continuous linear map

$$\phi^* : I(M, L) \to I(M', L') . \tag{2.2.19}$$

If ϕ is a submersion, this is a restriction of (2.1.18). In the case of a vector bundle E over M, we get

$$\phi^* : I(M, L; E) \to I(M', L'; \phi^* E) , \tag{2.2.20}$$

given by the $C^\infty(M)$-tensor product of the map (2.2.19) and the identity map on $C^\infty(M; E)$, using the versions of (2.1.6) and (2.1.20) for spaces of conormal distributions (see Sect. 2.1.6).

2.2.9 Push-Forward of Conormal Distributions

Let $\phi : M' \to M$ be a smooth submersion, and let $L \subset M$ and $L' \subset M'$ be regular submanifolds, which are closed subsets, such that $\phi(L') \subset L$ and the restriction $\phi : L' \to L$ is also a smooth submersion. Then (2.1.14) and (2.1.15) have continuous extensions

$$\phi_* : I^m_{\mathrm{c/cv}}(M', L'; \Omega_{\mathrm{fiber}}) \to I^{m+l/2-k/4}_{\mathrm{c/\cdot}}(M, L) \quad (m \in \mathbb{R}) , \tag{2.2.21}$$

where $k = \dim M' - \dim M$ and $l = \dim L' - \dim L$ [Sim90, Theorem 5.3.6], [Mel96, Proposition 6.7.2]. Taking inductive limits, we get a continuous linear map

$$\phi_* : I_{\mathrm{c/cv}}(M', L'; \Omega_{\mathrm{fiber}}) \to I_{\mathrm{c/\cdot}}(M, L) , \tag{2.2.22}$$

which is a restriction of (2.1.17). In the case of a vector bundle E over M, we get

$$\phi_* : I_{\mathrm{c/cv}}(M', L'; \phi^* E \otimes \Omega_{\mathrm{fiber}}) \to I_{\mathrm{c/\cdot}}(M, L; E) , \tag{2.2.23}$$

is given by the $C^\infty(M)$-tensor product of (2.2.22) and the identity map on $C^\infty(M; E)$, using the obvious versions of (2.1.6) and (2.1.21) for spaces of conormal distributions (see Sect. 2.1.6). The map (2.8.24) is also a restriction of (2.8.13).

2.3 Dual-Conormal Distributions

The dual space $I(M, L; E)'$ [Mel96, Chapter 6] also plays an important role in our work. Again, the case of $I(M, L)'$ is considered first; its extension for any vector bundle E can be made like in Sect. 2.2, and will be considered without further comment.

2.3.1 *Dual-Conormal Distributions when M Is Compact*

Consider the notation of Sects. 2.2.2 and 2.2.3, where M is assumed to be compact. The space of *dual-conormal distributions* of M at L (or of (M, L)) is [Mel96, Chapter 6]

$$I'(M, L) = I(M, L; \Omega)' . \tag{2.3.1}$$

Let also

$$I'^{(s)}(M, L) = I^{(-s)}(M, L; \Omega)' , \quad I'^{m}(M, L) = I^{-m}(M, L; \Omega)' . \tag{2.3.2}$$

$I'(M, L)$ is a complete Montel space, and every $I'^{(s)}(M, L)$ is bornological and barreled [ÁLKL24, Corollaries 5.1 and 5.2].

Transposing the versions of (2.2.4) and (2.2.7) with ΩM, we get continuous linear restriction maps, for $s' < s$ and $m < m'$,

$$I'^{(s')}(M, L) \leftarrow I'^{(s)}(M, L) , \quad I'^{m'}(M, L) \leftarrow I'^{m}(M, L) . \tag{2.3.3}$$

These maps form projective spectra (the Sobolev-order and symbol-order spectra), giving rise to $\varprojlim I'^{(s)}(M, L)$ as $s \uparrow +\infty$ and $\varprojlim I'^{m}(M, L)$ as $m \downarrow -\infty$. Similarly, from (2.2.5), we get continuous inclusions,

$$C^{-\infty}(M) \supset I'(M, L) \supset C^{\infty}(M) , \tag{2.3.4}$$

and (2.2.8) gives rise to continuous linear restriction maps

$$I'^{(-m+n/4-\epsilon)}(M, L) \leftarrow I'^{m}(M, L) \leftarrow I'^{(-m+n/4+\epsilon)}(M, L) , \tag{2.3.5}$$

for all $m \in \mathbb{R}$ and $\epsilon > 0$. We also have [ÁLKL24, Corollary 5.3]

$$I'(M, L) \equiv \varprojlim I'^{(s)}(M, L) \equiv \varprojlim I'^{m}(M, L) , \tag{2.3.6}$$

as $s \uparrow +\infty$ and $m \downarrow -\infty$, where the last equality follows from (2.3.5).

The left-hand-side maps of (2.3.3) have dense images, which follows from consequences of the Hahn-Banach theorem [NB11, Theorems 7.7.5 and 7.7.7 (c)], using that their transposes are the analogs of the inclusions (2.2.4) with ΩM by the reflexivity of the spaces $I^{(s)}(M, L; \Omega)$ (Sect. 2.2.2). Similarly, the inclusions (2.3.4) are dense.

2.3.2 Dual-Conormal Distributions when M Is Non-compact

If M is not supposed to be compact, the above concepts and properties can be extended as follows. We can similarly define the space $I'_K(M, L)$ of dual-conormal distributions supported in any compact $K \subset M$. Then define the LCHSs, $I'_{\mathrm{c}}(M, L) = \bigcup_K I'_K(M, L)$ like in (2.1.3), and $I'(M, L)$ like in (2.2.9) using $I'_{\mathrm{c}}(M, L)$ instead of $I_{\mathrm{c}}(M, L)$. These spaces satisfy a version of (2.3.1), interchanging arbitrary/compact support like in (2.1.5). $I'(M, L)$ is a complete Montel space, and (2.3.4) is also true. Similarly, we can define the spaces $I'^{(s)}_{\cdot/\mathrm{c}}(M, L)$ and $I'^{m}_{\cdot/\mathrm{c}}(M, L)$, which satisfy a version of (2.3.2) interchanging the support condition. Moreover (2.3.5) and (2.3.6) have obvious extensions.

If M is the domain of a given smooth submersion, the LCHS $I'_{\mathrm{cv}}(M; E)$ can be defined like $C^{-\infty}_{\mathrm{cv}}(M; E)$, using $I'_{\mathrm{c}}(M; E)$ instead of $C^{-\infty}_{\mathrm{c}}(M; E)$.

2.3.3 Conormal Distributions vs Dual-Conormal Distributions

Assume M is compact. Then [ÁLKL24, Theorem 8.11]

$$I(M, L) \cap I'(M, L) = C^{\infty}(M) .$$

2.3.4 Differential Operators on Dual-Conormal Distributional Sections

For any $A \in \mathrm{Diff}(M; E)$, the transpose of A^{t} on $I_{\mathrm{c}/\cdot}(M, L; E^* \otimes \Omega)$ (Sect. 2.2.7) is a continuous endomorphism A of $I'_{\cdot/\mathrm{c}}(M, L; E)$, which is a continuous extension of A on $C^{\infty}(M; E)$, and a restriction of A on $C^{-\infty}(M; E)$ (Sect. 2.1.7). By (2.2.15), if $A \in \mathrm{Diff}^m(M; E)$, we get induced continuous linear maps

$$A : I'^{(s)}_{\cdot/\mathrm{c}}(M, L; E) \to I'^{(s-m)}_{\cdot/\mathrm{c}}(M, L; E) , \tag{2.3.7}$$

If $A \in \mathrm{Diff}(M, L; E)$, the transpose of A^{t} of $I^{(-s)}_{\mathrm{c}/\cdot}(M, L; E^* \otimes \Omega)$ is a continuous endomorphism A of $I'^{(s)}_{\cdot/\mathrm{c}}(M, L; E)$.

2.3.5 Pull-Back of Dual-Conormal Distributions

With the notation and conditions of Sect. 2.2.9, transposing the compactly supported cases of (2.2.21) and (2.2.22) with ΩM, we get continuous linear maps

$$\begin{gathered} \phi^* : I'^m(M, L) \to I'^{m+l/2-k/4}(M', L') \quad (m \in \mathbb{R}) , \\ \phi^* : I'(M, L) \to I'(M', L') . \end{gathered} \tag{2.3.8}$$

In the case of a vector bundle E over M, like in (2.2.20), we get

$$\phi^* : I'(M, L; E) \to I'(M', L'; \phi^* E) . \tag{2.3.9}$$

The map (2.3.9) is an extension of (2.1.13) and a restriction of (2.1.18).

2.3.6 Push-Forward of Dual-Conormal Distributions

With the notation and conditions of Sect. 2.2.8, suppose ϕ is a submersion. Transposing the versions of (2.2.18) and (2.2.19) with ΩM, and using an analog of (2.1.16), we get continuous linear maps,

$$\begin{gathered} \phi_* : I'^m_{\mathrm{c}/\mathrm{cv}}(M', L' \otimes \Omega_{\mathrm{fiber}}) \to I'^{m-k/4}_{\mathrm{c}/\cdot}(M, L) \quad (m \in \mathbb{R}) , \\ \phi_* : I'_{\mathrm{c}/\mathrm{cv}}(M', L'; \Omega_{\mathrm{fiber}}) \to I'_{\mathrm{c}/\cdot}(M, L) . \end{gathered} \tag{2.3.10}$$

In the case of a vector bundle E over M, like in (2.2.23), we get

$$\phi_* : I'_{\mathrm{c}/\mathrm{cv}}(M', L'; \phi^* E \otimes \Omega_{\mathrm{fiber}}) \to I'_{\mathrm{c}/\cdot}(M, L; E) . \tag{2.3.11}$$

The map (2.3.10) is an extension of (2.1.14) and a restriction of (2.1.17).

2.4 Bounded Geometry

2.4.1 *Basic Notation*

The concepts recalled here become relevant when M is not compact. Equip M with a Riemannian metric g, and let ∇ denote its Levi-Civita connection, R its curvature tensor, and $\mathrm{inj}_M \geq 0$ its injectivity radius (the infimum of the injectivity radius at all points). If M is connected, we have an induced distance function d. If M is not connected, we can also define d taking $d(p, q) = \infty$ if p and q belong to different connected components. Observe that M is complete if $\mathrm{inj}_M > 0$. For $r > 0$, $p \in M$ and $S \subset M$, let $B(p, r)$ and $\overline{B}(p, r)$ denote the open and closed r-balls centered at p, and $\mathrm{Pen}(S, r)$ and $\overline{\mathrm{Pen}}(S, r)$ denote the open and closed r-penumbras of S (defined by the conditions $d(\cdot, S) < r$ and $d(\cdot, S) \leq r$, respectively). We may add the subscript "M" to this notation if needed, or a subscript "a" if we are referring to a family of Riemannian manifolds M_a.

2.4.2 *Manifolds and Vector Bundles of Bounded Geometry*

Recall that M is said to be of *bounded geometry* if $\mathrm{inj}_M > 0$ and $\sup |\nabla^m R| < \infty$ for every $m \in \mathbb{N}_0$. This concept has the following chart description.

Theorem 2.4.1 (Eichhorn [Eic91]; See also [Roe88, Sch96, Sch01]) *M is of bounded geometry if and only if, for some open ball $B \subset \mathbb{R}^n$ centered at 0, there are normal coordinates $y_p : V_p \to B$ at every $p \in M$ such that the corresponding Christoffel symbols Γ^i_{jk}, as a family of functions on B parametrized by i, j, k and p, lie in a bounded set of the Fréchet space $C^\infty(B)$. This equivalence holds as well replacing the Cristoffel symbols with the metric coefficients g_{ij}.*

Remark 2.4.2 Any non-connected Riemannian manifold of bounded geometry can be considered as a family of Riemannian manifolds (the connected components), which are of *equi-bounded geometry* in the sense that they satisfy the condition of bounded geometry with the same bounds.

Example 2.4.3 Typical examples of manifolds of bounded geometry are Lie groups with left invariant metrics, covering spaces of closed Riemannian manifolds and leaves of foliations on closed manifolds.

From now on in this section, assume M is of bounded geometry and consider the charts $y_p : V_p \to B$ given by Theorem 2.4.1. The radius of B will be denoted by r_0.

Proposition 2.4.4 (Schick [Sch96, Theorem A.22], [Sch01, Proposition 3.3]) *For every multi-index I, the function $|\partial_I(y_q y_p^{-1})|$ is bounded on $y_p(V_p \cap V_q)$, uniformly on $p, q \in M$.*

Proposition 2.4.5 (Shubin [Shu92, Appendix A1.1, Lemma 1.2]) *For any $0 < 2r \leq r_0$, there is a subset $\{p_k\} \subset M$ and some $N \in \mathbb{N}$ such that the balls $B(p_k, r)$ cover M, and every intersection of $N + 1$ sets $B(p_k, 2r)$ is empty.*

A vector bundle E of rank l over M is said to be of *bounded geometry* when it is equipped with a family of local trivializations over the charts (V_p, y_p), for small enough r_0, with corresponding defining cocycle $a_{pq} : V_p \cap V_q \to \mathrm{GL}(\mathbb{C}, l) \subset \mathbb{C}^{l \times l}$, such that, for all multi-index I, the function $|\partial_I(a_{pq}y_p^{-1})|$ is bounded on $y_p(V_p \cap V_q)$, uniformly on $p, q \in M$. When referring to local trivializations of a vector bundle of bounded geometry, we always mean that they satisfy this condition. If the corresponding defining cocycle is valued in $\mathrm{U}(l)$, then E is said to be of *bounded geometry* as a Hermitian vector bundle. Euclidean vector bundles of bounded geometry are similarly defined.

Example 2.4.6 The vector bundle E associated to the principal $\mathrm{O}(n)$-bundle P of orthonormal frames of M and any unitary representation of $\mathrm{O}(n)$ is of bounded geometry in a canonical way. In particular, this applies to $T_{\mathbb{C}}M$ and ΛM. If the representation is unitary, then bounded geometry holds as a Hermitian vector bundle. The same is true if we use any reduction Q of P with structural group $H \subset \mathrm{O}(n)$ and any unitary representation of H.

Example 2.4.7 Bounded geometry is preserved by operations of vector bundles induced by operations of vector spaces, like dual vector bundles, direct sums, tensor products, exterior products, densities, etc.

Example 2.4.8 Let E be a vector bundle E over a closed Riemannian manifold M, and let $\widetilde{M}$ be a covering of M. Then the lift $\widetilde{E}$ of E to $\widetilde{M}$ is of bounded geometry in a canonical way

2.4.3 *Uniform Spaces*

For every $m \in \mathbb{N}_0$, a function $u \in C^m(M)$ is said to be C^m*-uniformly bounded* if there is some $C_m \geq 0$ with $|\nabla^{m'} u| \leq C_m$ on M for all $m' \leq m$. These functions form the *uniform* C^m *space* $C^m_{\mathrm{ub}}(M)$, which is a Banach space with the norm $\|\cdot\|_{C^m_{\mathrm{ub}}}$ defined by the best constant C_m. As usual, we write $C_{\mathrm{ub}}(M) = C^0_{\mathrm{ub}}(M) = C(M) \cap L^\infty(M)$. Equivalently, we may take the norm $\|\cdot\|'_{C^m_{\mathrm{ub}}}$ defined by the best constant $C'_m \geq 0$ such that $|\partial_I(uy_p^{-1})| \leq C'_m$ on B for all $p \in M$ and $|I| \leq m$; in fact, it is enough to consider any subset of points p so that $\{V_p\}$ covers M [Sch96, Theorem A.22], [Sch01, Proposition 3.3]. The *uniform* C^∞ *space* is the Fréchet space $C^\infty_{\mathrm{ub}}(M) = \bigcap_m C^m_{\mathrm{ub}}(M)$, with the semi-norms $\|\cdot\|_{C^m_{\mathrm{ub}}}$ or $\|\cdot\|'_{C^m_{\mathrm{ub}}}$. It consists of the functions $u \in C^\infty(M)$ such that all functions uy_p^{-1} lie in a bounded set of $C^\infty(B)$.

The same definitions apply to functions with values in $\mathbb{C}^l$. Moreover the definition of uniform spaces with covariant derivative can be also considered for non-complete Riemannian manifolds.

Proposition 2.4.9 (Shubin [Shu92, Appendix A1.1, Lemma 1.3]; See also [Sch01, Proposition 3.2]) *Given r, $\{p_k\}$ and N like in Proposition 2.4.5 there is a partition of unity $\{f_k\}$ subordinated to the open covering $\{B(p_k, r)\}$, which is bounded in the Fréchet space $C^\infty_{\mathrm{ub}}(M)$.*

For a Hermitian vector bundle E of bounded geometry over M, the *uniform C^m space* $C^m_{\mathrm{ub}}(M; E)$ can be defined by introducing $\|\cdot\|'_{C^m_{\mathrm{ub}}}$ like the case of functions, using local trivializations of E to consider every uy_p^{-1} in $C^m(B, \mathbb{C}^l)$ for all $u \in C^m(M; E)$. Then, as above, we get the *uniform C^∞ space* $C^\infty_{\mathrm{ub}}(M; E)$, which consists of the sections $u \in C^\infty(M; E)$ such that all functions uy_p^{-1} define a bounded set of $C^\infty_{\mathrm{ub}}(B; \mathbb{C}^l)$. In particular, $\mathfrak{X}_{\mathrm{ub}}(M) := C^\infty_{\mathrm{ub}}(M; TM)$ is a $C^\infty_{\mathrm{ub}}(M)$-submodule and Lie subalgebra of $\mathfrak{X}(M)$.

The subset $\mathfrak{X}_{\mathrm{com}}(M) \subset \mathfrak{X}(M)$ of complete vector fields satisfies $\mathfrak{X}_{\mathrm{ub}}(M) \subset \mathfrak{X}_{\mathrm{com}}(M)$ [ÁLKL20, Proposition 3.8].

2.4.4 *Differential Operators of Bounded Geometry*

Like in Sect. 2.1.7, by using $\mathfrak{X}_{\mathrm{ub}}(M)$ and $C^\infty_{\mathrm{ub}}(M)$ instead of $\mathfrak{X}(M)$ and $C^\infty(M)$, we get the filtered subalgebra and $C^\infty_{\mathrm{ub}}(M)$-submodule $\mathrm{Diff}_{\mathrm{ub}}(M) \subset \mathrm{Diff}(M)$ of differential operators of *bounded geometry*. Observe that

$$C^m_{\mathrm{ub}}(M) = \{\, u \in C^m(M) \mid \mathrm{Diff}^m_{\mathrm{ub}}(M)\, u \subset L^\infty(M) \,\} \,. \tag{2.4.1}$$

The concept of $\mathrm{Diff}_{\mathrm{ub}}(M)$ can be extended to vector bundles of bounded geometry E and F over M by taking the $C^\infty_{\mathrm{ub}}(M)$-tensor product with $C^\infty_{\mathrm{ub}}(M; F \otimes E^*)$, obtaining the filtered $C^\infty_{\mathrm{ub}}(M)$-submodule $\mathrm{Diff}_{\mathrm{ub}}(M; E, F) \subset \mathrm{Diff}(M; E, F)$ (or $\mathrm{Diff}_{\mathrm{ub}}(M; E)$ if $E = F$). Bounded geometry of differential operators is preserved by compositions and by taking transposes, and by taking formal adjoints in the case of Hermitian vector bundles of bounded geometry; in particular, $\mathrm{Diff}_{\mathrm{ub}}(M; E)$ is a filtered subalgebra of $\mathrm{Diff}(M; E)$. Using local trivializations of E and F over the charts (V_p, y_p), we get a local description of any operator in $\mathrm{Diff}^m_{\mathrm{ub}}(M; E, F)$ by requiring its local coefficients to define a bounded subset of the Fréchet space $C^\infty(B, \mathbb{C}^{l'} \otimes \mathbb{C}^{l*})$, where l and l' are the ranks of E and F (Sect. 2.1.7). If E is a line bundle of bounded geometry, then [ÁLKL24, Eq. (2.24)]

$$\mathrm{Diff}^m_{\mathrm{ub}}(M; E) \equiv \mathrm{Diff}^m_{\mathrm{ub}}(M) \,. \tag{2.4.2}$$

Let $P_{\mathrm{ub}}(T^*M) \subset P(T^*M)$ be the graded subalgebra generated by $P^{(0)}_{\mathrm{ub}}(T^*M) \equiv C^\infty_{\mathrm{ub}}(M)$ and $P^{(1)}_{\mathrm{ub}}(T^*M) \equiv \mathfrak{X}_{\mathrm{ub}}(M)$, which is also a $C^\infty_{\mathrm{ub}}(M)$-submodule. Restricting (2.1.22), we get a short exact sequence with $\sigma_m : \mathrm{Diff}^m_{\mathrm{ub}}(M) \to P^{(m)}_{\mathrm{ub}}(T^*M)$. By taking the $C^\infty_{\mathrm{ub}}(M)$-tensor product with $C^\infty_{\mathrm{ub}}(M; F \otimes E^*)$, we get $P^{(m)}_{\mathrm{ub}}(T^*M; F \otimes E^*)$ and a short exact sequence with $\sigma_m : \mathrm{Diff}^m_{\mathrm{ub}}(M; E, F) \to P^{(m)}_{\mathrm{ub}}(T^*M; F \otimes E^*)$.

Using the norms $\|\cdot\|'_{C^m_{\mathrm{ub}}}$, it easily follows that every $A \in \mathrm{Diff}^m_{\mathrm{ub}}(M; E, F)$ defines bounded operators $A : C^{m+s}_{\mathrm{ub}}(M; E) \to C^s_{\mathrm{ub}}(M; F)$ ($s \in \mathbb{N}_0$), which induce a continuous linear map $A : C^\infty_{\mathrm{ub}}(M; E) \to C^\infty_{\mathrm{ub}}(M; F)$.

Example 2.4.10 In Example 2.4.6, the Levi-Civita connection ∇ induces a connection of bounded geometry on E, also denoted by ∇. In particular, ∇ itself is of bounded geometry on TM, and induces a connection ∇ of bounded geometry on ΛM. This holds as well for the connection on E induced by any other Riemannian connection of bounded geometry on TM.

Example 2.4.11 Bounded geometry of connections is preserved by taking the induced connections in the operations with vector bundles of bounded geometry indicated in Example 2.4.7.

Suppose E and F are Hermitian vector bundles of bounded geometry. Then any unitary connection ∇ of bounded geometry on E can be used to define an equivalent norm $\|\cdot\|_{C^m_{\mathrm{ub}}}$ on every Banach space $C^m_{\mathrm{ub}}(M; E)$, like in the case of $C^m_{\mathrm{ub}}(M)$.

It is said that $A \in \mathrm{Diff}^m(M; E, F)$ is *uniformly elliptic* if, given Hermitian metrics of bounded geometry on E and F, there is some $C \geq 1$ such that, for all $p \in M$ and $\xi \in T^*_pM$,

$$C^{-1}|\xi|^m \leq |\sigma_m(A)(p, \xi)| \leq C|\xi|^m . \tag{2.4.3}$$

This condition is independent of the choice of the Hermitian metrics of bounded geometry on E and F. Any $A \in \mathrm{Diff}^m_{\mathrm{ub}}(M; E, F)$ satisfies the second inequality.

Example 2.4.12 In Example 2.4.8, for any $A \in \mathrm{Diff}^m(M; E)$, its lift $\widetilde{A} \in \mathrm{Diff}^m(\widetilde{M}; \widetilde{E})$ is of bounded geometry in a canonical way. Moreover $\widetilde{A}$ is uniformly elliptic if A is elliptic.

2.4.5 *Sobolev Spaces of Manifolds of Bounded Geometry*

For any Hermitian vector bundle E of bounded geometry over M, any nonnegative symmetric uniformly elliptic $P \in \mathrm{Diff}^2_{\mathrm{ub}}(M; E)$ can be used to define the Sobolev space $H^s(M; E)$ ($s \in \mathbb{R}$) with a scalar product $\langle\cdot,\cdot\rangle_s$ (Sect. 2.1.11). Any choice of P defines the same Hilbertian space $H^s(M; E)$, which is a $C^\infty_{\mathrm{ub}}(M)$-module. In particular, $L^2(M; E)$ is the $C^\infty_{\mathrm{ub}}(M)$-tensor product of $L^2(M; \Omega^{1/2})$ and $C^\infty_{\mathrm{ub}}(M; E \otimes \Omega^{-1/2})$, and $H^s(M; E)$ is the $C^\infty_{\mathrm{ub}}(M)$-tensor product of $H^s(M)$ and $C^\infty_{\mathrm{ub}}(M; E)$. For instance, we may take $P = \nabla^*\nabla$ for any unitary connection ∇ of bounded geometry on E.

Example 2.4.13 In Example 2.4.8 and according to Example 2.4.12, $H^s(\widetilde{M};\widetilde{E})$ can be defined with the lift $\widetilde{P}$ of any nonnegative symmetric uniformly elliptic $P \in \mathrm{Diff}^2(M;E)$.

For $s \in \mathbb{N}_0$, the Sobolev space $H^s(M)$ can be also described with the scalar product

$$\langle u, v\rangle'_s = \sum_k \sum_{|I|\le s} \int_B f_k^2(x)\cdot \partial_I(u y_{p_k}^{-1})(x)\cdot \overline{\partial_I(v y_{p_k}^{-1})(x)}\,dx\ ,$$

using the partition of unity $\{f_k\}$ given by Proposition 2.4.5 [Sch96, Theorem A.22], [Sch01, Propositions 3.2 and 3.3], [Shu92, Appendices A1.2 and A1.3]. A similar scalar product $\langle\cdot,\cdot\rangle'_s$ can be defined for $H^s(M;E)$ with the help of local trivializations defining the bounded geometry of E. Every $A \in \mathrm{Diff}^m_{\mathrm{ub}}(M;E,F)$ defines bounded operators $A : H^{m+s}(M;E) \to H^s(M;F)$ $(s \in \mathbb{R})$, which induce continuous maps $A : H^{\pm\infty}(M;E) \to H^{\pm\infty}(M;F)$. For any almost everywhere positive $h \in C^\infty(M)$, we have $hH^m(M;E) = H^m(M;E)$ if and only if $h > 0$ and $h^{\pm 1} \in C^\infty_{\mathrm{ub}}(M)$.

If $m' > m + n/2$, then $H^{m'}(M;E) \subset C^m_{\mathrm{ub}}(M;E)$, continuously, and therefore $H^\infty(M;E) \subset C^\infty_{\mathrm{ub}}(M;E)$, continuously [Roe88, Proposition 2.8]. The Schwartz kernel mapping, $A \mapsto K_A$, defines a continuous linear map [Roe88, Proposition 2.9]

$$L(H^{-\infty}(M;E), H^\infty(M;F)) \to C^\infty_{\mathrm{ub}}(M; F \boxtimes (E^* \otimes \Omega))\ . \tag{2.4.4}$$

Remark 2.4.14 By (2.2.14), for any $A \in L(H^{-\infty}(M;E), H^\infty(M;F))$ and $r > 0$,

$$\operatorname{supp} K_A \subset \{\,(p,q) \in M^2 \mid d(p,q) \le r\,\}$$

if and only if $\operatorname{supp} Au \subset \overline{\mathrm{Pen}}(\operatorname{supp} u, r)$ for all $u \in H^{-\infty}(M;E)$.

Let $\mathcal{R}$ be the Fréchet space of rapidly decreasing functions on the real line. If $P \in \mathrm{Diff}^m_{\mathrm{ub}}(M;E)$ is uniformly elliptic and essentially self-adjoint, then the spectral theorem defines a continuous functional calculus

$$\mathcal{R} \to L(H^{-\infty}(M;E), H^\infty(M;E))\ , \quad \psi \mapsto \psi(P)\ .$$

Thus, by (2.4.4), the linear map

$$\mathcal{R} \to C^\infty_{\mathrm{ub}}(M; E \boxtimes (E^* \otimes \Omega))\ , \quad \psi \mapsto K_{\psi(P)}\ , \tag{2.4.5}$$

is continuous [Roe88, Proposition 2.10].

2.4.6 Maps of Bounded Geometry

For $a \in \{1,2\}$, let M_a be a Riemannian manifold of bounded geometry, of dimension n_a. Consider a normal chart $y_{a,p} : V_{a,p} \to B_a$ at every $p \in M_a$

satisfying the statement of Theorem 2.4.1. Let r_a denote the radius of B_a. For $0 < r \leq r_a$, let $B_{a,r} \subset \mathbb{R}^{n_a}$ denote the ball centered at the origin with radius r. We have $B_a(p,r) = y_{a,p}^{-1}(B_{a,r})$.

A smooth map $\phi : M_1 \to M_2$ is said to be of *bounded geometry* if, for some $0 < r < r_1$ and all $p \in M_1$, we have $\phi(B_1(p,r)) \subset V_{2,\phi(p)}$, and the compositions $y_{2,\phi(p)}\phi y_{1,p}^{-1}$ define a bounded set in the Fréchet space $C^\infty(B_{1,r}, \mathbb{R}^{n_2})$. This condition is preserved by the composition of maps. The set of smooth maps $M_1 \to M_2$ of bounded geometry is denoted by $C^\infty_{\mathrm{ub}}(M_1, M_2)$.

Let $\phi \in C^\infty_{\mathrm{ub}}(M_1, M_2)$. For every $m \in \mathbb{N}_0 \cup \{\infty\}$, using $\|\cdot\|'_{C^m_{\mathrm{ub}}}$ in the case where $m < \infty$ (Sect. 2.4.3) it follows that ϕ^* induces a continuous linear map [ÁLKL20, Eq. (19)]

$$\phi^* : C^m_{\mathrm{ub}}(M_2; \Lambda) \to C^m_{\mathrm{ub}}(M_1; \Lambda) . \tag{2.4.6}$$

Recall that ϕ is called *uniformly metrically proper* if, for any $s \geq 0$, there is some $t_s \geq 0$ so that, for all $p, q \in M_1$,

$$d_2(\phi(p), \phi(q)) \leq s \Rightarrow d_1(p,q) \leq t_s .$$

For all $m \in \mathbb{N}_0 \cup \{\infty\}$, if $\phi \in C^\infty_{\mathrm{ub}}(M_1, M_2)$ is uniformly metrically proper, then ϕ^* induces a continuous linear map [ÁLKL20, Eq. (21)]

$$\phi^* : H^m(M_2; \Lambda) \to H^m(M_1; \Lambda) . \tag{2.4.7}$$

If $\phi \in \mathrm{Diffeo}(M_1, M_2)$, and both ϕ and ϕ^{-1} are of bounded geometry, then ϕ is uniformly metrically proper. In this case, (2.4.7) can be continuously extended to Sobolev spaces of order $-m$.

The pull-back of a vector bundle of bounded geometry by a map of bounded geometry is of bounded geometry.

Homomorphisms of bounded geometry between vector bundles of bounded geometry have an obvious definition, but we will not use them.

2.4.7 *Smooth Families of Bounded Geometry*

Let T be a manifold, and let $\mathrm{pr}_1 : M \times T \to M$ denote the first factor projection. A section $u \in C^\infty(M \times T; \mathrm{pr}_1^* E)$ is called a *smooth family of smooth sections* of E *(parametrized by* T*)*, and we may use the notation $u = \{u_t \mid t \in T\}$, where $u_t = u(\cdot, t) \in C^\infty(M; E)$. Its *$T$-support* is $\overline{\{t \in T \mid u_t \neq 0\}}$. If the T-support is compact, then u is said to be *T-compactly supported*. It is said that u is *T-locally C^∞-uniformly bounded* if any $t \in T$ is in some chart (O, z) of T such that the maps $u(y_p \times z)^{-1}$ define a bounded subset of the Fréchet space $C^\infty(B \times z(O), \mathbb{C}^l)$, using local trivializations of E over the normal charts (V_p, y_p).

A *smooth family of differential operators*, $A = \{ A_t \mid t \in T \} \subset \mathrm{Diff}(M; E, F)$, can be defined by using smooth families of $\mathbb{C}$-valued functions, tangent vector fields and sections of $C^\infty(M; F \otimes E^*)$, like in Sect. 2.1.7. For this A, the T-*support* and the property of being T-*compactly supported* is defined like in the case of smooth families of sections. If the smooth families of functions, tangent vector fields and sections used to describe A are T-locally C^∞-uniformly bounded, then it is said that A is of T-*local bounded geometry* (cf. Sect. 2.4.4).

A smooth map $\phi : M_1 \times T \to M_2$ is called a *smooth family of smooth maps* $M_1 \to M_2$ (with *parameters* in T). It may be denoted by $\phi = \{ \phi^t \mid t \in T \}$, where $\phi^t = \phi(\cdot, t) : M_1 \to M_2$. It is said that ϕ is of T-*local bounded geometry* if every $t \in T$ is in some chart (O, z) of T such that, for some $0 < r < r_1$, we have $\phi(B_1(p,r) \times O) \subset V_{2,\phi(p)}$ for all $p \in M_1$, and the compositions $y_{2,\phi(p)}\phi(y_{1,p} \times z)^{-1}$, for $p \in M_1$, define a bounded subset of the Fréchet space $C^\infty(B_{1,r} \times z(O), \mathbb{R}^{n_2})$. The composition of smooth families of maps parametrized by T has the obvious sense and preserves the T-local bounded geometry condition. In particular, the $\mathbb{R}$-local bounded geometry condition makes sense for a flow $\phi = \{\phi^t\}$ on M. Given $X \in \mathfrak{X}_{\mathrm{com}}(M)$ with flow ϕ, we have $X \in \mathfrak{X}_{\mathrm{ub}}(M)$ if and only if ϕ is of $\mathbb{R}$-local bounded geometry [ÁLKL20, Proposition 3.18].

2.4.8 *Differential Complexes of Bounded Geometry*

With the notation of Sect. 2.1.14, assume that M, E and d are of bounded geometry (Sect. 2.4.2). Then we may also consider the topological complexes $(C^\infty_{\mathrm{ub}}(M; E), d)$ and $(H^{\pm\infty}(M; E), d)$ (Sects. 2.4.3 and 2.4.5).

(E, d) is said to be *uniformly elliptic* if D (or Δ) is uniformly elliptic (Sect. 2.4.4); this is equivalent to the obvious extension of (2.4.3) for (2.1.36). In this case, a version of (2.1.37) is true for $(H^{\pm\infty}(M; E), d)$, where the reduced cohomology is used instead of the cohomology, and the closures of the images of d, δ, D and Δ are used instead of their images.

2.5 Small b-Calculus

R. Melrose introduced b-calculus, a way to extend calculus to manifolds with boundary [Mel93, Mel96]. We will only use a part of it, called small b-calculus. For the sake of simplicity, we consider only compact manifolds with boundary, and the concepts and notation given here can be extended to the non-compact case like in Sect. 2.1, using compactly supported versions or local versions; some non-compact manifolds with boundary will be used in the paper. For the same reason, several kinds of section spaces and operators will be only defined in the case of functions or half-b-densities. Their extension to arbitrary vector bundles can be defined with tensor product expressions, like in Sect. 2.1. Most of these extensions will be used without further comments.

2.5.1 Some Notions of b-Geometry

Let M be a compact (smooth) n-manifold with boundary, whose interior is denoted by $\mathring{M}$. There exists a function $x \in C^\infty(M)$ so that $x \geq 0$, $\partial M = \{x = 0\}$ and $dx \neq 0$ on ∂M, which is called a *boundary-defining function*. Let ${}_+N\partial M \subset N\partial M$ be the inward-pointing subbundle of the normal bundle to the boundary. There is a unique trivialization $\nu \in C^\infty(\partial M; {}_+N\partial M)$ of ${}_+N\partial M$ so that $dx(\nu) = 1$. Take a collar neighborhood $T \equiv [0, \epsilon_0)_x \times \partial M_\varpi$ of ∂M. (In a product expression, every factor projection may be indicated as subscript of the corresponding factor.) Given coordinates $y = (y^1, \dots, y^{n-1})$ on some open $V \subset \partial M$, we get via ϖ coordinates $(x, y) = (x, y^1, \dots, y^{n-1})$, adapted (to ∂M), on the open subset $U \equiv [0, \epsilon_0) \times V \subset M$. There are vector bundles over M, ${}^{\mathrm{b}}TM$ and ${}^{\mathrm{b}}T^*M$, called *b-tangent* and *b-cotangent* bundles, which have the same restrictions to $\mathring{M}$ as TM and T^*M, and such that $x\partial_x, \partial_{y^1}, \dots, \partial_{y^{n-1}}$ and $x^{-1}dx, dy^1, \dots, dy^{n-1}$ extend to local frames around boundary points. This gives rise to versions of induced vector bundles, like ${}^{\mathrm{b}}\Omega^s M := \Omega^s({}^{\mathrm{b}}TM)$ ($s \in \mathbb{R}$) and ${}^{\mathrm{b}}\Omega M := {}^{\mathrm{b}}\Omega^1 M$. We have

$$C^\infty(M; \Omega^s) \equiv x^s C^\infty(M; {}^{\mathrm{b}}\Omega^s) . \tag{2.5.1}$$

Thus the integration operator $\int_M$ is defined on $xC^\infty(M; {}^{\mathrm{b}}\Omega)$, and induces a pairing between $C^\infty(M)$ and $xC^\infty(M; {}^{\mathrm{b}}\Omega)$.

At the points of ∂M, the local section $x\partial_x$ is independent of the choice of adapted local coordinates, spanning a trivial line subbundle ${}^{\mathrm{b}}N\partial M \subset {}^{\mathrm{b}}T_{\partial M}M$ with $T\partial M = {}^{\mathrm{b}}T_{\partial M}M/{}^{\mathrm{b}}N\partial M$. So ${}^{\mathrm{b}}\Omega^s_{\partial M}M \equiv \Omega^s\partial M \otimes \Omega^s({}^{\mathrm{b}}N\partial M)$, and a restriction map $C^\infty(M; {}^{\mathrm{b}}\Omega^s) \to C^\infty(\partial M; \Omega^s)$ is locally given by

$$u = a(x, y) \left|\frac{dx}{x} dy\right|^s \mapsto u|_{\partial M} = a(0, y)\, |dy|^s .$$

A Riemannian structure g on ${}^{\mathrm{b}}TM$ is called a *b-metric*. Locally,

$$g = a_0 \Big(\frac{dx}{x}\Big)^2 + 2 \sum_{j=1}^{n-1} a_{0j}\, \frac{dx}{x}\, dy^j + \sum_{j,k=1}^{n-1} a_{jk}\, dy^j\, dy^k ,$$

where a_0, a_{0j} and a_{jk} are C^∞ functions, provided that g is positive definite. If moreover $a_0 = 1 + O(x^2)$ and $a_{0j} = O(x)$ as $x \downarrow 0$, then g is called *exact*. In this case, the restriction of g to $\mathring{T} \equiv (0, \epsilon_0) \times \partial M$ is asymptotically cylindrical, and therefore the restriction of g to $\mathring{M}$ is a complete Riemannian metric. This restriction is of bounded geometry if it is *cylindrical* around the boundary; i.e., taking ϵ_0 small enough, we have $g = (\frac{dx}{x})^2 + h$ on $\mathring{T}$ for some Riemannian metric h on ∂M (considering $h \equiv \varpi^* h$).

2.5.2 Supported and Extendible Smooth Functions

Let $\breve{M}$ be any closed manifold containing M as submanifold of dimension n (for instance, $\breve{M}$ can be the double of M). Let $M' = \breve{M} \setminus \mathring{M}$, which is another compact submanifold with boundary of $\breve{M}$, of dimension n and with $\partial M' = M \cap M' = \partial M$.

The concepts, notation and conventions of Sect. 2.1.4 have straightforward extensions to manifolds with boundary, like the Fréchet space $C^\infty(M)$. Its elements are called *extendible functions* because the continuous linear restriction map

$$R : C^\infty(\breve{M}) \to C^\infty(M) \tag{2.5.2}$$

is surjective; in fact, there is a continuous linear extension map $E : C^\infty(M) \to C^\infty(\breve{M})$ [See64]. Since $C^\infty(\breve{M})$ and $C^\infty(M)$ are Fréchet spaces, the map (2.5.2) is open by the open mapping theorem, and therefore it is a surjective topological homomorphism. Its null space is $C^\infty_{M'}(\breve{M})$.

The Fréchet space of *supported functions* is the closed subspace of the smooth functions on M that vanish to all orders at the points of ∂M,

$$\dot{C}^\infty(M) = \bigcap_{m \geq 0} x^m C^\infty(M) \subset C^\infty(M) . \tag{2.5.3}$$

The extension by zero realizes $\dot{C}^\infty(M)$ as the closed subspace of functions on $\breve{M}$ supported in M,

$$\dot{C}^\infty(M) \equiv C^\infty_M(\breve{M}) \subset C^\infty(\breve{M}) . \tag{2.5.4}$$

By (2.5.3),

$$x^m \dot{C}^\infty(M) = \dot{C}^\infty(M) \quad (m \in \mathbb{R}) , \tag{2.5.5}$$

and therefore, by (2.5.1),

$$\dot{C}^\infty(M; {}^{\mathrm{b}}\Omega^s) \equiv \dot{C}^\infty(M; \Omega^s) \quad (s \in \mathbb{R}) . \tag{2.5.6}$$

We can similarly define Banach spaces $C^k(M)$ and $\dot{C}^k(M)$ ($k \in \mathbb{N}_0$) satisfying the analogs of (2.5.2)–(2.5.4), which in turn yield analogs of the first inclusions of (2.1.7), obtaining $C^\infty(M) = \bigcap_k C^k(M)$ and $\dot{C}^\infty(M) = \bigcap_k \dot{C}^k(M)$.

2.5.3 Supported and Extendible Distributions

The spaces of *supported* and *extendible distributions* on M are

$$\dot{C}^{-\infty}(M) = C^\infty(M; \Omega)' , \quad C^{-\infty}(M) = \dot{C}^\infty(M; \Omega)' .$$

These are barreled, ultrabornological, webbed, acyclic DF Montel spaces, and therefore complete, boundedly/compactly/sequentially retractive and reflexive [ÁLKL24, Proposition 6.1]. Transposing the version of (2.5.2) with ΩM, we get [Mel96, Proposition 3.2.1]

$$\dot{C}^{-\infty}(M) \equiv C^{-\infty}_{M}(\breve{M}) \subset C^{-\infty}(\breve{M}) . \tag{2.5.7}$$

Similarly, (2.5.4) and (2.5.3) give rise to continuous linear restriction maps

$$R : C^{-\infty}(\breve{M}) \to C^{-\infty}(M) , \tag{2.5.8}$$

$$R : \dot{C}^{-\infty}(M) \to C^{-\infty}(M) , \tag{2.5.9}$$

which are surjective topological homomorphisms [ÁLKL24, Proposition 6.2]. According to (2.5.7), the map (2.5.9) is a restriction of (2.5.8). There are continuous dense inclusions [Mel96, Lemma 3.2.1]

$$C^{\infty}_{\mathrm{c}}(\mathring{M}) \subset \dot{C}^{\infty}(M) \subset C^{\infty}(M) \subset \dot{C}^{-\infty}(M) , \tag{2.5.10}$$

the last one given by the integration pairing between $C^{\infty}(M)$ and $C^{\infty}(M; \Omega)$. The restriction of this pairing to $\dot{C}^{\infty}(M; \Omega)$ induces a continuous dense inclusion

$$C^{\infty}(M) \subset C^{-\infty}(M) . \tag{2.5.11}$$

Moreover (2.5.9) is the identity map on $C^{\infty}(M)$.

As before, from (2.5.5) and (2.5.6), we get

$$x^{m} C^{-\infty}(M) = C^{-\infty}(M) \quad (m \in \mathbb{R}) , \tag{2.5.12}$$

$$C^{-\infty}(M; {}^{\mathrm{b}}\Omega^{s}) \equiv C^{-\infty}(M; \Omega^{s}) \quad (s \in \mathbb{R}) . \tag{2.5.13}$$

The Banach spaces $C'^{-k}(M)$ and $\dot{C}'^{-k}(M)$ ($k \in \mathbb{N}_0$) are similarly defined. They satisfy the analogs of (2.5.7)–(2.5.13), and the analogs of the second inclusions of (2.1.7), obtaining $\bigcup_k C'^{-k}(M) = C^{-\infty}(M)$ and $\bigcup_k \dot{C}'^{-k}(M) = \dot{C}^{-\infty}(M)$.

2.5.4 Supported and Extendible Sobolev Spaces

The *supported Sobolev space* of order $s \in \mathbb{R}$ is the closed subspace of the elements supported in M,

$$\dot{H}^{s}(M) = H^{s}_{M}(\breve{M}) \subset H^{s}(\breve{M}) . \tag{2.5.14}$$

On the other hand, using the map (2.5.9), the *extendible Sobolev space* of order s is $H^s(M) = R(H^s(\breve{M}))$ with the inductive topology given by the linear map

$$R : H^s(\breve{M}) \to H^s(M) . \tag{2.5.15}$$

The null space of (2.5.15) is $H^s_{M'}(\breve{M})$. The analogs of (2.1.31)–(2.1.34) hold true in this setting using $\dot{C}^{\pm\infty}(M)$ and $C^{\pm\infty}(M)$. Furthermore the spaces $\dot{H}^s(M)$ and $H^s(M)$ form compact spectra of Hilbertian spaces.

The following properties are satisfied [Mel96, Proposition 3.5.1]. $C^\infty(M)$ is dense in $H^s(M)$, we have

$$\dot{H}^s(M) \equiv H^{-s}(M;\Omega)' , \quad H^s(M) \equiv \dot{H}^{-s}(M;\Omega)' , \tag{2.5.16}$$

and the map (2.5.9) has a continuous restriction

$$R : \dot{H}^s(M) \to H^s(M) , \tag{2.5.17}$$

which is surjective if $s \leq 1/2$, and injective if $s \geq -1/2$. In particular, $\dot{H}^0(M) \equiv H^0(M) \equiv L^2(M)$. The null space of (2.5.17) is $\dot{H}^s_{\partial M}(M)$.

2.5.5 The Space $\dot{C}^{-\infty}_{\partial M}(M)$

The indicated properties of (2.5.8) and (2.5.9) mean that we have short exact sequences in the category of continuous linear maps between LCSs (see also [Mel96, Proposition 3.3.1]),

$$\begin{gathered} 0 \to \dot{C}^{-\infty}(M') \xrightarrow{\iota} C^{-\infty}(\breve{M}) \xrightarrow{R} C^{-\infty}(M) \to 0 , \\ 0 \to \dot{C}^{-\infty}_{\partial M}(M) \xrightarrow{\iota} \dot{C}^{-\infty}(M) \xrightarrow{R} C^{-\infty}(M) \to 0 . \end{gathered} \tag{2.5.18}$$

From (2.5.7), we get

$$\dot{C}^{-\infty}_{\partial M}(M) \equiv C^{-\infty}_{\partial M}(\breve{M}) \subset C^{-\infty}(\breve{M}) . \tag{2.5.19}$$

The analogs of the second inclusion of (2.1.7), (2.1.31) and (2.1.33) hold true for the spaces $\dot{C}'^{-k}_{\partial M}(M)$ and $\dot{H}^s_{\partial M}(M)$. Thus the spaces $\dot{C}'^{-k}_{\partial M}(M)$ and $\dot{H}^s_{\partial M}(M)$ form spectra with the same union; the spectrum of spaces $\dot{H}^s_{\partial M}(M)$ is compact.

The following properties hold for $\dot{C}^{-\infty}_{\partial M}(M)$ [ÁLKL24, Proposition 6.4 and Corollary 6.5]: it is a limit subspace of the LF-space $\dot{C}^{-\infty}(M)$; and it is barreled, ultrabornological, webbed acyclic DF Montel space, and therefore complete, reflexive and boundedly/compactly/sequentially retractive. A description of $\dot{C}^{-\infty}_{\partial M}(M)$ will be indicated in Remark 2.6.1.

2.5.6 Differential Operators Acting on $C^{-\infty}(M)$ and $\dot{C}^{-\infty}(M)$

The notions of Sect. 2.1.7 also have straightforward extensions to manifolds with boundary. The action of any $A \in \mathrm{Diff}(M)$ on $C^{\infty}(M)$ preserves $\dot{C}^{\infty}(M)$, giving rise to extended continuous actions of A on $C^{-\infty}(M)$ and $\dot{C}^{-\infty}(M)$. They fit into commutative diagrams

$$\begin{array}{ccc} \dot{C}^{-\infty}(M) & \xrightarrow{A} & \dot{C}^{-\infty}(M) \\ {\scriptstyle R}\downarrow & & \downarrow{\scriptstyle R} \\ C^{-\infty}(M) & \xrightarrow{A} & C^{-\infty}(M) \end{array} \qquad \begin{array}{ccc} C^{-\infty}(M) & \xrightarrow{A} & C^{-\infty}(M) \\ {\scriptstyle \iota}\uparrow & & \uparrow{\scriptstyle \iota} \\ C^{\infty}(M) & \xrightarrow{A} & C^{\infty}(M)\,. \end{array} \tag{2.5.20}$$

However the analogous diagram

$$\begin{array}{ccc} \dot{C}^{-\infty}(M) & \xrightarrow{A} & \dot{C}^{-\infty}(M) \\ {\scriptstyle \iota}\uparrow & & \uparrow{\scriptstyle \iota} \\ C^{\infty}(M) & \xrightarrow{A} & C^{\infty}(M) \end{array} \tag{2.5.21}$$

may not be commutative. Using the notation $u \mapsto u_{\mathrm{c}}$ for the injection $C^{\infty}(M) \subset \dot{C}^{-\infty}(M)$ of (2.5.10), we have $A(u_{\mathrm{c}}) - (Au)_{\mathrm{c}} \in C^{-\infty}_{\partial M}(M)$ for all $u \in C^{\infty}(M)$ [Mel96, Eq. (3.4.8)].

From (2.5.2) and its version for vector fields, we get a surjective restriction map

$$\mathrm{Diff}(\breve{M}) \to \mathrm{Diff}(M)\,, \quad \breve{A} \mapsto \breve{A}|_M\,. \tag{2.5.22}$$

For any $\breve{A} \in \mathrm{Diff}(\breve{M})$ with $\breve{A}|_M = A$, we have the commutative diagrams

$$\begin{array}{ccc} C^{-\infty}(\breve{M}) & \xrightarrow{\breve{A}} & C^{-\infty}(\breve{M}) \\ {\scriptstyle R}\downarrow & & \downarrow{\scriptstyle R} \\ C^{-\infty}(M) & \xrightarrow{A} & C^{-\infty}(M)\,, \end{array} \qquad \begin{array}{ccc} C^{-\infty}(\breve{M}) & \xrightarrow{\breve{A}} & C^{-\infty}(\breve{M}) \\ {\scriptstyle \iota}\uparrow & & \uparrow{\scriptstyle \iota} \\ \dot{C}^{-\infty}(M) & \xrightarrow{A} & \dot{C}^{-\infty}(M)\,, \end{array} \tag{2.5.23}$$

where the left-hand side square extends the left-hand side square of (2.5.20).

If $A \in \mathrm{Diff}^m(M)$ ($m \in \mathbb{N}_0$), its actions on $\dot{C}^{-\infty}(M)$ and $C^{-\infty}(M)$ define continuous linear maps,

$$A : \dot{H}^s(M) \to \dot{H}^{s-m}(M)\,, \quad A : H^s(M) \to H^{s-m}(M)\,. \tag{2.5.24}$$

The maps (2.5.17) and (2.5.24) fit into a commutative diagram given by the left-hand side square of (2.5.20).

2.5.7 Differential Operators Tangent to the Boundary

The concepts of Sect. 2.2 can be generalized to the case with boundary when $L = \partial M$ [Mel96, Chapter 6] (see also [Mel93, Section 4.9]), giving rise to the Lie subalgebra and $C^\infty(M)$-submodule $\mathfrak{X}_{\mathrm{b}}(M) \subset \mathfrak{X}(M)$ of vector fields tangent to ∂M, called *b-vector fields*. We have $\mathfrak{X}_{\mathrm{b}}(M) \equiv C^\infty(M; {}^{\mathrm{b}}TM)$. Using $\mathfrak{X}_{\mathrm{b}}(M)$ like in Sect. 2.1.7, we get the filtered $C^\infty(M)$-submodule and filtered subalgebra $\mathrm{Diff}_{\mathrm{b}}(M) \subset \mathrm{Diff}(M)$ of *b-differential operators*; they are the operators $A \in \mathrm{Diff}(M)$ such that (2.5.21) is commutative [Mel96, Exercise 3.4.20]. The definition of $\mathrm{Diff}_{\mathrm{b}}(M)$ can be extended to arbitrary vector bundles like in Sect. 2.1.7. The condition of being tangent to the boundary is closed by taking transposes and formal adjoints. The restriction map (2.5.22) satisfies

$$\mathrm{Diff}(\breve{M}, \partial M)|_M = \mathrm{Diff}_{\mathrm{b}}(M) \,. \tag{2.5.25}$$

For all $a \in \mathbb{R}$ and $k \in \mathbb{N}_0$, we have [Mel96, Eqs. (4.2.7) and (4.2.8)]

$$\mathrm{Diff}_{\mathrm{b}}^k(M)\, x^a = x^a\, \mathrm{Diff}_{\mathrm{b}}^k(M) \,. \tag{2.5.26}$$

$\mathrm{Diff}(M)$ is spanned by ∂_x and $\mathrm{Diff}_{\mathrm{b}}(M)$ as algebra, and therefore

$$\mathrm{Diff}^k(M)\, x^a \subset x^{a-k}\, \mathrm{Diff}^k(M) \,. \tag{2.5.27}$$

2.5.8 Conormal Distributions at the Boundary

The spaces of *supported* and *extendible conormal distributions* at the boundary of Sobolev order $s \in \mathbb{R}$ are the $C^\infty(M)$-modules and LCSs,

$$\dot{\mathcal{A}}^{(s)}(M) = \{\, u \in \dot{C}^{-\infty}(M) \mid \mathrm{Diff}_{\mathrm{b}}(M)\, u \subset \dot{H}^s(M) \,\} \,,$$
$$\mathcal{A}^{(s)}(M) = \{\, u \in C^{-\infty}(M) \mid \mathrm{Diff}_{\mathrm{b}}(M)\, u \subset H^s(M) \,\} \,,$$

with the topologies defined like in (2.1.1), which are totally reflexive Fréchet spaces [ÁLKL24, Proposition 6.6]. They satisfy the analogs of the continuous inclusions (2.2.4), giving rise to the filtered $C^\infty(M)$-modules and LCSs of *supported* and *extendible conormal distributions* at the boundary,

$$\dot{\mathcal{A}}(M) = \bigcup_s \dot{\mathcal{A}}^{(s)}(M) \,, \quad \mathcal{A}(M) = \bigcup_s \mathcal{A}^{(s)}(M) \,, \tag{2.5.28}$$

which are barreled, ultrabornological and webbed [ÁLKL24, Corollary 6.7]. By definition, there are continuous inclusions

$$\dot{\mathcal{A}}(M) \subset \dot{C}^{-\infty}(M)\,, \quad \mathcal{A}(M) \subset C^{-\infty}(M)\,. \tag{2.5.29}$$

Thus $\dot{\mathcal{A}}(M)$ and $\mathcal{A}(M)$ are Hausdorff. We have

$$\bigcap_s \dot{\mathcal{A}}^{(s)}(M) = \dot{C}^{\infty}(M)\,, \quad \bigcap_s \mathcal{A}^{(s)}(M) = C^{\infty}(M)\,, \tag{2.5.30}$$

obtaining continuous dense inclusions [Mel96, Proposition 4.1.1 and Lemma 4.6.1]

$$\dot{C}^{\infty}(M) \subset \dot{\mathcal{A}}(M)\,, \quad C^{\infty}(M) \subset \mathcal{A}(M), \dot{\mathcal{A}}(M)\,. \tag{2.5.31}$$

By (2.5.31) and the density of the inclusions (2.5.10) and (2.5.11), it follows that the inclusions (2.5.29) are also dense. On the other hand, by elliptic regularity, we get continuous inclusions [Mel96, Eq. (4.1.4)]

$$\dot{\mathcal{A}}(M)|_{\mathring{M}}, \mathcal{A}(M) \subset C^{\infty}(\mathring{M})\,. \tag{2.5.32}$$

The maps (2.5.17) restrict to continuous linear maps

$$R : \dot{\mathcal{A}}^{(s)}(M) \to \mathcal{A}^{(s)}(M)\,, \tag{2.5.33}$$

which are surjective for $s \leq 1/2$ and injective for $s \geq -1/2$. If $s = 0$, then (2.5.33) is a TVS-isomorphism because $\dot{H}^0(M) \equiv H^0(M)$. The maps (2.5.33) induce a surjective topological homomorphism [Mel96, Proposition 4.1.1], [ÁLKL24, Proposition 6.8]

$$R : \dot{\mathcal{A}}(M) \to \mathcal{A}(M)\,, \tag{2.5.34}$$

which is the identity on $C^{\infty}(M)$.

2.5.9 The Spaces $x^m L^{\infty}(M)$

For $m \in \mathbb{R}$, consider the weighted space $x^m L^{\infty}(M)$ (Sect. 2.1.12). There is a continuous inclusion

$$x^m L^{\infty}(M) \subset C^{-\infty}(M)\,.$$

For $m' < m$, we also have a continuous inclusion

$$x^m L^{\infty}(M) \subset x^{m'} L^{\infty}(M)\,, \tag{2.5.35}$$

and $C^{\infty}_{\mathrm{c}}(\mathring{M})$ is dense in $x^m L^{\infty}(M)$ with the topology of $x^{m'} L^{\infty}(M)$ [ÁLKL24, Proposition 6.10].

2.5.10 Filtration of $\mathcal{A}(M)$ by Bounds

For every $m \in \mathbb{R}$, let

$$\mathcal{A}^m(M) = \{ u \in C^{-\infty}(M) \mid \mathrm{Diff}_\mathrm{b}(M)\, u \subset x^m L^\infty(M) \} .$$

This is another $C^\infty(M)$-module and Fréchet space with the topology like in (2.1.1). By (2.5.35), there is a continuous inclusion

$$\mathcal{A}^m(M) \subset \mathcal{A}^{m'}(M) \quad (m' < m) . \tag{2.5.36}$$

Moreover there are continuous inclusions [Mel96, Proof of Proposition 4.2.1]

$$\mathcal{A}^{(s)}(M) \subset \mathcal{A}^m(M) \subset \mathcal{A}^{(\min\{m,0\})}(M) \quad (m < s - n/2 - 1) . \tag{2.5.37}$$

Hence

$$\mathcal{A}(M) = \bigcup_m \mathcal{A}^m(M) . \tag{2.5.38}$$

Despite of defining the same LF-space, the filtrations of $\mathcal{A}(M)$ given by the spaces $\mathcal{A}^{(s)}(M)$ and $\mathcal{A}^m(M)$ are not equivalent because, in contrast with (2.5.30),

$$\dot{C}^\infty(M) = \bigcap_m \mathcal{A}^m(M) .$$

The following is true [ÁLKL24, Corollaries 6.14–6.16 and 6.39 and Remark 6.41]: the topologies of $\mathcal{A}(M)$ and $C^\infty(\mathring{M})$ coincide on every $\mathcal{A}^m(M)$ (however the second inclusion of (2.5.32) is not a TVS-embedding); $C^\infty_\mathrm{c}(\mathring{M})$ is dense in every $\mathcal{A}^m(M)$, and therefore in every $\mathcal{A}^{(s)}(M)$ and $\mathcal{A}(M)$; and $\mathcal{A}(M)$ is an acyclic Montel space, and therefore complete, boundedly/compactly/sequentially retractive and reflexive.

2.5.11 $\dot{\mathcal{A}}(M)$ and $\mathcal{A}(M)$ vs $I(\breve{M}, \partial M)$

The restriction maps (2.5.15) define continuous linear maps

$$R : I^{(s)}(\breve{M}, \partial M) \to \mathcal{A}^{(s)}(M) ,$$

which induce a surjective topological homomorphism [ÁLKL24, Proposition 6.18]

$$R : I(\breve{M}, \partial M) \to \mathcal{A}(M) . \tag{2.5.39}$$

The null space of (2.5.39) is $I_{M'}(\breve{M}, \partial M)$. There are TVS-identities

$$\dot{\mathcal{A}}^{(s)}(M) \equiv I_M^{(s)}(\breve{M}, \partial M) , \tag{2.5.40}$$

inducing a TVS-isomorphism [ÁLKL24, Corollary 6.20]

$$\dot{\mathcal{A}}(M) \xrightarrow{\cong} I_M(\breve{M}, \partial M) . \tag{2.5.41}$$

Moreover $I_M(\breve{M}, \partial M)$ is a limit subspace of the LF-space $I(\breve{M}, \partial M)$ [ÁLKL24, Proposition 6.19].

2.5.12 Filtration of $\dot{\mathcal{A}}(M)$ by the Symbol Order

Like in (2.5.40), let

$$\dot{\mathcal{A}}^m(M) = I_M^m(\breve{M}, \partial M) \subset I^m(\breve{M}, \partial M) \quad (m \in \mathbb{R}) , \tag{2.5.42}$$

which are closed subspaces satisfying the analogs of (2.2.7) and (2.2.8). Thus

$$\dot{\mathcal{A}}(M) = \bigcup_m \dot{\mathcal{A}}^m(M) , \quad \dot{C}^\infty(M) = \bigcap_m \dot{\mathcal{A}}^m(M) ,$$

and the TVS-isomorphism (2.5.41) is also compatible with the symbol filtration. $\dot{\mathcal{A}}(M)$ is an acyclic Montel space, and therefore complete, boundedly/compactly/sequentially retractive and reflexive [ÁLKL24, Corollary 6.22].

2.5.13 The Space $\mathcal{K}(M)$

The condition of being supported in ∂M defines the LCHSs and $C^\infty(M)$-modules

$$\mathcal{K}^{(s)}(M) = \dot{\mathcal{A}}^{(s)}_{\partial M}(M) , \quad \mathcal{K}^m(M) = \dot{\mathcal{A}}^m_{\partial M}(M) , \quad \mathcal{K}(M) = \dot{\mathcal{A}}_{\partial M}(M) .$$

These are the null spaces of the corresponding restrictions of the map (2.5.34) to $\dot{\mathcal{A}}^{(s)}(M)$, $\dot{\mathcal{A}}^m(M)$ and $\dot{\mathcal{A}}(M)$. They satisfy the analogs of (2.2.4), (2.2.7) and (2.2.8), obtaining $\bigcup_s \mathcal{K}^{(s)}(M) = \bigcup_m \mathcal{K}^m(M)$.

The properties of (2.5.34) mean that the following sequence is exact in the category of continuous linear maps between LCSs:

$$0 \to \mathcal{K}(M) \xrightarrow{\iota} \dot{\mathcal{A}}(M) \xrightarrow{R} \mathcal{A}(M) \to 0 . \tag{2.5.43}$$

It is called the *conormal sequence at the boundary*. We have

$$\mathcal{K}^{(s)}(M) = \{\, u \in \dot{C}^{-\infty}_{\partial M}(M) \mid \mathrm{Diff}_{\mathrm{b}}(M)\, u \subset \dot{H}^{s}_{\partial M}(M) \,\} ,$$

with the topology defined like in (2.1.1). The following properties hold [ÁLKL24, Propositions 6.24 and 6.25 and Corollaries 6.26–6.28]: every $\mathcal{K}^{(s)}(M)$ is a totally reflexive Fréchet space; $\mathcal{K}(M)$ is a limit subspace of the LF-space $\dot{\mathcal{A}}(M)$; and $\mathcal{K}(M)$ is barreled, ultrabornological, webbed and an acyclic Montel space, and therefore complete, boundedly/compactly/sequentially retractive and reflexive.

The TVS-isomorphism (2.5.41) restricts to a TVS-identity

$$\mathcal{K}(M) \equiv I_{\partial M}(\breve{M}, \partial M) , \tag{2.5.44}$$

which in turn restricts to identities between the LCHSs defining the Sobolev-order and symbol-order filtrations, according to (2.5.40) and (2.5.42).

A description of $\mathcal{K}^{(s)}(M)$ and $\mathcal{K}(M)$ will be indicated in Remark 2.6.3.

2.5.14 *Action of* Diff(M) *on* $\dot{\mathcal{A}}(M)$, $\mathcal{A}(M)$ *and* $\mathcal{K}(M)$

Any $A \in \mathrm{Diff}(M)$ defines continuous endomorphisms A of $\dot{\mathcal{A}}(M)$, $\mathcal{A}(M)$ and $\mathcal{K}(M)$. If $A \in \mathrm{Diff}^k(M)$, these maps also satisfy the analogs of (2.2.15). If $A \in \mathrm{Diff}_{\mathrm{b}}(M)$, then it defines continuous endomorphisms of $\dot{\mathcal{A}}^{(s)}(M)$, $\mathcal{A}^{(s)}(M)$, $\mathcal{A}^{m}(M)$ and $\mathcal{K}^{(s)}(M)$. All of these maps are restrictions of the endomorphisms A of $\dot{C}^{-\infty}(M)$, $C^{-\infty}(M)$ and $C^{\infty}(\mathring{M})$, and extensions of the endomorphisms A of $\dot{C}^{\infty}(M)$ and $C^{\infty}(M)$.

2.5.15 *Partial Extension Maps*

Given linear subspaces, $X \subset \mathcal{A}(M)$ and $Y \subset \dot{\mathcal{A}}(M)$, a map $E : X \to Y$ is called a *partial extension map* if $R(Y) \subset X$ and $RE = 1$ on X. The surjectivity of (2.5.34) is given by the following result. Its proof is recalled here because it will play an important role in our work.

Proposition 2.5.1 (Cf. [Mel96, Section 4.4]) *For all $m \in \mathbb{R}$, there is a continuous linear partial extension map $E_m : \mathcal{A}^{m}(M) \to \dot{\mathcal{A}}^{(s)}(M)$, where $s = 0$ if $m \geq 0$, and $m > s \in \mathbb{Z}^-$ if $m < 0$. For $m \geq 0$, $E_m : \mathcal{A}^{m}(M) \to \dot{\mathcal{A}}^{(0)}(M)$ is a continuous inclusion map.*

Proof First, let us consider the non-compact n-manifold with boundary $\mathbb{R}^n_1 := [0,\infty) \times \mathbb{R}^{n-1}$, whose double is $\mathbb{R}^n$. Consider the canonical coordinates on $\mathbb{R}^n_1$ given by the factor projections, $x : \mathbb{R}^n_1 \to [0,\infty)$ and $y : \mathbb{R}^n_1 \to \mathbb{R}^{n-1}$. We use the obvious generalization to the non-compact case of the spaces of extendible and supported conormal distributions at the boundary, of Sobolev order s, whose definitions involve $H^s_{\mathrm{loc}}(\mathbb{R}^n_1)$ and $\dot{H}^s_{\mathrm{loc}}(\mathbb{R}^n_1)$ like in Sect. 2.2.4.

For $m \geq 0$, since $x^m L^\infty(\mathbb{R}^n_1) \subset L^2_{\mathrm{loc}}(\mathbb{R}^n_1)$, continuously, we get $\mathcal{A}^m(\mathbb{R}^n_1) \subset \dot{\mathcal{A}}^{(0)}(\mathbb{R}^n_1)$, continuously. This also follows from (2.5.37) using that $\dot{\mathcal{A}}^{(0)}(\mathbb{R}^n_1) \equiv \mathcal{A}^{(0)}(\mathbb{R}^n_1)$. Thus E_m must be the inclusion map in this case.

Now fix $m < 0$. For $0 < \delta \leq 1$ such that $m + \delta \leq 0$ if $m \neq -1$, and $m + \delta < 0$ if $m = -1$, we have a continuous linear map $J : \mathcal{A}^m(\mathbb{R}^n_1) \to \mathcal{A}^{m+\delta}(\mathbb{R}^n_1)$ defined by

$$Ju(x, y) = \int_1^x u(\xi, y)\, d\xi \,. \tag{2.5.45}$$

So, for $-m < -s =: N \in \mathbb{N}$, we get the continuous linear maps (see Sect. 2.5.14)

$$\mathcal{A}^m(\mathbb{R}^n_1) \xrightarrow{J^N} \mathcal{A}^0(\mathbb{R}^n_1) \xrightarrow{E_0} \dot{\mathcal{A}}^{(0)}(\mathbb{R}^n_1) \xrightarrow{\partial_x^N} \dot{\mathcal{A}}^{(s)}(\mathbb{R}^n_1)\,,$$

whose composition is the desired extension E_m. For all $u \in \mathcal{A}^m(\mathbb{R}^n_1)$, we have

$$\partial\mathbb{R}^n_1 \cap \operatorname{supp} E_m u \subset \{0\} \times y(\operatorname{supp} u)\,. \tag{2.5.46}$$

Consider now a compact manifold with boundary M. Cover ∂M with a finite collection of adapted charts $(U_j, (x_j, y_j))$, and let $\{\lambda_j, \mu\}$ be a partition of unity subordinated to the open covering $\{U_j, \mathring{M}\}$ of M. By the case of $\mathbb{R}^n_1$, we directly get $\mathcal{A}^m(U_j) \subset \dot{\mathcal{A}}^{(0)}(U_j)$, continuously, if $m \geq 0$. By (2.5.46), if $m < 0$ and $-m < N \in \mathbb{N}$, we get a continuous linear partial extension map $E_{m,j} : \mathcal{A}^m(U_j) \to \dot{\mathcal{A}}^{(-N)}(U_j)$, which preserves the condition of being compactly supported. Then the result follows with $E_m : \mathcal{A}^m(M) \to \dot{\mathcal{A}}^{(s)}(M)$ defined by $E_m u = \mu u + \sum_j E_{m,j}(\lambda_j u)$. □

Remark 2.5.2 Consider the case where $m < 0$ in the proof of Proposition 2.5.1. Taking a collar neighborhood of the boundary, $T \equiv [0, \epsilon)_x \times \partial M_\varpi$, we can use adapted charts $(U_j \equiv [0, \epsilon) \times V_j, (x, y_j))$ defined by charts (V_j, y_j) of ∂M, like in Sect. 2.5.1. Then the operators $\partial_x \in \mathrm{Diff}(U_j)$ can be combined to define an operator $\partial_x \in \mathrm{Diff}(T)$, which indeed is the derivative operator on $C^\infty(T) \equiv C^\infty([0, \epsilon), C^\infty(\partial M))$. On the other hand, by integrating from ϵ to x, like in (2.5.45), we get a continuous linear map $J : \mathcal{A}^m(T) \to \mathcal{A}^{m+\delta}(T)$; in fact, this defines a continuous endomorphism J of $C^\infty(\mathring{T})$. In this way, a continuous linear extension map $E_{m,T} : \mathcal{A}^m(T) \to \dot{\mathcal{A}}^{(s)}(T)$ can be defined like in the case of $\mathbb{R}^n_1$. Then $E_m : \mathcal{A}^m(M) \to \dot{\mathcal{A}}^{(s)}(M)$ can be given by $E_m u = \mu u + E_{m,T}(\lambda u)$, where $\{\lambda, \mu\}$ is a partition of unity subordinated to the open covering $\{T, \mathring{M}\}$ of M.

Remark 2.5.3 A version of Proposition 2.5.1 with a vector bundle E over M can be achieved by taking the $C^\infty(M)$-tensor product with the identity map on $C^\infty(M; E \otimes E^*)$. We can also adapt the proof as follows. With the notation of Remark 2.5.2, there is an identity $E_T \equiv \varpi^* E_{\partial M} \equiv [0, \epsilon) \times E_{\partial M}$ over T, which induces trivializations $E_{U_j} \equiv [0,\epsilon) \times E_{V_j} \equiv [0,\epsilon) \times V_j \times \mathbb{C}^l$ over domains $U_j \equiv [0,\epsilon) \times V_j$. Like in Remark 2.5.2, these local trivializations can be used to define $\partial_x \in \mathrm{Diff}^1(T; E)$, which is considered as the derivative operator on $C^\infty([0,\epsilon), C^\infty(\partial M; E)) \equiv C^\infty(T; E)$. As usual, integration by parts shows that

$$\partial_x^{\mathrm{t}} = -\partial_x \in \mathrm{Diff}^1(T; E^* \otimes \Omega) . \tag{2.5.47}$$

If $E = \Lambda M$, then $\partial_x \in \mathrm{Diff}^1(T; \Lambda)$ is the Lie derivative with respect to $\partial_x \in \mathfrak{X}(T)$.

Remark 2.5.4 By (2.5.46), all steps of the proof of Proposition 2.5.1 have obvious compactly supported versions. This also applies to Remarks 2.5.2 and 2.5.3.

Given m and s satisfying the conditions of Proposition 2.5.1, let us denote by $E_{m,s}$ the partial extension map constructed in the proof of Proposition 2.5.1. This notation will make it easier to analyze its dependence on m and s in the following result.

Proposition 2.5.5 *Let $s' \le s$ and $m' \le m$ such that the maps $E_{m,s}$, $E_{m,s'}$ and $E_{m',s'}$ are defined. Then $E_{m,s}u = E_{m',s'}u$ for all $u \in \mathcal{A}^m(M)$.*

Proof According to the proof of Proposition 2.5.1, it is enough to consider the case of $\mathbb{R}^n_1$.

If $m' \ge 0$, there is nothing to prove.

In the case $m < 0$, we have $s, s' \in \mathbb{Z}^-$ with $m' > s > s'$. Let $N = -s$, $N' = -s'$ and $k = s - s' = N' - N$ in $\mathbb{Z}^+$. Since $\mathcal{A}^0(\mathbb{R}^n_1) \subset L^\infty(\mathbb{R}^n_1)$, the composition

$$\mathcal{A}^0(\mathbb{R}^n_1) \xrightarrow{J^k} \mathcal{A}^0(\mathbb{R}^n_1) \hookrightarrow \dot{\mathcal{A}}^{(0)}(\mathbb{R}^n_1) \xrightarrow{\partial_x^k} \dot{\mathcal{A}}^{(-k)}(\mathbb{R}^n_1)$$

is equal to the inclusion map $\mathcal{A}^0(\mathbb{R}^n_1) \hookrightarrow \dot{\mathcal{A}}^{(-k)}(\mathbb{R}^n_1)$. So, for all $u \in \mathcal{A}^m(\mathbb{R}^n_1)$, since $J^N u \in \mathcal{A}^0(\mathbb{R}^n_1)$, we have

$$E_{m',s'}u = E_{m,s'}u = \partial_x^{N'} J^{N'} u = \partial_x^N \partial_x^k J^k J^N u = \partial_x^N J^N u = E_{m,s}u .$$

In the case $m' < 0 \le m$, we have $s = 0$ and $m' > s' \in \mathbb{Z}^-$. Then the result follows with a similar argument using $k = -s' \in \mathbb{Z}^+$. □

Corollary 2.5.6 *For all s and m such that the map $E_{m,s}$ is defined, we have $E_{m,s}u = u$ for all $u \in C^\infty_{\mathrm{c}}(\mathring{M})$.*

Proof Use that $C^\infty_{\mathrm{c}}(\mathring{M}) \subset \mathcal{A}^m(M)$ and apply Proposition 2.5.5. □

Remark 2.5.7 The proof of Proposition 2.5.5 can be also applied to maps $E_{m,s}$ with $m \ge 0$ and $s \in \mathbb{Z}^-$, defined E_m like in the case $m < 0$. Including these maps, the map $E_{m,s'}$ of the statement is always defined under the other assumptions.

Remark 2.5.8 Proposition 2.5.5 and Corollary 2.5.6 are also true with the definitions of E_m given in Remarks 2.5.2 and 2.5.3, with similar proofs.

2.5.16 L^2 and L^∞ Half-b-Densities

We have

$$L^2(M; {}^{\mathrm{b}}\Omega^{\frac{1}{2}}) \equiv x^{-\frac{1}{2}} L^2(M; \Omega^{\frac{1}{2}}) , \tag{2.5.48}$$

$$L^\infty(M; {}^{\mathrm{b}}\Omega^{\frac{1}{2}}) \equiv x^{-\frac{1}{2}} L^\infty(M; \Omega^{\frac{1}{2}}) , \tag{2.5.49}$$

where (2.5.48) holds as Hilbert spaces, and (2.5.49) holds as LCHSs endowed with a family of equivalent Banach space norms [ÁLKL24, Eqs. (6.50) and (6.52)].

Equip M with a b-metric g (Sect. 2.5.1), and endow $\mathring{M}$ with the restriction of g, also denoted by g. With the corresponding Euclidean/Hermitean structures on $\Omega^{1/2}\mathring{M}$ and ${}^{\mathrm{b}}\Omega^{1/2}M$, we get $L^\infty(\mathring{M}; \Omega^{\frac{1}{2}}) \equiv L^\infty(M; {}^{\mathrm{b}}\Omega^{\frac{1}{2}})$ as Banach spaces.

2.5.17 b-Sobolev Spaces

For $m \in \mathbb{N}_0$, the *b-Sobolev spaces* of *order* $\pm m$ are the $C^\infty(M)$-modules and Hilbertian spaces defined by the following analogs of (2.1.29) and (2.1.30):

$$H_{\mathrm{b}}^m(M; {}^{\mathrm{b}}\Omega^{\frac{1}{2}}) = \{ u \in L^2(M; {}^{\mathrm{b}}\Omega^{\frac{1}{2}}) \mid \mathrm{Diff}_{\mathrm{b}}^m(M; {}^{\mathrm{b}}\Omega^{\frac{1}{2}})\, u \subset L^2(M; {}^{\mathrm{b}}\Omega^{\frac{1}{2}}) \} ,$$

$$\mathrm{Diff}_{\mathrm{b}}^m(M; {}^{\mathrm{b}}\Omega^{\frac{1}{2}})\, L^2(M; {}^{\mathrm{b}}\Omega^{\frac{1}{2}}) = H_{\mathrm{b}}^{-m}(M; {}^{\mathrm{b}}\Omega^{\frac{1}{2}}) = H_{\mathrm{b}}^m(M; {}^{\mathrm{b}}\Omega^{\frac{1}{2}})' .$$

Any finite set of $C^\infty(M)$-generators of $\mathrm{Diff}_{\mathrm{b}}^m(M; {}^{\mathrm{b}}\Omega^{1/2})$ defines a scalar product on $H_{\mathrm{b}}^{\pm m}(M; {}^{\mathrm{b}}\Omega^{1/2})$. The intersections/unions of the spaces $H_{\mathrm{b}}^m(M; {}^{\mathrm{b}}\Omega^{1/2})$ $(m \in \mathbb{Z})$ are denoted by $H_{\mathrm{b}}^{\pm\infty}(M; {}^{\mathrm{b}}\Omega^{1/2})$. In particular, $H_{\mathrm{b}}^\infty(M; {}^{\mathrm{b}}\Omega^{1/2}) = \mathcal{A}^{(0)}(M; {}^{\mathrm{b}}\Omega^{1/2})$.

2.5.18 Weighted b-Sobolev Spaces

We will also use the *weighted b-Sobolev space* $x^a H_{\mathrm{b}}^m(M; {}^{\mathrm{b}}\Omega^{1/2})$ $(a \in \mathbb{R})$, another Hilbertian space defined like in Sect. 2.1.12. We have [ÁLKL24, Section 6.19]

$$\bigcap_{a,m} x^a H_{\mathrm{b}}^m(M; {}^{\mathrm{b}}\Omega^{\frac{1}{2}}) = \dot{C}^\infty(M; {}^{\mathrm{b}}\Omega^{\frac{1}{2}}) .$$

2.5.19 Action of $\mathrm{Diff}_{\mathrm{b}}^m(M)$ on Weighted b-Sobolev Spaces

We have

$$\mathrm{Diff}_{\mathrm{b}}^m(M; {}^{\mathrm{b}}\Omega^{\frac{1}{2}}) \equiv \mathrm{Diff}_{\mathrm{b}}^m(M) \equiv \mathrm{Diff}_{\mathrm{b}}^m(M; \Omega^{\frac{1}{2}}) , \tag{2.5.50}$$

like in (2.1.23). Moreover any $A \in \mathrm{Diff}^k_{\mathrm{b}}(M; {}^{\mathrm{b}}\Omega^{1/2})$ defines continuous linear maps [Mel93, Lemma 5.14]

$$A : x^a H^m_{\mathrm{b}}(M; {}^{\mathrm{b}}\Omega^{\frac{1}{2}}) \to x^a H^{m-k}_{\mathrm{b}}(M; {}^{\mathrm{b}}\Omega^{\frac{1}{2}}) ,$$

which induce a continuous endomorphism A of $x^a H^{\pm\infty}_{\mathrm{b}}(M; {}^{\mathrm{b}}\Omega^{1/2})$.

2.5.20 A Description of $\mathcal{A}(M)$

In this subsection, unless the contrary is indicated, assume the following properties:

(A) $\mathring{M}$ is of bounded geometry with g.
(B) The collar neighborhood T of ∂M can be chosen so that:

(a) every $A \in \mathfrak{X}(\partial M)$ has an extension $A' \in \mathfrak{X}_{\mathrm{b}}(T)$ such that A' is ϖ-projectable to A, and $A'|_{\mathring{T}}$ is orthogonal to the ϖ-fibers; and
(b) $\mathfrak{X}_{\mathrm{ub}}(\mathring{M})|_{\mathring{T}}$ is $C^\infty_{\mathrm{ub}}(\mathring{M})|_{\mathring{T}}$-generated by $x\partial_x$ and the restrictions $A'|_{\mathring{T}}$ of the vector fields A' of (a), for $A \in \mathfrak{X}(\partial M)$.

For instance, (A) and (B) are true if $\mathring{T}$ is cylindrical with g (Sect. 2.5.1). The following properties hold [ÁLKL24, Corollaries 6.32, 6.34, 6.35, 6.37, 6.38 and 6.40 and Propositions 6.33 and 6.36]: the restriction to $\mathring{M}$ defines a continuous injection $C^\infty(M) \subset C^\infty_{\mathrm{ub}}(\mathring{M})$ (thus $C^\infty_{\mathrm{ub}}(\mathring{M})$ becomes a $C^\infty(M)$-module); as $C^\infty_{\mathrm{ub}}(\mathring{M})$-modules,

$$\mathrm{Diff}^m_{\mathrm{ub}}(\mathring{M}) \equiv \mathrm{Diff}^m_{\mathrm{b}}(M) \otimes_{C^\infty(M)} C^\infty_{\mathrm{ub}}(\mathring{M}) ,$$

$$\mathrm{Diff}^m_{\mathrm{ub}}(\mathring{M}; \Omega^{\frac{1}{2}}) \equiv \mathrm{Diff}^m_{\mathrm{b}}(M; {}^{\mathrm{b}}\Omega^{\frac{1}{2}}) \otimes_{C^\infty(M)} C^\infty_{\mathrm{ub}}(\mathring{M}) ;$$

as $C^\infty(M)$-modules and Hilbertian spaces or LCHSs, for $m \in \mathbb{Z}$,

$$H^m(\mathring{M}; \Omega^{1/2}) \equiv H^m_{\mathrm{b}}(M; {}^{\mathrm{b}}\Omega^{1/2}) , \qquad H^m(\mathring{M}) = x^{-1/2} H^m_{\mathrm{b}}(M) ,$$

$$H^{\pm\infty}(\mathring{M}; \Omega^{1/2}) \equiv H^{\pm\infty}_{\mathrm{b}}(M; {}^{\mathrm{b}}\Omega^{1/2}) , \quad H^{\pm\infty}(\mathring{M}) = x^{-1/2} H^{\pm\infty}_{\mathrm{b}}(M) ;$$

as $C^\infty(M)$-modules and LCHSs, for $m \in \mathbb{R}$,

$$\mathcal{A}^m(M; \Omega^{1/2}) \equiv x^{m+1/2} H^\infty_{\mathrm{b}}(M; {}^{\mathrm{b}}\Omega^{1/2}) ,$$

$$\mathcal{A}^m(M) \equiv x^m H^\infty_{\mathrm{b}}(M) \equiv x^{m+1/2} H^\infty(\mathring{M}) , \tag{2.5.51}$$

$$\mathcal{A}(M) \equiv \bigcup_m x^m H^\infty_{\mathrm{b}}(M) = \bigcup_m x^m H^\infty(\mathring{M}) . \tag{2.5.52}$$

Actually, the first identities of (2.5.51) and (2.5.52) are independent of g, and therefore they hold true without the assumptions (A) and (B) [ÁLKL24, Remark 6.41].

2.5.21 Dual-Conormal Distributions at the Boundary

Consider the LCHSs [Hör85, Section 18.3], [Mel96, Chapter 4],

$$\mathcal{K}'(M) = \mathcal{K}(M;\Omega)' , \quad \mathcal{A}'(M) = \dot{\mathcal{A}}(M;\Omega)' , \quad \dot{\mathcal{A}}'(M) = \mathcal{A}(M;\Omega)' ,$$

which are complete Montel spaces [ÁLKL24, Proposition 6.42]. The elements of $\mathcal{A}'(M)$ (resp., $\dot{\mathcal{A}}'(M)$) will be called *extendible* (resp., *supported*) *dual-conormal distributions* at the boundary. Consider also the LCHSs

$$\mathcal{K}'^{(s)}(M) = \mathcal{K}^{(-s)}(M;\Omega)' , \quad \mathcal{K}'^{m}(M) = \mathcal{K}^{-m}(M;\Omega)' ,$$

and, similarly, define $\mathcal{A}'^{(s)}(M)$, $\mathcal{A}'^{m}(M)$, $\dot{\mathcal{A}}'^{(s)}(M)$ and $\dot{\mathcal{A}}'^{m}(M)$. The spaces $\mathcal{K}'^{(s)}(M)$, $\mathcal{A}'^{(s)}(M)$ and $\dot{\mathcal{A}}'^{(s)}(M)$ are bornological and barreled [ÁLKL24, Corollary 6.43]. The transpositions of the analogs of (2.2.4) and (2.2.7) for the spaces $\mathcal{K}^{(s)}(M;\Omega)$, $\mathcal{K}^{m}(M;\Omega)$, $\dot{\mathcal{A}}^{(s)}(M;\Omega)$ and $\dot{\mathcal{A}}^{m}(M;\Omega)$ are continuous linear restriction maps

$$\mathcal{K}'^{(s')}(M) \to \mathcal{K}'^{(s)}(M) , \quad \mathcal{K}'^{m}(M) \to \mathcal{K}'^{m'}(M) ,$$
$$\mathcal{A}'^{(s')}(M) \to \mathcal{A}'^{(s)}(M) , \quad \mathcal{A}'^{m}(M) \to \mathcal{A}'^{m'}(M) ,$$

for $s < s'$ and $m < m'$. These maps form projective spectra, giving rise to projective limits. The spaces $\mathcal{K}'^{(s)}(M)$, $\mathcal{K}'^{m}(M)$, $\mathcal{A}'^{(s)}(M)$ and $\mathcal{A}'^{m}(M)$ satisfy the analogs of (2.3.5) and (2.3.6) [ÁLKL24, Corollary 6.44].

Similarly, transposing the analogs of (2.2.4) and (2.5.36) for the spaces $\mathcal{A}(M,\Omega M)$, we get continuous inclusions

$$\dot{\mathcal{A}}'^{(s)}(M) \supset \dot{\mathcal{A}}'^{(s')}(M) , \quad \dot{\mathcal{A}}'^{m}(M) \supset \dot{\mathcal{A}}'^{m'}(M) ,$$

for $s < s'$ and $m < m'$. The version of (2.5.37) with ΩM yields continuous inclusions

$$\dot{\mathcal{A}}'^{(s)}(M) \supset \dot{\mathcal{A}}'^{m}(M) \supset \dot{\mathcal{A}}'^{(\max\{m,0\})}(M) \quad (m > s + n/2 + 1) . \tag{2.5.53}$$

We also have [ÁLKL24, Corollary 6.44]

$$\dot{\mathcal{A}}'(M) = \bigcap_s \dot{\mathcal{A}}'^{(s)}(M) = \bigcap_m \dot{\mathcal{A}}'^{m}(M) , \tag{2.5.54}$$

where the last equality is a consequence of (2.5.53).

Transposing the versions of (2.5.3), (2.5.29), and (2.5.31) with ΩM, we get continuous inclusions [Mel96, Section 4.6]

$$C^\infty(M) \subset \mathcal{A}'(M) \subset C^{-\infty}(M), \dot{C}^{-\infty}(M) , \tag{2.5.55}$$

$$\dot{C}^\infty(M) \subset \dot{\mathcal{A}}'(M) \subset \dot{C}^{-\infty}(M), C^{-\infty}(M) , \tag{2.5.56}$$

and $R : \dot{C}^{-\infty}(M) \to C^{-\infty}(M)$ restricts to the identity map on $\mathcal{A}'(M)$ and $\dot{\mathcal{A}}'(M)$. The first inclusion of (2.5.56) is dense; in fact, $C^\infty_{\mathrm{c}}(\mathring{M})$ is dense in every $\dot{\mathcal{A}}'^m(M)$, and therefore in $\dot{\mathcal{A}}'(M)$ [ÁLKL24, Corollary 6.50 and Remark 6.51].

2.5.22 Dual-Conormal Sequence at the Boundary

Transposing maps in the version of (2.5.43) with ΩM, we get the sequence

$$0 \leftarrow \mathcal{K}'(M) \xleftarrow{R'} \mathcal{A}'(M) \xleftarrow{\iota'} \dot{\mathcal{A}}'(M) \leftarrow 0 , \tag{2.5.57}$$

where $R' = \iota^{\mathrm{t}}$ and $\iota' = R^{\mathrm{t}}$. It is called the *dual-conormal sequence at the boundary* of M, which is exact in the category of continuous linear maps between LCSs [ÁLKL24, Proposition 6.45].

2.5.23 $\dot{\mathcal{A}}(M)$ and $\mathcal{A}(M)$ vs $\mathcal{A}'(M)$

Using (2.5.29), (2.5.31), and (2.5.55), we have [Hör85, Proposition 18.3.24], [Mel96, Theorem 4.6.1]

$$\dot{\mathcal{A}}(M) \cap \mathcal{A}'(M) = C^\infty(M) . \tag{2.5.58}$$

2.5.24 A Description of $\dot{\mathcal{A}}'(M)$

If (A) and (B) are true, then, for $m \in \mathbb{R}$ [ÁLKL24, Corollaries 6.48 and 6.49],

$$\dot{\mathcal{A}}'^m(M) \equiv x^m H_{\mathrm{b}}^{-\infty}(M) = x^{m-\frac{1}{2}} H^{-\infty}(\mathring{M}) , \tag{2.5.59}$$

$$\dot{\mathcal{A}}'(M) \equiv \bigcap_m x^m H_{\mathrm{b}}^{-\infty}(M) = \bigcap_m x^m H^{-\infty}(\mathring{M}) . \tag{2.5.60}$$

The first identities of (2.5.59) and (2.5.60) are independent of g, and hold without the assumptions (A) and (B).

2.5.25 *Action of* Diff(M) *on* $\mathcal{A}'(M)$, $\dot{\mathcal{A}}'(M)$ *and* $\mathcal{K}'(M)$

Any $A \in \operatorname{Diff}(M)$ induces continuous linear endomorphisms A of $\mathcal{A}'(M)$, $\dot{\mathcal{A}}'(M)$ and $\mathcal{K}'(M)$ [Mel96, Proposition 4.6.1], which are the transposes of A^{t} on $\dot{\mathcal{A}}(M;\Omega)$, $\mathcal{A}(M;\Omega)$ and $\mathcal{K}(M;\Omega)$ (Sects. 2.1.5 and 2.5.14). If $A \in \operatorname{Diff}^k(M)$, these maps satisfy the analogs of (2.3.7). If $A \in \operatorname{Diff}_{\mathrm{b}}(M)$, it induces continuous endomorphisms of $\mathcal{A}'^{(s)}(M)$, $\mathcal{A}'^{m}(M)$, $\dot{\mathcal{A}}'^{(s)}(M)$ and $\mathcal{K}'^{(s)}(M)$.

2.5.26 *The b-Stretched Product*

Let $Y_1, \dots, Y_r$ be the connected components of ∂M. Consider the submanifold $B := \bigcup_{j=1}^{r} Y_j^2$ of the C^∞ manifold with corners M^2. Its inward-pointing spherical normal bundle is $S_+NB = {}_+NB/\mathbb{R}^+$, where $\mathbb{R}^+$ acts on ${}_+NB$ by multiplication. The *b-stretched product* M^2_{b} is the compact smooth manifold with corners obtained from M^2 by blowing-up B [Mel93, Sections 4.1 and 4.2], [Mel96, Chapter 4], with corresponding surjective smooth blow-down map $\beta_{\mathrm{b}} : M^2_{\mathrm{b}} \to M^2$; namely, $M^2_{\mathrm{b}} = S_+NB \sqcup (M^2 \setminus B)$, and β_{b} is the combination of the projection $S_+NB \to B$ and the identity map on $M^2 \setminus B$. The topology and C^∞ structure of M^2_{b} can be described as follows.

For any C^1 curve $\chi : [0,1] \to M^2$ with $\chi(0) \in B$ and $\chi((0,1]) \subset M^2 \setminus B$, let $\tilde{\chi} : [0,1] \to M^2_{\mathrm{b}}$ be the lift of χ so that $\tilde{\chi}(0)$ is defined by $\chi'(0)$. Then a subset $U \subset M^2_{\mathrm{b}}$ is open if it has open intersections with S_+NB and $M^2 \setminus B$, and, for any such curve χ with $\tilde{\chi}([0,1]) \subset U$, we have $\eta([0,1]) \subset U$ for all C^1 curve $\eta : [0,1] \to M^2$ of the same type as χ and C^1-close enough to χ.

Let x and x' denote the lifts to M^2 of the boundary-defining function x from the left and right factors. The C^∞ function

$$\tau := \frac{x - x'}{x + x'} : M^2 \setminus (\partial M)^2 \to [-1, 1]$$

has a continuous extension τ to the open neighborhood $S_+NB \sqcup (M^2 \setminus (\partial M)^2)$ of S_+NB in M^2_{b}. Then $C^\infty(M^2_{\mathrm{b}})$ is locally generated by τ and $\beta_{\mathrm{b}}^* C^\infty(M^2)$.

The manifold with corners M^2_{b} has three boundary hypersurfaces,

$$\mathrm{ff} = \beta_{\mathrm{b}}^{-1}(B)\,, \quad \mathrm{lb} = \overline{\beta_{\mathrm{b}}^{-1}(\partial M \times \mathring{M})}\,, \quad \mathrm{rb} = \overline{\beta_{\mathrm{b}}^{-1}(\mathring{M} \times \partial M)}\,,$$

called the *front face*, and the *left* and *right boundaries*. They satisfy $\mathrm{lb} \cap \mathrm{rb} = \emptyset$. Another embedded, compact submanifold of M^2_{b} is the *b-diagonal*, $\Delta_{\mathrm{b}} = \overline{\beta_{\mathrm{b}}^{-1}(\Delta \setminus B)}$, where $\Delta \subset M^2$ is the diagonal. We have $\Delta_{\mathrm{b}} \pitchfork \mathrm{ff}$, $\Delta_{\mathrm{b},0} := \Delta_{\mathrm{b}} \cap \mathrm{ff} = \partial \Delta_{\mathrm{b}}$ and $\Delta_{\mathrm{b}} \cap \mathrm{lb} = \Delta_{\mathrm{b}} \cap \mathrm{rb} = \emptyset$. Moreover $\beta_{\mathrm{b}} : \Delta_{\mathrm{b}} \to \Delta \equiv M$ is a diffeomorphism, where the last identity is given by the diagonal map.

Let also $x = \beta_b^* x$ and $x' = \beta_b^* x'$ on M_b^2. Thus $r = x + x'$ is a defining function of ff in M_b^2 (in the same sense as in Sect. 2.5.1 for ∂M). For adapted local coordinates (x, y), the lifts y and y' of y to open subsets of M^2 and M_b^2 are defined like x and x'. Then (r, τ, y, y') or (x, τ, y, y') are local coordinates of M_b^2 around points of ff, the submanifold Δ_b is locally described by the conditions $\tau = 0$ and $y = y'$, and $\Delta_{b,0}$ is locally described by the conditions $r = \tau = 0$ and $y = y'$. Other local coordinates (r, s, y, y') or (x, s, y, y') of M_b^2 around points of ff are defined using the function

$$s := \frac{1+\tau}{1-\tau} = \frac{x}{x'} : M_b^2 \setminus \mathrm{rb} \to (0, \infty) .$$

With the obvious extensions to manifolds with corners of some concepts of Sects. 2.5.1 and 2.5.7, we get the following [Mel93, Section 4.5]. First,

$$\begin{aligned} {}^{\mathrm{b}}TM^2 &\equiv ({}^{\mathrm{b}}TM)^2 , & {}^{\mathrm{b}}TM_b^2 &\equiv \beta_b^*({}^{\mathrm{b}}TM^2) , \\ \mathfrak{X}_b(M^2) &\equiv C^\infty(M^2; {}^{\mathrm{b}}TM^2) , & \mathfrak{X}_b(M_b^2) &\equiv C^\infty(M^2; {}^{\mathrm{b}}TM_b^2) . \end{aligned}$$

Second, any vector field in $\mathfrak{X}(M^2, B)$ can be lifted to a vector field in $\mathfrak{X}(M^2, \mathrm{ff})$; in particular, the lifts to M^2 of $\mathfrak{X}_b(M)$, from the left and right factors, generate $\mathfrak{X}_b(M^2)$ over $C^\infty(M^2)$. Third, there is a lifting map $\beta_b^* : \mathfrak{X}_b(M^2) \to \mathfrak{X}_b(M_b^2)$, whose image spans $\mathfrak{X}_b(M_b^2)$ over $C^\infty(M_b^2)$. It induces a lifting map $\beta_b^* : \mathrm{Diff}_b^m(X^2) \to \mathrm{Diff}_b^m(M_b^2)$, whose image spans $\mathrm{Diff}_b^m(M_b^2)$ over $C^\infty(M_b^2)$ [Mel93, Exercise 4.11]. For instance, using the above local coordinates, the lift of $x\partial_x$ is $\frac{1}{2}(1+\tau)r\partial_r + \frac{1}{2}(1-\tau^2)\partial_\tau$. Finally, the lift to M_b^2 of $\mathfrak{X}_b(M)$ from the left factor of M^2 is a Lie subalgebra of $\mathfrak{X}_b(M_b^2)$ transverse to Δ_b, giving rise to natural isomorphisms $N\Delta_b \cong {}^{\mathrm{b}}TM$ and $N^*\Delta_b \cong {}^{\mathrm{b}}T^*M$ [Mel93, Lemmas 4.5 and 4.6]. Thus there is a canonical isomorphism ${}^{\mathrm{b}}\Omega^{1/2}(M_b^2)|_{\Delta_b} \cong {}^{\mathrm{b}}\Omega M$ (cf. [Mel93, Eq. (4.125)]).

2.5.27 *The b-Pseudodifferential Operators*

A refinement of the Schwartz kernel theorem gives a bijection [Mel93, Lemma 4.20]

$$L(\dot{C}^\infty(M; {}^{\mathrm{b}}\Omega^{\frac{1}{2}}), C^{-\infty}(M; {}^{\mathrm{b}}\Omega^{\frac{1}{2}})) \to C^{-\infty}(M_b^2; \beta_b^*({}^{\mathrm{b}}\Omega^{\frac{1}{2}} \boxtimes {}^{\mathrm{b}}\Omega^{\frac{1}{2}})) ,$$

$$A \mapsto \kappa_A , \quad \langle Au, v\rangle = \langle \kappa_A, \beta_b^*(v \otimes u)\rangle , \quad u, v \in \dot{C}^\infty(M; {}^{\mathrm{b}}\Omega^{\frac{1}{2}}) .$$

The concept of conormal distributional sections can be also extended to submanifolds whose boundary is given by a transverse intersection with the boundary faces, like $\Delta_b \subset M_b^2$. Then a continuous linear map $A : \dot{C}^\infty(M; {}^{\mathrm{b}}\Omega^{1/2}) \to C^{-\infty}(M; {}^{\mathrm{b}}\Omega^{1/2})$ is called a *b-pseudodifferential operator* of *order* at most $m \in \mathbb{R}$

if $\kappa_A \in I^m(M_b^2, \Delta_b; \beta_b^*({}^b\Omega^{1/2} \boxtimes {}^b\Omega^{1/2}))$ and κ_A vanishes to all orders at $\mathrm{lb} \cup \mathrm{rb}$ [Mel93, Definition 4.22]. Such operators form a $C^\infty(M_b^2)$-module $\Psi_b^m(M; {}^b\Omega^{1/2})$, obtaining the filtered $C^\infty(M_b^2)$-module $\Psi_b(M; {}^b\Omega^{1/2}) = \bigcup_m \Psi_b^m(M; {}^b\Omega^{1/2})$. Moreover the submodule $\Psi_b^{-\infty}(M; {}^b\Omega^{1/2}) := \bigcap_m \Psi_b^m(M; {}^b\Omega^{1/2})$ (resp., $\mathrm{Diff}_b(M; {}^b\Omega^{1/2})$) of $\Psi_b(M; {}^b\Omega^{1/2})$ consists of the operators $A \in \Psi_b(M; {}^b\Omega^{1/2})$ with smooth κ_A (resp., $\operatorname{supp} \kappa_A \subset \Delta_b$). The obvious generalization of the definition of principal symbol, like in Sect. 2.2.3, now gives the *principal b-symbol* exact sequence,

$$0 \to \Psi_b^{m-1}(M; {}^b\Omega^{\frac{1}{2}}) \hookrightarrow \Psi_b^m(M; {}^b\Omega^{\frac{1}{2}}) \xrightarrow{{}^b\sigma_m} S^{(m)}({}^bT^*M; {}^b\Omega^{\frac{1}{2}}) \to 0\,.$$

The principal b-symbol is used to define *b-ellipticity* like ellipticity in the case of pseudodifferential operators (Sect. 2.1.9).

Omitting ${}^b\Omega^{1/2}$, if $A \in \Psi_b^{-\infty}(M)$ and $\kappa := \kappa_A$ is supported in the domain of a chart (x, s, y, y'), then we can write $\kappa = \kappa'(x, s, y, y')\, s^{-1} ds\, dy'$ because κ is rapidly decreasing as $s \downarrow 0$ and as $s \uparrow +\infty$, obtaining

$$Au(x, y) = \int_{\partial M} \int_0^\infty \kappa'(x, s, y, y')\, u\Big(\frac{x}{s}, y'\Big) \frac{ds}{s}\, dy'\,, \tag{2.5.61}$$

for all $u \in \dot{C}^\infty(M)$ supported in the domain of the chart (x, y).

Any $A \in \Psi_b^m(M; {}^b\Omega^{1/2})$ defines continuous endomorphisms A of $\dot{C}^\infty(M; {}^b\Omega^{1/2})$ and $C^{\pm\infty}(M; {}^b\Omega^{1/2})$ [Mel93, Propositions 4.29 and 4.34 and Exercise 4.33]. In this sense, $\Psi_b(M; {}^b\Omega^{1/2})$ becomes a filtered algebra with the operation of composition [Mel93, Propositions 5.20], and the principal b-symbol map is multiplicative.

2.5.28 The Indicial Family

Let $A \in \Psi_b^m(M; {}^b\Omega^{1/2})$ ($m \in \mathbb{R}$) and write $\kappa = \kappa_A$. Roughly speaking, the *indicial family* of A is an entire family, $I_\nu(A, \lambda) \in \Psi^m(\partial M; \Omega^{1/2})$ ($\lambda \in \mathbb{C}$), depending on the trivialization ν of ${}_+N\partial M$ (Sect. 2.5.1), defined by taking the "fiberwise" Mellin transform of certain conormal distributional section defined by $\kappa|_{\mathrm{ff}}$. Thus $I_\nu(A, \lambda) = 0$ for all λ just when $\kappa|_{\mathrm{ff}} = 0$.

The indicial family can be also described as follows. For $z \in \mathbb{C}$ and $m \in \mathbb{R}$, the mapping $A \mapsto x^{-z} A x^z$ defines an automorphism of $\Psi_b^m(M; {}^b\Omega^{1/2})$ [Mel93, Proposition 5.7]. Hence every $A \in \Psi_b(M; {}^b\Omega^{1/2})$ defines a continuous endomorphism A of $x^k C^\infty(M; {}^b\Omega^{1/2})$ ($k \in \mathbb{N}$). Therefore a continuous endomorphism A_∂ of $C^\infty(\partial M; \Omega^{1/2})$ is well defined by $A_\partial v = Au|_{\partial M}$ if $u \in C^\infty(M; {}^b\Omega^{1/2})$ with $u|_{\partial M} = v$ [Mel93, Eq. (5.31)]. Then [Mel93, Proposition 5.8]

$$I_\nu(A, \lambda) = (x^{-i\lambda} A x^{i\lambda})_\partial\,. \tag{2.5.62}$$

We will only use the indicial family in the following cases, where ${}^{\mathrm{b}}\Omega^{1/2}$ is omitted for the sake of simplicity. First, if $A \in \Psi_{\mathrm{b}}^{-\infty}(M)$ and κ is supported in the domain of a chart (x, s, y, y'), as described in (2.5.61), then $I_\nu(A, \lambda) \in \Psi_{\mathrm{b}}^{-\infty}(\partial M; \Omega^{1/2})$ is given by

$$K_{I_\nu(A,\lambda)}(y, y') = \int_0^\infty s^{-i\lambda}\kappa'(0, s, y, y')\frac{ds}{s} . \tag{2.5.63}$$

The support condition can be obviously removed by using a partition of unity or a collar neighborhood of ∂M. Second, if $A \in \mathrm{Diff}_{\mathrm{b}}^m(M)$ ($m \in \mathbb{N}_0$) is locally given by

$$A = \sum_{j+|I|\leq m} a_{j,I}(x, y)(xD_x)^j D_y^I ,$$

using adapted local coordinates (x, y), then

$$I_\nu(A, \lambda) = \sum_{j+|I|\leq m} a_{j,I}(0, y)\lambda^j D_y^I . \tag{2.5.64}$$

The indicial family is multiplicative [Mel93, Corollary of Proposition 5.20], and compatible with the operation of taking formal adjoints of b-differential operators (Cf. [Mel93, Eq. (4.112)]).

2.5.29 *The b-Integral*

The *b-integral* is the linear map

$${}^{\nu}\!\!\int = {}^{\nu}\!\!\int_M : C^1(M; {}^{\mathrm{b}}\Omega) \to \mathbb{C} ,$$

depending on ν, defined by [Mel93, Lemma 4.62]

$${}^{\nu}\!\!\int u = \lim_{\epsilon\downarrow 0}\left(\int_{x\geq\epsilon} u + \ln\epsilon \int_{\partial M} u|_L\right) ,$$

using a boundary-defining function x with $dx(\nu) = 1$. Another trivialization $\mu \in C^\infty(M; {}_+N\partial M)$ is of the form $\mu = a\nu$ for some $0 < a \in C^\infty(\partial M)$, and

$${}^{\mu}\!\!\int u - {}^{\nu}\!\!\int u = \int_{\partial M} \ln a \cdot u|_{\partial M} .$$

Lemma 2.5.9 *${}^{\nu}\!\!\int$ is continuous with the C^1 topology on $C^1(M; {}^{\mathrm{b}}\Omega)$.*

Proof Consider a chart (V, y) of ∂M, and the adapted local coordinates (x, y) on $U \equiv [0, \epsilon_0)_x \times V \subset M$ $(\epsilon_0 > 0)$. Since ${}^{\nu}\!\!\int \equiv \int_M$ on $C^1_c(\mathring{M}; {}^b\Omega) \equiv C^1_c(\mathring{M}; \Omega)$, it is easy to see that it is enough to prove the continuity of ${}^{\nu}\!\!\int$ on $C^1_c(U; {}^b\Omega)$. Every $u \in C^1_c(U; {}^b\Omega)$ is of the form $u(x, y) = a(x, y)\,|x^{-1}dx\,dy|$ for some $a \in C^1_c(U)$. Then

$$ {}^{\nu}\!\!\int u = \ln \epsilon_0 \cdot \int_V a(0, y)\,dy + \lim_{\epsilon \downarrow 0} \left(\int_V \int_\epsilon^{\epsilon_0} (a(x, y) - a(0, y)) \frac{dx}{x}\,dy \right) . $$

Hence

$$ \left| {}^{\nu}\!\!\int u \right| \le \operatorname{vol} V \cdot \left(\ln \epsilon_0 \cdot \max_{y \in V} |a(0, y)| + \epsilon_0 \cdot \max_{(\xi, y) \in U} |\partial_x a(\xi, y)| \right) . $$

□

Corollary 2.5.10 *Let T and T' be collar neighborhoods of ∂M in M with $\overline{T'} \subset T$. For any sequence u_k in $C^1(M; {}^b\Omega)$, if $u_k|_T \to 0$ in $C^1(T; {}^b\Omega)$ and $\lim_k \int_{M \setminus T'} u_k = a \in \mathbb{C}$, then $\lim_k {}^{\nu}\!\!\int u_k = a$.*

Proof Let $\{\lambda, \mu\}$ be a smooth partition of unity of M subordinated to the open covering $\{T, M \setminus \overline{T'}\}$. Then $\lim_k \lambda u_k = 0$ in $C^1(M; {}^b\Omega)$, obtaining $\lim_k {}^{\nu}\!\!\int \lambda u_k = 0$ by Lemma 2.5.9. Moreover $\lim_k \int_{M \setminus T'} \lambda u_k = 0$. Therefore

$$ \lim_k {}^{\nu}\!\!\int u_k = \lim_k {}^{\nu}\!\!\int \mu u_k = \lim_k \int_{M \setminus T'} \mu u_k = \lim_k \int_{M \setminus T'} u_k = a . $$

□

Remark 2.5.11 Consider a collar neighborhood of ∂M in M of the form $T \equiv [0, \epsilon)_x \times \partial M_\varpi$, and the intermediate space

$$ \begin{aligned} C^1(T; {}^b\Omega) &\equiv C^1([0, \epsilon), C^1(\partial M; {}^b\Omega_{\partial M})) \\ &\subset C^{0,1}_\varpi(T; {}^b\Omega) := C^1([0, \epsilon), C^0(\partial M; {}^b\Omega_{\partial M})) \\ &\subset C^0(T; {}^b\Omega) \equiv C^0([0, \epsilon), C^0(\partial M; {}^b\Omega_{\partial M})) . \end{aligned} $$

Then ${}^{\nu}\!\!\int_M$ is actually defined on

$$ \{ u \in C^0(M; {}^b\Omega) \mid u|_T \in C^{0,1}_\varpi(T; {}^b\Omega) \} , $$

and the proof of Lemma 2.5.9 shows that it is continuous with the obvious topology defined by the topologies of $C^0(M; {}^b\Omega)$ and $C^{0,1}_\varpi(T; {}^b\Omega)$. So Corollary 2.5.10 is true with the weaker condition $u_k|_T \to 0$ in $C^{0,1}_\varpi(T; {}^b\Omega)$.

2.5.30 The b-Trace

Any $A \in \Psi_{\mathrm{b}}^{-\infty}(M; {}^{\mathrm{b}}\Omega^{1/2})$ is of trace class if and only if $A \in r\Psi_{\mathrm{b}}^{-\infty}(M; {}^{\mathrm{b}}\Omega^{1/2})$ (i.e., $\kappa_A|_{\mathrm{ff}} = 0$). The *b-trace* ${}^{\mathrm{b}}\mathrm{Tr} : \Psi_{\mathrm{b}}^{-\infty}(M; {}^{\mathrm{b}}\Omega^{1/2}) \to \mathbb{C}$ is an extension of the trace $\mathrm{Tr} : r\Psi_{\mathrm{b}}^{-\infty}(M; {}^{\mathrm{b}}\Omega^{1/2}) \to \mathbb{C}$ given by

$$ {}^{\mathrm{b}}\mathrm{Tr}\, A = {}^{\nu}\!\!\int_M \kappa_A|_{\Delta_{\mathrm{b}}}\,, $$

using the canonical isomorphism ${}^{\mathrm{b}}\Omega^{1/2}(M_{\mathrm{b}}^2)|_{\Delta_{\mathrm{b}}} \cong {}^{\mathrm{b}}\Omega M$ (Sect. 2.5.26). If $A, B \in \Psi_{\mathrm{b}}^{-\infty}(M; {}^{\mathrm{b}}\Omega^{1/2})$, then [Mel93, Proposition 5.9]

$$ {}^{\mathrm{b}}\mathrm{Tr}[A, B] = -\frac{1}{2\pi i}\int_{-\infty}^{+\infty} \mathrm{Tr}(\partial_\lambda I_\nu(A, \lambda)\, I_\nu(B, \lambda))\, d\lambda\,. \tag{2.5.65} $$

This equality also holds if $A \in \mathrm{Diff}_{\mathrm{b}}(M; {}^{\mathrm{b}}\Omega^{1/2})$ and $B \in \Psi_{\mathrm{b}}^{-\infty}(M; {}^{\mathrm{b}}\Omega^{1/2})$ [Mel93, Lemma 5.10].

If E is a $\mathbb{Z}_2$-graded Hermitian vector bundle over M with degree involution w ($\mathsf{w}u = (-1)^k u$ for $u \in E^k$ and $k \in \mathbb{Z}_2$), and $A \in \Psi_{\mathrm{b}}^{-\infty}(M; E)$ is homogeneous of degree zero, then its *b-supertrace* is ${}^{\mathrm{b}}\mathrm{Str}\, A = {}^{\mathrm{b}}\mathrm{Tr}(A\mathsf{w})$. This notion extends the supertrace $\mathrm{Str}(B)$ of any homogeneous operator $B \in r\Psi_{\mathrm{b}}^{-\infty}(M; E)$ of degree zero.

2.6 Conormal Sequence

In this section and the next one, for the sake of simplicity, we only consider submanifolds of codimension one because that is the only case we use. However, the results can be extended to submanifolds of arbitrary codimension with more work.

2.6.1 Cutting Along a Submanifold

Again, for brevity reasons, we consider only the case of a closed manifold and the trivial line bundle. Like in other sections, the spaces of distributions we are going to define have obvious extensions to non-compact manifolds and arbitrary vector bundles, taking compact support or no support conditions, and taking regular submanifolds that are closed subspaces. We will consider those types of extensions without further comment.

Let M be a closed connected manifold, and $L \subset M$ be a (possibly non-connected) regular closed submanifold of codimension one. $M \setminus L$ may have several connected components. First assume also that L is transversely oriented. Then, like in the case with boundary (Sect. 2.5.1), there is some real-valued smooth function

x on some tubular neighborhood T of L in M, with projection $\varpi : T \to L$, so that $L = \{x = 0\}$ and $dx \neq 0$ on L. Any function x satisfying these conditions is called a *defining function* of L on T. We can suppose $T \equiv (-\epsilon, \epsilon)_x \times L_\varpi$, for some $\epsilon > 0$. (If M and L were not compact, and L were a regular submanifold that is a closed subset, then the tubular neighborhood would have a more involved expression, using a smooth positive function $\epsilon(y)$ on L instead of a fixed positive number ϵ.) For any atlas $\{V_j, y_j\}$ of L, we get an atlas of T of the form $\{U_j \equiv (-\epsilon, \epsilon) \times V_j, (x, y)\}$, whose charts are adapted to L. The corresponding local vector fields $\partial_x \in \mathfrak{X}(U_j)$ can be combined to define a vector field $\partial_x \in \mathfrak{X}(T)$; we can consider ∂_x as the derivative operator on $C^\infty(T) \equiv C^\infty((-\epsilon, \epsilon), C^\infty(L))$. For every j, $\operatorname{Diff}(U_j, L \cap U_j)$ is spanned by $x\partial_x, \partial_j^1, \dots, \partial_j^{n-1}$ using the operations of $C^\infty(U_j)$-module and algebra, where $\partial_j^k = \partial/\partial y_j^k$. Using $T \equiv (-\epsilon, \epsilon)_x \times L$, any $A \in \operatorname{Diff}(L)$ induces an operator $1 \otimes A \in \operatorname{Diff}(T, L)$, such that $(1 \otimes A)(u(x)v(y)) = u(x)\,(Av)(y)$ for $u \in C^\infty(-\epsilon, \epsilon)$ and $v \in C^\infty(L)$. This defines a canonical injection $\operatorname{Diff}(L) \equiv 1 \otimes \operatorname{Diff}(L) \subset \operatorname{Diff}(T, L)$ so that $(1 \otimes A)|_L = A$. (This also shows the surjectivity of (2.2.2) in this case.) Moreover $\operatorname{Diff}(T)$ (resp., $\operatorname{Diff}(T, L)$) is spanned by ∂_x (resp., $x\partial_x$) and $1 \otimes \operatorname{Diff}(L)$ using the operations of $C^\infty(T)$-module and algebra. Clearly,

$$[\partial_x, 1 \otimes \operatorname{Diff}(L)] = 0\,, \quad [\partial_x, x\partial_x] = \partial_x\,, \tag{2.6.1}$$

yielding

$$[\partial_x, \operatorname{Diff}^k(T, L)] \subset \operatorname{Diff}^k(T, L) + \operatorname{Diff}^{k-1}(T, L)\,\partial_x\,. \tag{2.6.2}$$

$\operatorname{Diff}^k(T, L)$ and $\operatorname{Diff}^k(T)$ satisfy the obvious versions of (2.5.26) and (2.5.27).

For a vector bundle E over M, there is an identity $E_T \equiv (-\epsilon, \epsilon) \times E_L$ over $T \equiv (-\epsilon, \epsilon) \times L$, which can be used to define $\partial_x \in \operatorname{Diff}^1(T; E)$. With this interpretation of ∂_x and using tensor products like in (2.1.6), the vector bundle versions of the concepts and properties of this section are straightforward.

Let $\boldsymbol{M}$ be the smooth manifold with boundary defined by "cutting" M along L; i.e., modifying M only on the tubular neighborhood $T \equiv (-\epsilon, \epsilon) \times L$, which is replaced with $\boldsymbol{T} = ((-\epsilon, 0] \sqcup [0, \epsilon)) \times L$ in the obvious way. ($\boldsymbol{M}$ is the blowing-up $[M, L]$ of M along L [Mel96, Chapter 5].) Thus $\partial\boldsymbol{M} \equiv L \sqcup L$ because L is transversely oriented, and $\mathring{\boldsymbol{M}} \equiv M \setminus L$. A canonical projection $\pi : \boldsymbol{M} \to M$ is defined as the combination of the identity map $\mathring{\boldsymbol{M}} \to M \setminus L$ and the map $\boldsymbol{T} \to T$ given by the product of the canonical projection $(-\epsilon, 0] \sqcup [0, \epsilon) \to (-\epsilon, \epsilon)$ and id_L. This projection realizes M as a quotient space of $\boldsymbol{M}$ by the equivalence relation defined by the homeomorphism $h \equiv h_0 \times \mathrm{id}$ of $\partial\boldsymbol{M} \equiv \partial\boldsymbol{T} = (\{0\} \sqcup \{0\}) \times L$, where h_0 switches the two points of $\{0\} \sqcup \{0\}$. Moreover $\pi : \boldsymbol{M} \to M$ is a local embedding of a compact manifold with boundary to a closed manifold of the same dimension.

Like in Sect. 2.1.6, we have the continuous linear pull-back map

$$\pi^* : C^\infty(M) \to C^\infty(\boldsymbol{M})\,, \tag{2.6.3}$$

which is clearly injective. The transpose of the version of (2.6.3) with ΩM and $\Omega \boldsymbol{M} \equiv \boldsymbol{\pi}^*\Omega M$ is the continuous linear push-forward map

$$\boldsymbol{\pi}_* : \dot{C}^{-\infty}(\boldsymbol{M}) \to C^{-\infty}(M) , \tag{2.6.4}$$

which is a surjective topological homomorphism [ÁLKL24, Proposition 7.4].

After distinguishing a connected component L_0 of L, let $\widetilde{M}$ and $\widetilde{L}$ be the quotients of $\boldsymbol{M}\sqcup\boldsymbol{M} \equiv \boldsymbol{M}\times\mathbb{Z}_2$ and $\partial\boldsymbol{M}\sqcup\partial\boldsymbol{M} \equiv \partial\boldsymbol{M}\times\mathbb{Z}_2$ by the equivalence relation generated by $(p, a) \sim (h(p), a)$ if $\boldsymbol{\pi}(p) \in L \setminus L_0$ and $(p, a) \sim (h(p), a + 1)$ if $\boldsymbol{\pi}(p) \in L_0$ ($p \in \boldsymbol{\pi}^{-1}(L) = \partial\boldsymbol{M}$ in both cases). Let us remark that $\widetilde{M}$ may not be homeomorphic to the double of $\boldsymbol{M}$, which is the quotient of $\boldsymbol{M} \times \mathbb{Z}_2$ by the equivalence relation generated by $(p, 0) \sim (p, 1)$, for $p \in \partial\boldsymbol{M}$. Note that $\widetilde{M}$ is a closed connected manifold and $\widetilde{L}$ is a closed regular submanifold. Moreover the quotient $\widetilde{T}$ of $\boldsymbol{T} \sqcup \boldsymbol{T}$ becomes a tubular neighborhood of $\widetilde{L}$ in $\widetilde{M}$. The combination $\boldsymbol{\pi} \sqcup \boldsymbol{\pi} : \boldsymbol{M} \sqcup \boldsymbol{M} \to M$ induces a twofold covering map $\tilde{\pi} : \widetilde{M} \to M$, whose restrictions to $\widetilde{L}$ and $\widetilde{T}$ are trivial twofold coverings of L and T, respectively; i.e., $\widetilde{L} \equiv L \sqcup L$ and $\widetilde{T} \equiv T \sqcup T$. The group of deck transformations of $\tilde{\pi} : \widetilde{M} \to M$ is $\{\mathrm{id}, \sigma\}$, where $\sigma : \widetilde{M} \to \widetilde{M}$ is induced by the map $\sigma_0 : \boldsymbol{M} \times \mathbb{Z}_2 \to \boldsymbol{M} \times \mathbb{Z}_2$ defined by switching the elements of $\mathbb{Z}_2$. The composition of the injection $\boldsymbol{M} \to \boldsymbol{M} \times \mathbb{Z}_2$, $p \mapsto (p, 0)$, with the quotient map $\boldsymbol{M} \sqcup \boldsymbol{M} \to \widetilde{M}$ is a smooth embedding $\boldsymbol{M} \to \widetilde{M}$. This will be considered as an inclusion map of a regular submanifold with boundary, obtaining $\partial\boldsymbol{M} \equiv \widetilde{L}$.

Since $\tilde{\pi}$ is a twofold covering map, we have continuous linear maps (Sect. 2.1.6)

$$\begin{gathered}\tilde{\pi}_* : C^{\infty}(\widetilde{M}) \to C^{\infty}(M) , \quad \tilde{\pi}^* : C^{\infty}(M) \to C^{\infty}(\widetilde{M}) , \\ \tilde{\pi}^* : C^{-\infty}(M) \to C^{-\infty}(\widetilde{M}) , \quad \tilde{\pi}_* : C^{-\infty}(\widetilde{M}) \to C^{-\infty}(M) ,\end{gathered} \tag{2.6.5}$$

both pairs of maps satisfying

$$\tilde{\pi}_*\tilde{\pi}^* = 2 , \quad \tilde{\pi}^*\tilde{\pi}_* = A_\sigma , \tag{2.6.6}$$

where $A_\sigma : C^{\pm\infty}(\widetilde{M}) \to C^{\pm\infty}(\widetilde{M})$ is given by $A_\sigma u = u + \sigma_* u$. Using the continuous linear restriction and inclusion maps given by (2.5.2) and (2.5.7), we get the commutative diagrams

$$\begin{array}{ccccccc} C^{\infty}(\widetilde{M}) & \xrightarrow{R} & C^{\infty}(\boldsymbol{M}) & \quad & \dot{C}^{-\infty}(\boldsymbol{M}) & \xrightarrow{\iota} & C^{-\infty}(\widetilde{M}) \\ {\scriptstyle\tilde{\pi}^*}\uparrow & & \uparrow{\scriptstyle\boldsymbol{\pi}^*} & & {\scriptstyle\boldsymbol{\pi}_*}\downarrow & & \downarrow{\scriptstyle\tilde{\pi}_*} \\ C^{\infty}(M) & = & C^{\infty}(M) , & & C^{-\infty}(M) & = & C^{-\infty}(M) , \end{array} \tag{2.6.7}$$

the second one is the transpose of the density-bundles version of the first one.

2.6.2 Lift of Differential Operators from M to $\widetilde{M}$

For any $A \in \mathrm{Diff}(M)$, let $\widetilde{A} \in \mathrm{Diff}(\widetilde{M})$ denote its lift via the covering map $\tilde{\pi} : \widetilde{M} \to M$. The action of $\widetilde{A}$ on $C^{\pm\infty}(\widetilde{M})$ corresponds to the action of A on $C^{\pm\infty}(M)$ via $\tilde{\pi}^* : C^{\pm\infty}(M) \to C^{\pm\infty}(\widetilde{M})$ and $\tilde{\pi}_* : C^{\pm\infty}(\widetilde{M}) \to C^{\pm\infty}(M)$. According to (2.5.22), $\widetilde{A}|_{\boldsymbol{M}} \in \mathrm{Diff}(\boldsymbol{M})$ is the lift of A via the local embedding $\pi : \boldsymbol{M} \to M$, sometimes also denoted by $\widetilde{A}$. The action of $\widetilde{A}$ on $C^{\infty}(\boldsymbol{M})$ (resp., $C^{-\infty}(\boldsymbol{M})$) corresponds to the action of A on $C^{\infty}(M)$ (resp., $C^{-\infty}(M)$) via $\pi^* : C^{\infty}(M) \to C^{\infty}(\boldsymbol{M})$ (resp., $\pi_* : C^{-\infty}(\boldsymbol{M}) \to C^{-\infty}(M)$). If $A \in \mathrm{Diff}(M, L)$, then $\widetilde{A} \in \mathrm{Diff}(\widetilde{M}, \widetilde{L})$ and $\widetilde{A}|_{\boldsymbol{M}} \in \mathrm{Diff}_{\mathrm{b}}(\boldsymbol{M})$ by (2.5.25).

2.6.3 The Spaces $C^{\pm\infty}(M, L)$

Consider the closed subspaces,

$$C^{\infty}(M, L) \subset C^{\infty}(M)\,, \quad C^{k}(M, L) \subset C^{k}(M) \quad (k \in \mathbb{N}_0)\,, \tag{2.6.8}$$

consisting of functions that vanish to all orders at the points of L in the first case, and that vanish up to order k at the points of L in the second case. Then let

$$C^{-\infty}(M, L) = C^{\infty}(M, L; \Omega)'\,, \quad C'^{-k}(M, L) = C^{k}(M, L; \Omega)'\,.$$

$C^{-\infty}(M, L)$ is a barreled, ultrabornological, webbed, acyclic DF Montel space, and therefore complete, boundedly/compactly/sequentially retractive and reflexive [ÁLKL24, Corollary 7.1]. Note that (2.6.3) restricts to TVS-isomorphisms

$$\pi^* : C^{\infty}(M, L) \xrightarrow{\cong} \dot{C}^{\infty}(\boldsymbol{M})\,, \quad \pi^* : C^{k}(M, L) \xrightarrow{\cong} \dot{C}^{k}(\boldsymbol{M})\,. \tag{2.6.9}$$

Taking the transposes of its versions with density bundles, it follows that (2.6.4) restricts to TVS-isomorphisms

$$\pi_* : C^{-\infty}(\boldsymbol{M}) \xrightarrow{\cong} C^{-\infty}(M, L)\,, \quad \pi_* : C'^{-k}(\boldsymbol{M}) \xrightarrow{\cong} C'^{-k}(M, L)\,. \tag{2.6.10}$$

So the spaces $C^{\infty}(M, L)$, $C^{k}(M, L)$, $C^{-\infty}(M, L)$ and $C'^{-k}(M, L)$ satisfy the analogs of (2.1.7) and (2.1.8).

On the other hand, there are Hilbertian spaces $H^{r}(M, L)$ ($r > n/2$) and $H'^{s}(M, L)$ ($s \in \mathbb{R}$), continuously included in $C^{0}(M, L)$ and $C^{-\infty}(M, L)$, resp., such that the second map of (2.6.9) for $k = 0$ and the first map of (2.6.10) restrict to TVS-isomorphisms

$$\pi^* : H^{r}(M, L) \xrightarrow{\cong} \dot{H}^{r}(\boldsymbol{M})\,, \quad \pi_* : H^{s}(\boldsymbol{M}) \xrightarrow{\cong} H'^{s}(M, L)\,. \tag{2.6.11}$$

For $s = 0$, the second TVS-isomorphism of (2.6.11) becomes

$$\pi_* : L^2(\boldsymbol{M}) \xrightarrow{\cong} L^2(M) . \tag{2.6.12}$$

By (2.5.16),

$$H'^{-r}(M, L) \equiv H^r(M, L; \Omega)' , \quad H^r(M, L) \equiv H'^{-r}(M, L; \Omega)' . \tag{2.6.13}$$

Now, the second identity of (2.6.13) can be used to extend the definition of $H^r(M, L)$ for all $r \in \mathbb{R}$.

Alternatively, we may also use trace theorems [Ada75, Theorem 7.53 and 7.58] to define $H^m(M, L)$ for $m \in \mathbb{Z}^+$, and then use the first identity of (2.6.13) to define $H'^{-m}(M, L)$.

From (2.6.3), (2.6.4), (2.6.11) and the analogs of (2.1.32)–(2.1.34) mentioned in Sect. 2.5.4, we get

$$C^\infty(M, L) = \bigcap_r H^r(M, L) , \quad C^{-\infty}(M, L) = \bigcup_s H'^s(M, L) , \tag{2.6.14}$$

as well as a continuous inclusion and a continuous linear surjection,

$$C^\infty(\boldsymbol{M}) \subset \bigcap_s H'^s(M, L) , \quad C^{-\infty}(\boldsymbol{M}) \leftarrow \bigcup_r H^r(M, L) . \tag{2.6.15}$$

By (2.6.13) and (2.6.14),

$$C^\infty(M, L) = C^{-\infty}(M, L; \Omega)' . \tag{2.6.16}$$

The transpose of the version of the first inclusion of (2.6.8) with ΩM is a surjective topological homomorphism [ÁLKL24, Proposition 7.4]

$$R : C^{-\infty}(\boldsymbol{M}) \to C^{-\infty}(M, L) , \tag{2.6.17}$$

whose restriction to $C^\infty(\boldsymbol{M})$ is the identity. It can be also described as the composition

$$C^{-\infty}(\boldsymbol{M}) \xrightarrow{\tilde{\pi}^*} C^{-\infty}(\widetilde{M}) \xrightarrow{R} C^{-\infty}(\boldsymbol{M}) \xrightarrow{\pi_*} C^{-\infty}(M, L) .$$

The canonical pairing between $C^\infty(\boldsymbol{M})$ and $C^\infty(M, L; \Omega)$ defines a continuous dense inclusion

$$C^\infty(\boldsymbol{M}) \subset C^{-\infty}(M, L) \tag{2.6.18}$$

such that (2.6.17) is the identity on $C^\infty(M)$. We also get commutative diagrams

$$\begin{array}{ccc} C^\infty(M) & \xleftarrow{\iota} & \dot C^\infty(M) \\ {\scriptstyle \pi^*}\uparrow & & \uparrow{\scriptstyle \cong\ \pi^*} \\ C^\infty(M) & \xleftarrow{\iota} & C^\infty(M,L)\,, \end{array} \qquad \begin{array}{ccc} \dot C^{-\infty}(M) & \xrightarrow{R} & C^{-\infty}(M) \\ {\scriptstyle \pi_*}\downarrow & & \downarrow{\scriptstyle \cong\ \pi_*} \\ C^{-\infty}(M) & \xrightarrow{R} & C^{-\infty}(M,L)\,, \end{array} \tag{2.6.19}$$

the second one is the transpose of the density-bundles version of the first one.

2.6.4 The Space $C^{-\infty}_L(M)$

The condition of being supported in L define closed subspaces,

$$C^{-\infty}_L(M) \subset C^{-\infty}(M)\,, \quad C'^{-k}_L(M) \subset C'^{-k}(M)\,, \quad H^s_L(M) \subset H^s(M)\,,$$

which are the null spaces of restrictions of (2.6.17). These spaces satisfy continuous inclusions analogous to (2.1.7), (2.1.31), and (2.1.33). The following properties hold [ÁLKL24, Corollaries 7.2 and 7.3]: $C^{-\infty}_L(M)$ is a limit subspace of the LF-space $C^{-\infty}(M)$; and it is a barreled, ultrabornological, webbed, acyclic DF Montel space, and therefore complete, boundedly/compactly/sequentially retractive and reflexive.

According to (2.5.19) and Sect. 2.6.1, we have [ÁLKL24, Eq. (7.19)]

$$\dot C^{-\infty}_{\partial M}(M) \equiv C^{-\infty}_L(M) \oplus C^{-\infty}_L(M)\,, \tag{2.6.20}$$

The maps (2.6.4) and (2.6.5) have restrictions

$$\pi_* = \tilde\pi_* : \dot C^{-\infty}_{\partial M}(M) \to C^{-\infty}_L(M)\,, \quad \tilde\pi^* : C^{-\infty}_L(M) \to \dot C^{-\infty}_{\partial M}(M)\,. \tag{2.6.21}$$

Using (2.6.20), these maps are given by $\pi_*(u,v) = u + v$ and $\tilde\pi^* u = (u,u)$.

Moreover the right-hand side diagram of (2.6.19) can be completed to get the commutative diagram

$$\begin{array}{ccccccccc} 0 \to & \dot C^{-\infty}_{\partial M}(M) & \xrightarrow{\iota} & \dot C^{-\infty}(M) & \xrightarrow{R} & C^{-\infty}(M) & \to 0 \\ & {\scriptstyle \pi_*}\downarrow & & {\scriptstyle \pi_*}\downarrow & & \downarrow{\scriptstyle \cong\ \pi_*} & \\ 0 \to & C^{-\infty}_L(M) & \xrightarrow{\iota} & C^{-\infty}(M) & \xrightarrow{R} & C^{-\infty}(M,L) & \to 0\,. \end{array} \tag{2.6.22}$$

The bottom row of this diagram is exact in the category of continuous linear maps between LCSs by the properties of (2.6.17).

2.6.5 A Description of $C_L^{-\infty}(M)$

According to (2.2.10) and Sects. 2.1.7 and 2.6.1, we have TVS-isomorphisms

$$\partial_x^m : C^{-\infty}(L; \Omega^{-1}NL) \xrightarrow{\cong} \partial_x^m C^{-\infty}(L; \Omega^{-1}NL) \subset C_L^{-\infty}(M) , \tag{2.6.23}$$

for $m \in \mathbb{N}_0$, inducing TVS-isomorphisms [ÁLKL24, Proposition 7.7]

$$\bigoplus_{m=0}^{\infty} C^{-\infty}(L; \Omega^{-1}NL) \xrightarrow{\cong} C_L^{-\infty}(M) , \tag{2.6.24}$$

$$\bigoplus_{m=0}^{k} C^{m-k}(L; \Omega^{-1}NL) \xrightarrow{\cong} C_L^{\prime -k}(M) \quad (k \in \mathbb{N}_0) . \tag{2.6.25}$$

Remark 2.6.1 (See [Mel96, Exercise 3.3.18]) In Sect. 2.5.5, for any compact manifold with boundary M, the analogs of (2.6.24) and (2.6.25) for $\dot{C}_{\partial M}^{-\infty}(M)$ follows from their application to $C_{\partial M}^{-\infty}(\breve{M})$.

2.6.6 Action of Diff(M) on $C^{-\infty}(M, L)$ and $C_L^{-\infty}(M)$

For every $A \in \mathrm{Diff}(M)$, A^{t} preserves $C^{\infty}(M, L; \Omega)$, and therefore A induces a continuous linear map $A = A^{tt}$ on $C^{-\infty}(M, L)$. By locality, it restricts to a continuous endomorphism A of $C_L^{-\infty}(M)$.

2.6.7 The Space $J(M, L)$

According to Sects. 2.5.8 and 2.6.3, there is a LCHS $J(M, L)$, with a dense continuous inclusion

$$J(M, L) \subset C^{-\infty}(M, L) , \tag{2.6.26}$$

so that (2.6.10) restricts to a TVS-isomorphism

$$\pi_* : \mathcal{A}(M) \xrightarrow{\cong} J(M, L) , \tag{2.6.27}$$

where $\mathcal{A}(M)$ is defined in (2.5.28). By (2.5.32), there is a continuous inclusion

$$J(M, L) \subset C^{\infty}(M \setminus L) .$$

We also get spaces $J^{(s)}(M, L)$ and $J^m(M, L)$ ($s, m \in \mathbb{R}$) corresponding to $\mathcal{A}^{(s)}(\boldsymbol{M})$ and $\mathcal{A}^m(\boldsymbol{M})$ via (2.6.27). Let $\boldsymbol{x}$ be an extension of $|x|$ to M so that it is positive and smooth on $M \setminus L$. Its lift $\boldsymbol{\pi}^*\boldsymbol{x}$ is a boundary-defining function of $\boldsymbol{M}$, also denoted by $\boldsymbol{x}$. Using the first map of (2.6.10) and second map of (2.6.11), and according to Sect. 2.6.2, we can also describe

$$J^{(s)}(M, L) = \{u \in C^{-\infty}(M, L) \mid \mathrm{Diff}(M, L)\, u \subset H'^{s}(M, L)\}\,, \tag{2.6.28}$$

$$J^m(M, L) = \{u \in C^{-\infty}(M, L) \mid \mathrm{Diff}(M, L)\, u \subset \boldsymbol{x}^m L^\infty(\boldsymbol{M})\}\,,$$

with topologies like in (2.1.1). These spaces satisfy the analogs of (2.2.4), (2.5.28) and (2.5.36)–(2.5.38). Using (2.6.15) and (2.6.28), we get a continuous dense inclusion [ÁLKL24, Eq. (7.28) and Corollary 7.14]

$$C^\infty(M) \subset J(M, L)\,. \tag{2.6.29}$$

In fact, $C^\infty_{\mathrm{c}}(M \setminus L)$ is dense in every $J^{(s)}(M, L)$ and $J^m(M, L)$, and therefore in $J(M, L)$ [ÁLKL24, Corollaries 7.14 and 7.17 and the analog of Remark 6.41 for $J(M, L)$]. Moreover the following properties hold [ÁLKL24, Corollaries 7.11–7.13 and 7.15]: every $J^{(s)}(M, L)$ is a totally reflexive Fréchet space; $J(M, L)$ is barreled, ultrabornological, webbed and an acyclic Montel space, and therefore complete, boundedly/compactly/sequentially retractive and reflexive; and the topologies of $J(M, L)$ and $C^\infty(M \setminus L)$ coincide on every $J^m(M, L)$.

2.6.8 A Description of $J(M, L)$

Take a b-metric $\boldsymbol{g}$ on $\boldsymbol{M}$ satisfying (A) and (B) (Sect. 2.5.20), and consider its restriction to $\mathring{\boldsymbol{M}}$. Consider also the boundary-defining function $\boldsymbol{x}$ of $\boldsymbol{M}$ (Sect. 2.6.7). Taking $m \in \mathbb{R}$, we have TVS-isomorphisms [ÁLKL24, Corollaries 7.16 and 7.18]

$$J^m(M, L) \cong \boldsymbol{x}^m H^\infty_{\mathrm{b}}(\boldsymbol{M}) \equiv \boldsymbol{x}^{m+1/2} H^\infty(\mathring{\boldsymbol{M}})\,, \tag{2.6.30}$$

$$J(M, L) \cong \bigcup_m \boldsymbol{x}^m H^\infty_{\mathrm{b}}(\boldsymbol{M}) = \bigcup_m \boldsymbol{x}^m H^\infty(\mathring{\boldsymbol{M}})\,. \tag{2.6.31}$$

The first isomorphisms of (2.6.30) and (2.6.31) are independent of $\boldsymbol{g}$; thus they hold without assuming (A) and (B) [ÁLKL24, the analog of Remark 6.41 for $J(M, L)$].

2.6.9 $I(M, L)$ vs $\dot{\mathcal{A}}(\boldsymbol{M})$ and $J(M, L)$

According to Sects. 2.2.8 and 2.2.9, we have the continuous linear maps

$$\tilde{\pi}^* : I(M, L) \to I(\widetilde{M}, \widetilde{L})\,, \quad \tilde{\pi}_* : I(\widetilde{M}, \widetilde{L}) \to I(M, L)\,, \tag{2.6.32}$$

which are restrictions of the maps (2.6.5), and therefore they satisfy (2.6.6). These maps are compatible with the symbol and Sobolev filtrations because $\tilde{\pi} : \widetilde{M} \to M$ is a covering map (Sects. 2.2.8 and 2.2.9).

Since (2.5.41) gives a TVS-embedding $\dot{\mathcal{A}}(\boldsymbol{M}) \subset I(\widetilde{M}, \widetilde{L})$, which preserves the Sobolev-order and symbol-order filtrations, the map $\tilde{\pi}_*$ of (2.6.32) has the restriction

$$\pi_* : \dot{\mathcal{A}}(\boldsymbol{M}) \to I(M, L) . \tag{2.6.33}$$

By (2.6.12) and according to Sect. 2.6.2, the map (2.6.33) restricts to a TVS-isomorphism

$$\pi_* : \dot{\mathcal{A}}^{(0)}(\boldsymbol{M}) \xrightarrow{\cong} I^{(0)}(M, L) . \tag{2.6.34}$$

On the other hand, the map (2.6.17) restricts to a continuous linear map

$$R : I(M, L) \to J(M, L) , \tag{2.6.35}$$

which is the identity on $C^\infty(M)$, and can be also described as the composition

$$I(M, L) \xrightarrow{\tilde{\pi}^*} I(\widetilde{M}, \widetilde{L}) \xrightarrow{R} \mathcal{A}(\boldsymbol{M}) \xrightarrow{\pi_*} J(M, L) .$$

Both (2.6.33) and (2.6.35) are surjective topological homomorphisms [ÁLKL24, Proposition 7.29], and therefore $C^\infty(M)$ is dense in $J(M, L)$ [ÁLKL24, Corollary 7.32].

2.6.10 The Space $K(M, L)$

Like in Sect. 2.5.13, the condition of being supported in L defines the LCHSs and $C^\infty(M)$-modules

$$K^{(s)}(M, L) = I_L^{(s)}(M, L) , \quad K^m(M, L) = I_L^m(M, L) , \quad K(M, L) = I_L(M, L) .$$

These are closed subspaces of $I^{(s)}(M, L)$, $I_L^m(M, L)$ and $I(M, L)$, respectively; more precisely, they are the null spaces of the corresponding restrictions of the map (2.6.35). The identity (2.6.20) restricts to a TVS-identity

$$\mathcal{K}(\boldsymbol{M}) \equiv K(M, L) \oplus K(M, L) . \tag{2.6.36}$$

Furthermore the maps (2.6.21) induce continuous linear maps

$$\pi_* : \mathcal{K}(\boldsymbol{M}) \to K(M, L) , \quad \tilde{\pi}^* : K(M, L) \to \mathcal{K}(\boldsymbol{M}) . \tag{2.6.37}$$

Using (2.6.36), these maps are given by $\pi_*(u,v) = u + v$ and $\tilde{\pi}^* u = (u,u)$.

By (2.5.40) and (2.5.42), $K^{(s)}(M,L)$ and $K^m(M,L)$ satisfy analogs of (2.6.36), using $\mathcal{K}^{(s)}(\boldsymbol{M})$ and $\mathcal{K}^m(\boldsymbol{M})$. The following properties hold true [ÁLKL24, Corollaries 7.19–7.21 and 7.23]: $K(M,L)$ is a limit subspace of the LF-space $I(M,L)$; every $K^{(s)}(M,L)$ is a totally reflexive Fréchet space; moreover it is barreled, ultrabornological and webbed, and therefore so is $K(M,L)$; and $K(M,L)$ is an acyclic Montel space, and therefore complete, boundedly/compactly/sequentially retractive and reflexive.

Example 2.6.2 With the notation of Sect. 2.1.9, $\mathrm{Diff}(M) \equiv K(M^2,\Delta)$ becomes a filtered $C^\infty(M^2)$-submodule of $\Psi(M)$, with the order filtration corresponding to the symbol filtration. In this way, $\mathrm{Diff}(M)$ also becomes a LCHS satisfying the above properties. If M is compact, it is also a filtered subalgebra of $\Psi(M)$.

2.6.11 A Description of $K(M,L)$

By (2.2.17) and (2.2.15), for $s < -1/2$, every isomorphism (2.6.23) restricts to a TVS-isomorphism

$$\partial_x^m : C^\infty(L;\Omega^{-1}NL) \xrightarrow{\cong} \partial_x^m C^\infty(L;\Omega^{-1}NL) \subset K^{(s-m)}(M,L)\,, \tag{2.6.38}$$

Then (2.6.24) restricts to a TVS-isomorphisms [ÁLKL24, Proposition 7.26]

$$\bigoplus_{m=0}^{\infty} C^\infty(L;\Omega^{-1}NL) \xrightarrow{\cong} K(M,L)\,, \tag{2.6.39}$$

$$\bigoplus_{m<-s-\frac{1}{2}} C^\infty(L;\Omega^{-1}NL) \xrightarrow{\cong} K^{(s)}(M,L) \quad (s < -1/2)\,. \tag{2.6.40}$$

Remark 2.6.3 In Sect. 2.5.13, for any compact manifold with boundary M, the analogs of (2.6.39) and (2.6.40) for $\mathcal{K}(M)$ follows from their application to $K(\breve{M},\partial M)$ using (2.5.44).

2.6.12 The Conormal Sequence

The diagram (2.6.22) has the restriction

$$\begin{array}{ccccccccc} 0 \to & \mathcal{K}(\boldsymbol{M}) & \xrightarrow{\iota} & \dot{\mathcal{A}}(\boldsymbol{M}) & \xrightarrow{R} & \mathcal{A}(\boldsymbol{M}) & \to 0 \\ & {\scriptstyle\pi_*}\downarrow & & {\scriptstyle\pi_*}\downarrow & & \cong\downarrow{\scriptstyle\pi_*} & \\ 0 \to & K(M,L) & \xrightarrow{\iota} & I(M,L) & \xrightarrow{R} & J(M,L) & \to 0\,. \end{array} \tag{2.6.41}$$

The bottom row of (2.6.41) is exact in the category of continuous linear maps between LCSs [ÁLKL24, Corollary 7.30]; it will be called the *conormal sequence* of M at L (or of (M, L)).

The surjectivity of (2.6.35) can be realized with the partial extension maps given by the following consequence of Proposition 2.5.1, whose proof is recalled by its relevance in Chaps. 5 and 6.

Corollary 2.6.4 ([ÁLKL24, Corollary 7.31]) *For all $m \in \mathbb{R}$, there is a continuous linear partial extension map $E_m : J^m(M, L) \to I^{(s)}(M, L)$, where $s = 0$ if $m \geq 0$, and $m > s \in \mathbb{Z}^-$ if $m < 0$. For $m \geq 0$, $E_m : J^m(M, L) \to I^{(0)}(M)$ is a continuous inclusion map.*

Proof By the commutativity of (2.6.41) and using Proposition 2.5.1, we can define $E_m : J^m(M, L) \to I^{(s)}(M, L)$ as the composition

$$J^m(M, L) \xrightarrow{\pi_*^{-1}} \mathcal{A}^m(\boldsymbol{M}) \xrightarrow{E_m} \dot{\mathcal{A}}^{(s)}(\boldsymbol{M}) \xrightarrow{\pi_*} I^{(s)}(M, L) \,.$$

The last assertion follows from Propositions 2.5.1 and 2.5.5 and (2.6.34). □

According to this proof, Remarks 2.5.2 to 2.5.4, 2.5.7, and 2.5.8, Proposition 2.5.5, and Corollary 2.5.6 have obvious versions for the maps given by Corollary 2.6.4.

2.6.13 *Action of* Diff(M) *on the Conormal Sequence*

According to Sect. 2.2.7, every $A \in \mathrm{Diff}(M)$ defines a continuous linear map A on $I(M, L)$, which preserves $K(M, L)$, and induces a continuous linear map A on $J(M, L)$. This map satisfies the analog of (2.2.15).

The map A on $J(M, L)$ can be also described as a restriction of A on $C^{-\infty}(M, L)$ (Sect. 2.6.6). On the other hand, according to Sect. 2.5.14, the lift $\widetilde{A} \in \mathrm{Diff}(\boldsymbol{M})$ defines continuous linear maps on the top spaces of (2.6.41) which correspond to the operators defined by A on the bottom spaces via the maps π_*. If $A \in \mathrm{Diff}(M, L)$, then it defines continuous endomorphisms A of $J^{(s)}(M, L)$ and $J^m(M, L)$.

2.6.14 *Pull-Back Maps on the Conormal Sequence*

Consider the notation and conditions of Sect. 2.2.8. By the exactness of the conormal sequences of (M, L) and (M', L') in the category of continuous linear maps between LCSs, the map (2.2.19) induces continuous linear maps,

$$\phi^* : K(M, L) \to K(M', L') \,, \tag{2.6.42}$$

$$\phi^* : J(M, L) \to J(M', L') . \tag{2.6.43}$$

The map (2.6.42) is the restriction of (2.2.19), which is well defined because the map (2.2.19) can be locally defined, and (2.6.43) is the induced map in the quotient. These maps are compatible with the maps ι and R of the conormal sequences, and satisfy the analog of (2.2.20).

2.6.15 *Push-Forward Maps on the Conormal Sequence*

Consider the notation and conditions of Sect. 2.2.9. Like in Sect. 2.6.14, the map (2.2.22) induces continuous linear maps,

$$\phi_* : K(M', L'; \Omega_{\text{fiber}}) \to K(M, L) , \tag{2.6.44}$$

$$\phi_* : J(M', L'; \Omega_{\text{fiber}}) \to J(M, L) . \tag{2.6.45}$$

They are also compatible with the maps ι and R of the conormal sequences, and satisfy the analog of (2.2.23).

2.6.16 *Case where L Is Not Transversely Orientable*

If L is not transversely orientable, we still have a tubular neighborhood T of L in M, but there is no defining function x of L in T trivializing the projection $\varpi : T \to L$. We can cut M along L as well to produce a bounded compact manifold, $\boldsymbol{M}$, with a projection $\pi : \boldsymbol{M} \to M$ and a boundary collar $\boldsymbol{T}$ over T.

Using a boundary-defining function $\boldsymbol{x}$ of $\boldsymbol{M}$, we get the same definitions, properties and descriptions of $C^{\pm\infty}(M, L)$ and $J(M, L)$ (Sects. 2.6.3, 2.6.7, and 2.6.8). $C_L^{-\infty}(M)$ and $K(M, L)$ also have the same definitions (Sects. 2.6.4 and 2.6.10). However (2.6.20) and (2.6.36) are not true because the covering map $\pi : \partial\boldsymbol{M} \to L$ is not trivial, and the descriptions given in (2.6.24), (2.6.25), (2.6.39), and (2.6.40) need a slight modification. This problem can be solved as follows.

Let $\check{\pi} : \check{L} \to L$ denote the twofold covering of transverse orientations of L, and let $\check{\sigma}$ denote its deck transformation different from the identity. Since the lift of NL to $\check{L}$ is trivial, $\check{\pi}$ on $\check{L} \equiv \{0\} \times \check{L}$ can be extended to a twofold covering $\check{\pi} : \check{T} := (-\epsilon, \epsilon)_x \times \check{L} \to T$, for some $\epsilon > 0$. Its deck transformation different from the identity is an extension of $\check{\sigma}$ on $\check{L} \equiv \{0\} \times \check{L}$, also denoted by $\check{\sigma}$. Then $\check{L}$ is transversely oriented in $\check{T}$; i.e., its normal bundle $N\check{L}$ is trivial. Thus $C_{\check{L}}^{-\infty}(\check{T})$ and $K(\check{T}, \check{L})$ satisfy (2.6.20), (2.6.24), (2.6.25), (2.6.36), (2.6.39), and (2.6.40). Since $N\check{L} \equiv \check{\pi}^*NL$, the map $\check{\sigma}$ lifts to a homomorphism of $N\check{L}$, which induces a homomorphism of $\Omega^{-1}NL$ also denoted by $\check{\sigma}$. Let L_{-1} be the union of non-transversely oriented connected components of L, and L_1 the union of its transversely oriented components. Correspondingly, let $\check{L}_{\pm 1} = \check{\pi}^{-1}(L_{\pm 1})$ and $\check{T}_{\pm 1} = (-\epsilon, \epsilon) \times \check{L}_{\pm 1}$. Since $\check{\sigma}^* x = \pm x$ on $T_{\pm 1}$, the isomorphisms (2.6.24), (2.6.25), (2.6.39) and (2.6.40) become true in this case by replacing $C^r(L; \Omega^{-1} NL)$ ($r \in \mathbb{Z} \cup \{\pm\infty\}$) with the direct sum of the spaces

$$\{ u \in C^r(L_{\pm 1}; \Omega^{-1} N L_{\pm 1}) \mid \check{\sigma}^* u = \pm u \} .$$

The other results about $C_L^{-\infty}(M)$ and $K(M, L)$ (Sects. 2.6.4, 2.6.5, 2.6.10, and 2.7.3) can be obtained by using these extensions of (2.6.24), (2.6.25), (2.6.39), and (2.6.40) instead of (2.6.20) and (2.6.36). Sections 2.6.12 to 2.6.15 also have straightforward extensions.

2.7 Dual-Conormal Sequence

2.7.1 *The Spaces $K'(M, L)$ and $J'(M, L)$*

Consider the notation of Sect. 2.6 assuming that L is transversely oriented; the extension to the non-transversely orientable case can be made with the procedure of Sect. 2.6.16. Like in Sects. 2.3.1 and 2.5.21, let

$$K'(M, L) = K(M, L; \Omega)' , \quad J'(M, L) = J(M, L; \Omega)' .$$

By (2.6.27) and (2.6.36),

$$\mathcal{K}'(M) \equiv K'(M, L) \oplus K'(M, L) , \quad \dot{\mathcal{A}}'(M) \equiv J'(M, L) . \tag{2.7.1}$$

Let also

$$\begin{cases} K'^{(s)}(M, L) = K^{(-s)}(M, L; \Omega)' , & K'^{m}(M, L) = K^{-m}(M, L; \Omega)' , \\ J'^{(s)}(M, L) = J^{(-s)}(M, L; \Omega)' , & J'^{m}(M, L) = J^{-m}(M, L; \Omega)' , \end{cases} \tag{2.7.2}$$

which satisfy the analog of (2.7.1). Like in Sect. 2.5.21, for $s < s'$ and $m < m'$, we get continuous linear restriction maps

$$K'^{(s')}(M, L) \to K'^{(s)}(M, L) , \quad K'^{m}(M, L) \to K'^{m'}(M, L) ,$$

and continuous injections

$$J'^{(s')}(M, L) \subset J'^{(s)}(M, L) , \quad J'^{m'}(M, L) \subset J'^{m}(M, L) ,$$

forming projective spectra. By (2.7.1) and its analog for the spaces (2.7.2), and according to Sect. 2.5.21, we get that the spaces $K'^{(s)}(M, L)$ and $K'^{m}(M, L)$ satisfy the analogs of (2.3.5) and (2.3.6), and the spaces $J'^{(s)}(M, L)$ and $J'^{m}(M, L)$ satisfy the analogs of (2.5.53) and (2.5.54) [ÁLKL24, Corollary 8.3]. Furthermore, $K'(M, L)$ and $J'(M, L)$ are complete Montel spaces [ÁLKL24, Corollary 8.1], and $K'^{(s)}(M, L)$ and $J'^{(s)}(M, L)$ are bornological and barreled [ÁLKL24, Corollary 8.2].

Like in Sect. 2.5.21, the versions of (2.6.16), (2.6.26), and (2.6.29) with ΩM induce continuous inclusions

$$C^{-\infty}(M) \supset J'(M, L) \supset C^{\infty}(M, L) . \tag{2.7.3}$$

2.7.2 Description of $J'(M, L)$

With the notation and conditions of Sect. 2.6.8, we have the following [ÁLKL24, Corollaries 8.4 and 8.5]:

$$J'^{m}(M, L) \cong \boldsymbol{x}^{m} H_{\mathrm{b}}^{-\infty}(\boldsymbol{M}) = \boldsymbol{x}^{m-\frac{1}{2}} H^{-\infty}(\mathring{\boldsymbol{M}}) , \tag{2.7.4}$$

$$J'(M, L) \cong \bigcap_{m} \boldsymbol{x}^{m} H_{\mathrm{b}}^{-\infty}(\boldsymbol{M}) = \bigcap_{m} \boldsymbol{x}^{m} H^{-\infty}(\mathring{\boldsymbol{M}}) . \tag{2.7.5}$$

Actually, the first isomorphisms of (2.7.4) and (2.7.5) are independent of g, and hold true without the assumptions (A) and (B). Furthermore $C_{\mathrm{c}}^{\infty}(M \setminus L)$ is dense in every $J'^{m}(M, L)$ and in $J'(M, L)$ [ÁLKL24, Corollary 8.6]. Therefore, the right-hand side inclusion of (2.7.3) is also dense.

2.7.3 Description of $K'(M, L)$

The transposes of the versions of (2.6.39) and (2.6.40) with ΩM are TVS-isomorphisms [ÁLKL24, Corollary 8.7],

$$K'(M, L) \xrightarrow{\cong} \prod_{m=0}^{\infty} C^{-\infty}(L) , \tag{2.7.6}$$

$$K'^{(s)}(M, L) \xrightarrow{\cong} \prod_{m<s-1/2} C^{-\infty}(L) \quad (s > 1/2) , \tag{2.7.7}$$

because

$$C^{\infty}(L; \Omega^{-1} NL \otimes \Omega M)' = C^{\infty}(L; \Omega)' = C^{-\infty}(L) .$$

2.7.4 Dual-Conormal Sequence

The transpose of the density-bundles version of (2.6.41) is the commutative diagram

$$\begin{array}{ccccc} 0 \leftarrow \mathcal{K}'(\boldsymbol{M}) & \xleftarrow{R'} & \mathcal{A}'(\boldsymbol{M}) & \xleftarrow{\iota'} & \dot{\mathcal{A}}'(\boldsymbol{M}) \leftarrow 0 \\ \pi^*\uparrow & & \pi^*\uparrow & & \pi^*\uparrow\cong \\ 0 \leftarrow K'(M,L) & \xleftarrow{R'} & I'(M,L) & \xleftarrow{\iota'} & J'(M,L) \leftarrow 0\,, \end{array} \tag{2.7.8}$$

where $R' = \iota^{\mathrm{t}}$ and $\iota' = R^{\mathrm{t}}$. Its bottom row is exact in the category of continuous linear maps between LCSs [ÁLKL24, Proposition 8.8], and is called the *dual-conormal sequence* of M at L (or of (M, L)).

2.7.5 Action of Diff(M) on the Dual-Conormal Sequence

With the notation of Sect. 2.6.13, consider the actions of A^{t} and $\widetilde{A}^{\mathrm{t}}$ on the bottom and top spaces of the version of (2.6.41) with ΩM and $\Omega \boldsymbol{M}$. Taking transposes again, we get induced actions of A and $\widetilde{A}$ on the bottom and top spaces of (2.7.8), which correspond one another via the linear maps π^*. These maps satisfy the analogs of (2.3.7).

2.7.6 Pull-Back Maps on the Dual-Conormal Sequence

Consider the notation and conditions of Sect. 2.3.5 (the same as in Sect. 2.2.9). Like in Sect. 2.3.5, transposing the compactly supported case of the analog of (2.2.23) for (2.6.44) and (2.6.45) with $E = \Omega M$, we get continuous linear maps,

$$\phi^* : K'(M, L) \to K'(M', L')\,, \tag{2.7.9}$$

$$\phi^* : J'(M, L) \to J'(M', L')\,. \tag{2.7.10}$$

They are compatible with the maps ι' and R' of the dual-conormal sequences, and satisfy the analog of (2.3.9).

2.7.7 Push-Forward Maps on the Dual-Conormal Sequence

Consider the notation and conditions of Sect. 2.3.6 (the same as in Sect. 2.2.8). Like in Sect. 2.3.6, transposing the analogs of (2.2.20) for (2.8.20) and (2.8.21) with

$E = \Omega M$, and using an analog of (2.1.16), we get continuous linear maps,

$$\phi_* : K'(M', L'; \Omega_{\text{fiber}}) \to K'(M, L) , \tag{2.7.11}$$

$$\phi_* : J'(M', L'; \Omega_{\text{fiber}}) \to J'(M, L) . \tag{2.7.12}$$

They are compatible with the maps ι' and R' of the dual-conormal sequences, and satisfy the analog of (2.3.11).

2.8 Currents

Here, again, the manifold M may not be compact, and $L \subset M$ is a regular submanifold that is a closed subset. When using $J(M, L; \Lambda)$ or $K(M, L; \Lambda)$, it is also assumed that L is of codimension one.

2.8.1 *Differential Forms and Currents*

Consider the space $C^\infty(M; \Lambda)$ of smooth differential forms, and the space $C^{-\infty}(M; \Lambda)$ of currents. The most typical example of elliptic complex is given by the de Rham derivative d on $C^\infty(M; \Lambda)$, giving rise to the de Rham cohomology $H^\bullet(M)$. The extension of d to $C^{-\infty}(M; \Lambda)$ is another topological complex, which produces isomorphic cohomology [dR84]. We typically consider cohomology with complex coefficients without further comment; real cohomology classes are only considered in a few cases, where it is indicated; the same applies to other cohomologies that will be considered. The basic properties of $(C^{\pm\infty}(M; \Lambda), d)$ and $H^\bullet(M)$ can be seen in [dR84, BT82]; for instance, the general properties of elliptic complexes apply in this setup (Sect. 2.1.14). Some properties will be seen in Sect. 2.9 with more generality.

A Riemannian metric g on M defines a Hermitian structure on ΛM, also denoted by g. Then we have the additional operators δ (the *de Rham coderivative*), D and Δ (the *Laplacian*) of Sect. 2.1.14. If needed, the subscript "M" may be added to this notation, and to other similar notation.

We may also consider the de Rham complex with coefficients in a flat vector bundle $\mathcal{F}$, $d = d^{\mathcal{F}}$ on $C^\infty(M; \Lambda \otimes \mathcal{F})$. As above, g and a Hermitian structure on $\mathcal{F}$ induce additional operators $\delta = \delta^{\mathcal{F}}$, $D = D^{\mathcal{F}}$ and $\Delta = \Delta^{\mathcal{F}}$.

For any $V \in \mathfrak{X}(M)$, let ι_V and $\mathcal{L}_V$ denote the corresponding inner product and Lie derivative on $C^\infty(M; \Lambda)$. For $\eta = V^\flat \in C^\infty(M; \Lambda^1)$, we write $\eta\lrcorner = -(\eta\wedge)^* = -\iota_V$. Let w be the degree involution on ΛM. For the bundle of Clifford algebras of T^*M, we have the identity $\mathrm{Cl}(T^*M) \equiv \Lambda_{\mathbb{R}} M$ defined by the symbol of filtered algebras. Via this identity, the left Clifford multiplication by η is $c(\eta) = \eta\wedge + \eta\lrcorner$, and the composition of w with the right Clifford multiplication by η is $\hat{c}(\eta) = \eta\wedge - \eta\lrcorner$.

2.8.2 Product of Differential Forms and Currents

The exterior product of smooth differential forms has continuous extensions,[1]

$$C^{\pm\infty}(M;\Lambda)\otimes C^{\mp\infty}(M;\Lambda)\to C^{-\infty}(M;\Lambda)\;. \tag{2.8.1}$$

For example, with the notation of Sect. 2.2.6, assuming that M and L are oriented, it easily follows that, for $\alpha\in C^{\infty}(M;\Lambda)$ and $\beta\in C^{\infty}(L;\Lambda\otimes\Omega^{-1}NL)$,

$$\alpha\wedge\delta_L^{\beta}=\delta_L^{\alpha|_L\wedge\beta}\;,\quad \delta_L^{\beta}\wedge\alpha=\delta_L^{\beta\wedge\alpha|_L}\;. \tag{2.8.2}$$

2.8.3 Currents on Oriented Manifolds

Assume M is oriented. The orientation induces a canonical identity $\Omega M\equiv\Lambda^n M$. Then, for every degree k, the non-degenerate pairing $\Lambda^k M\otimes\Lambda^{n-k}M\to\Lambda^n M$ defined by the wedge product induces a canonical identity

$$(\Lambda^k M)^*\otimes\Omega M\equiv\Lambda^{n-k}M\;. \tag{2.8.3}$$

By (2.8.3), the space (2.1.5) becomes

$$C^{-\infty}_{\cdot/\mathrm{c}}(M;\Lambda^k)\equiv C^{\infty}_{\mathrm{c}/\cdot}(M;\Lambda^{n-k})' \tag{2.8.4}$$

in this case. This identity corresponds to a pairing

$$C^{\pm\infty}_{\cdot/\mathrm{c}}(M;\Lambda^k)\otimes C^{\mp\infty}_{\mathrm{c}/\cdot}(M;\Lambda^{n-k})\to\mathbb{C}\;,$$

which will be denoted with parentheses to distinguish it from the scalar product. This pairing can be given by the composition of (2.8.1) and the extension

$$C^{-\infty}_{\mathrm{c}}(M;\Lambda^n)\to\mathbb{C}\;,\quad \alpha\mapsto(\alpha,1)\;,$$

of $\int_M : C^{\infty}_{\mathrm{c}}(M;\Lambda M)\to\mathbb{C}$.

2.8.4 Hodge Operator on Oriented Manifolds

Continuing with the assumption of orientation, let $\star$ on ΛM denote the $\mathbb{C}$-linear extension of the Hodge operator $\star$ on the real forms, which is unitary, and let $\bar\star$ denote its $\mathbb{C}$-antilinear extension. These operators are determined by the conditions, for $\alpha,\beta\in C^{\infty}(M;\Lambda)$,

$$\alpha\wedge\overline{\star\beta}=g(\alpha,\beta)\,\mathrm{dvol}=\alpha\wedge\bar\star\beta\;,$$

[1] This holds with more generality under conditions on the wavefront set [Hör71, Theorem 8.2.10], but we will not use it.

where $\mathrm{dvol} = \star 1$ is the volume form. Recall that, on $C^\infty(M; \Lambda^k)$,

$$\begin{cases} \star^2 = (-1)^{nk+k}\,, \quad \delta = (-1)^{nk+n+1} \star d \star\,, \quad \eta\lrcorner = (-1)^{nk+n+1} \star \eta\wedge\star\,, \\ d\star = (-1)^k \star \delta\,, \quad \delta\star = (-1)^{k+1} \star d\,, \quad \Delta\star = \star\Delta\,, \\ \eta\wedge\star = (-1)^k \star \eta\lrcorner\,, \quad \eta\lrcorner\star = (-1)^{k+1} \star \eta\wedge\,. \end{cases} \tag{2.8.5}$$

The equalities (2.8.5) are also true with $\bar\star$, and can be extended to $C^{-\infty}(M; \Lambda)$.

For all $\alpha \in C^\infty(M; \Lambda^k)$ and $\beta \in C^\infty_{\mathrm{c}}(M; \Lambda^{n-k})$,

$$\alpha \wedge \beta = (-1)^{kn+k} \alpha \wedge \bar\star^2 \beta = (-1)^{kn+k} g(\alpha, \bar\star\beta)\,\mathrm{dvol}\,,$$

yielding

$$(\alpha, \beta) = (-1)^{kn+k} \langle \alpha, \bar\star\beta \rangle\,. \tag{2.8.6}$$

2.8.5 Pull-Back and Push-Forward of Currents

Given a smooth map $\phi : M' \to M$, recall that its tangent map $T\phi = \phi_* : TM' \to TM$ defines a homomorphism $\phi_* : TM' \to \phi^* TM$, which induces a homomorphism

$$\phi^* : \phi^* \Lambda M \to \Lambda M'\,. \tag{2.8.7}$$

Then recall that the pull-back homomorphism

$$\phi^* : C^\infty(M; \Lambda) \to C^\infty(M'; \Lambda) \tag{2.8.8}$$

can be given as the composition

$$C^\infty(M; \Lambda) \xrightarrow{\phi^*} C^\infty(M'; \phi^* \Lambda M) \xrightarrow{\phi^*} C^\infty(M'; \Lambda)\,, \tag{2.8.9}$$

where the first map ϕ^* is given by (2.1.13), and the second map ϕ^* is induced by (2.8.7).

Now, suppose ϕ is a submersion and its vertical subbundle $\mathcal{V}$ is oriented. Let $\pi_{\mathrm{top}} : \Lambda\mathcal{V} \to \Lambda^{\mathrm{top}}\mathcal{V}$ denote the canonical projection. The orientation of $\mathcal{V}$ gives a canonical identity $\Omega_{\mathrm{fiber}} \equiv \Lambda^{\mathrm{top}}\mathcal{V}$. So

$$\phi^* \Lambda M \otimes \Omega_{\mathrm{fiber}} \equiv \phi^* \Lambda M \otimes \Lambda^{\mathrm{top}}\mathcal{V} \subset \phi^* \Lambda M \otimes \Lambda\mathcal{V} \equiv \Lambda M'\,.$$

Moreover, $\pi_{\mathrm{top}} : \Lambda\mathcal{V} \to \Lambda^{\mathrm{top}}\mathcal{V}$ induces a projection

$$\pi_{\mathrm{top}} : \Lambda M' \to \phi^* \Lambda M \otimes \Omega_{\mathrm{fiber}}\,. \tag{2.8.10}$$

The push-forward homomorphism or *integration along the fibers* [BT82, Section I.6],

$$\phi_* : C^{\infty}_{c/cv}(M'; \Lambda) \to C^{\infty}_{c/\cdot}(M; \Lambda) , \tag{2.8.11}$$

can be described as the composition

$$C^{\infty}_{c/cv}(M'; \Lambda) \xrightarrow{\pi_{top}} C^{\infty}_{c/cv}(M'; \phi^*\Lambda M \otimes \Omega_{fiber}) \xrightarrow{\phi_*} C^{\infty}_{c/\cdot}(M; \Lambda) , \tag{2.8.12}$$

where π_{top} is induced by (2.8.10), and ϕ_* is given by (2.1.14) with $E = \Lambda M$.

We also get the push-forward and pull-back maps on currents,

$$\phi_* : C^{-\infty}_{c/cv}(M'; \Lambda) \to C^{-\infty}_{c/\cdot}(M; \Lambda) , \tag{2.8.13}$$

$$\phi^* : C^{-\infty}(M; \Lambda) \to C^{-\infty}(M'; \Lambda) , \tag{2.8.14}$$

given by the compositions

$$C^{-\infty}_{c/cv}(M'; \Lambda) \xrightarrow{\pi_{top}} C^{-\infty}_{c/cv}(M'; \phi^*\Lambda M \otimes \Omega_{fiber}) \xrightarrow{\phi_*} C^{\infty}_{c/\cdot}(M; \Lambda) , \tag{2.8.15}$$

$$C^{-\infty}(M; \Lambda) \xrightarrow{\phi^*} C^{-\infty}(M'; \phi^*\Lambda M) \xrightarrow{\phi^*} C^{-\infty}(M'; \Lambda) , \tag{2.8.16}$$

where ϕ_* and the first map ϕ^* are given by (2.1.17)–(2.1.19) with $E = \Lambda M$, and π_{top} is induced by (2.8.10). The notation f_ϕ is also used for ϕ_*, or f_F if ϕ is a trivial bundle with typical fiber F.

Proposition 2.8.1 *The compactly supported case of* (2.8.13) *is the transpose of* (2.8.8), *and* (2.8.14) *is the transpose of the compactly supported case of* (2.8.11).

Proof By passing to double covers of orientations, we can assume M and M' are oriented, and therefore we can use (2.8.4). By the density of the space of smooth forms in the space of currents (Sect. 2.1.4), it is enough to check the statement on smooth forms, where it is given by [BT82, Proposition 6.15 (b)]: for $\alpha \in C^{\infty}(M; \Lambda)$ and $\beta \in C^{\infty}_{c}(M'; \Lambda)$,

$$(\phi^*\alpha, \beta) = \int_{M'} \phi^*\alpha \wedge \beta = \int_M \alpha \wedge \phi_*\beta = (\alpha, \phi_*\beta) .$$

□

2.8.6 Homotopy Operators

Recall that any smooth homotopy, $H : M' \times I \to M$ ($I = [0, 1]$), induces a continuous homotopy operator $\mathsf{h} : C^{\infty}(M'; \Lambda) \to C^{\infty}(M; \Lambda)$ (a linear map, which is homogeneous of degree -1, and satisfies $H_1^* - H_0^* = \mathsf{h}d + d\mathsf{h}$, where $H_t = H(\cdot, t) : M' \to M$). For instance, we can take h equal to the composition [BT82, Section 4]

$$C^\infty(M; \Lambda) \xrightarrow{H^*} C^\infty(M' \times I; \Lambda) \xrightarrow{f_I} C^\infty(M'; \Lambda) . \tag{2.8.17}$$

2.8.7 Pull-Back of Conormal Currents

With the notations and conditions of Sect. 2.2.8, the map (2.8.8) has a continuous extension

$$\phi^* : I(M, L; \Lambda) \to I(M', L; \Lambda) , \tag{2.8.18}$$

which can be given as the composition

$$I(M, L; \Lambda) \xrightarrow{\phi^*} I(M', L'; \phi^* \Lambda M) \xrightarrow{\phi^*} I(M', L; \Lambda) , \tag{2.8.19}$$

where the first map ϕ^* is given by (2.2.20) with $E = \Lambda M$, and the second map ϕ^* is induced by (2.8.7). If ϕ is a smooth submersion with oriented vertical subbundle, then (2.8.18) is also a restriction of (2.8.14).

Similarly, when L is of codimension one, there are continuous homomorphisms,

$$\phi^* \cdot K(M, L; \Lambda) \to K(M', L'; \Lambda) , \tag{2.8.20}$$

$$\phi^* : J(M, L; \Lambda) \to J(M', L'; \Lambda) , \tag{2.8.21}$$

which can be given as the compositions

$$K(M, L; \Lambda) \xrightarrow{\phi^*} K(M', L'; \phi^* \Lambda M) \xrightarrow{\phi^*} K(M', L; \Lambda) , \tag{2.8.22}$$

$$J(M, L; \Lambda) \xrightarrow{\phi^*} J(M', L'; \phi^* \Lambda M) \xrightarrow{\phi^*} J(M', L; \Lambda) , \tag{2.8.23}$$

where the first maps ϕ^* are given by the analogs of (2.2.20) with $E = \Lambda M$ for (2.6.42) and (2.6.43), and the second maps ϕ^* are induced by (2.8.7).

2.8.8 Push-Forward of Conormal Currents

With the notations and conditions of Sect. 2.2.9, assume also that the vertical subbundle of ϕ is oriented. Then the push-forward homomorphism (2.8.11) has a continuous extension

$$\phi_* : I_{c/cv}(M', L'; \Lambda) \to I_{c/\cdot}(M, L; \Lambda) , \tag{2.8.24}$$

which can be described as the composition

$$I_{c/cv}(M', L'; \Lambda) \xrightarrow{\pi_{top}} I_{c/cv}(M', L'; \phi^*\Lambda M \otimes \Omega_{fiber}) \xrightarrow{\phi_*} I_{c/\cdot}(M, L; \Lambda) , \tag{2.8.25}$$

where π_{top} is induced by (2.8.10), and ϕ_* is given by (2.2.23) with $E = \Lambda M$. The map (2.8.24) is also a restriction of (2.8.13).

Similarly, if L is of codimension one, there are continuous homomorphisms,

$$\phi_* : K_{c/cv}(M', L'; \Lambda) \to K_{c/\cdot}(M, L; \Lambda) , \tag{2.8.26}$$

$$\phi_* : J_{c/cv}(M', L'; \Lambda) \to J_{c/\cdot}(M, L; \Lambda) . \tag{2.8.27}$$

which can be described as the compositions

$$K_{c/cv}(M', L'; \Lambda) \xrightarrow{\pi_{top}} K_{c/cv}(M', L'; \phi^*\Lambda M \otimes \Omega_{fiber}) \xrightarrow{\phi_*} K_{c/\cdot}(M, L; \Lambda) , \tag{2.8.28}$$

$$J_{c/cv}(M', L'; \Lambda) \xrightarrow{\pi_{top}} J_{c/cv}(M', L'; \phi^*\Lambda M \otimes \Omega_{fiber}) \xrightarrow{\phi_*} J_{c/\cdot}(M, L; \Lambda) , \tag{2.8.29}$$

where the maps π_{top} are induced by (2.8.10), and the maps ϕ_* are given by the analogs of (2.2.23) with $E = \Lambda M$ for (2.6.44) and (2.6.45).

2.8.9 Pull-Back of Dual-Conormal Currents

Consider the notations and conditions of Sect. 2.3.5 (the same as in Sect. 2.2.9). The map (2.8.8) has a continuous extension

$$\phi^* : I'(M, L; \Lambda) \to I'(M', L'; \Lambda) , \tag{2.8.30}$$

which can be given as the composition

$$I'(M, L; \Lambda) \xrightarrow{\phi^*} I'(M', L'; \phi^*\Lambda M) \xrightarrow{\phi^*} I'(M', L'; \Lambda) ,$$

using (2.3.9) like in (2.8.19). The map (2.8.30) is also a restriction of (2.8.14).

Similarly, when L is of codimension one, there are continuous homomorphisms,

$$\phi^* : K'(M, L; \Lambda) \to K'(M', L'; \Lambda) , \tag{2.8.31}$$

$$\phi^* : J'(M, L; \Lambda) \to J'(M', L'; \Lambda) , \tag{2.8.32}$$

which can be given as the compositions

$$K'(M, L; \Lambda) \xrightarrow{\phi^*} K'(M', L'; \phi^* \Lambda M) \xrightarrow{\phi^*} K'(M', L'; \Lambda) ,$$

$$J'(M, L; \Lambda) \xrightarrow{\phi^*} J'(M', L'; \phi^* \Lambda M) \xrightarrow{\phi^*} J'(M', L'; \Lambda) ,$$

using the analogs of (2.3.9) for (2.7.9) and (2.7.10) like in (2.8.22) and (2.8.23).

2.8.10 Push-Forward of Dual-Conormal Currents

With the notations and conditions of Sect. 2.3.6, assume also that the vertical subbundle of ϕ is oriented. Then the map (2.8.11) has a continuous extension

$$\phi_* : I'_{c/cv}(M', L'; \Lambda) \to I'_{c/\cdot}(M, L; \Lambda) , \tag{2.8.33}$$

which can be described as the composition

$$I'_{c/cv}(M', L'; \Lambda) \xrightarrow{\pi_{top}} I'_{c/cv}(M', L'; \phi^* \Lambda M \otimes \Omega_{fiber}) \xrightarrow{\phi_*} I'_{c/\cdot}(M, L; \Lambda) ,$$

using (2.3.11) like in (2.8.25). The map (2.8.33) is also a restriction of (2.8.13).

Similarly, if L is of codimension one, there are continuous homomorphisms,

$$\phi_* : K'_{c/cv}(M', L'; \Lambda) \to K'_{c/\cdot}(M, L; \Lambda) , \tag{2.8.34}$$

$$\phi_* : J'_{c/cv}(M', L'; \Lambda) \to J'_{c/\cdot}(M, L; \Lambda) , \tag{2.8.35}$$

which can be described as the compositions

$$K'_{c/cv}(M', L'; \Lambda) \xrightarrow{\pi_{top}} K'_{c/cv}(M', L'; \phi^* \Lambda M \otimes \Omega_{fiber}) \xrightarrow{\phi_*} K'_{c/\cdot}(M, L; \Lambda) ,$$

$$J'_{c/cv}(M', L'; \Lambda) \xrightarrow{\pi_{top}} J'_{c/cv}(M', L'; \phi^* \Lambda M \otimes \Omega_{fiber}) \xrightarrow{\phi_*} J'_{c/\cdot}(M, L; \Lambda) ,$$

using the analogs of (2.3.11) for (2.7.11) and (2.7.12) like in (2.8.28) and (2.8.29).

Proposition 2.8.2 *The compact-support cases of* (2.8.33)–(2.8.35) *are transposes of* (2.8.18), (2.8.20) *and* (2.8.21)*; and* (2.8.30)–(2.8.32) *are transposes of the compact-support cases of* (2.8.24), (2.8.26), *and* (2.8.27).

Proof We have the commutative diagrams

$$
\begin{array}{ccccccc}
I_{\mathrm{c}}(M', L'; \Lambda) & \xrightarrow{\phi_*} & I_{\mathrm{c}}(M, L; \Lambda) & \qquad & I'(M', L'; \Lambda) & \xleftarrow{\phi^*} & I'(M, L; \Lambda) \\
\uparrow & & \uparrow & & \downarrow & & \downarrow \\
C^\infty_{\mathrm{c}}(M'; \Lambda) & \xrightarrow{\phi_*} & C^\infty_{\mathrm{c}}(M; \Lambda) & & C^{-\infty}(M'; \Lambda) & \xleftarrow{\phi^*} & C^{-\infty}(M; \Lambda) \,,
\end{array}
$$

where the vertical arrows are continuous dense inclusions given by (2.2.5) and (2.3.4) with ΛM. By Proposition 2.8.1, the transpose of the first diagram is

$$
\begin{array}{ccc}
I'(M', L'; \Lambda) & \xleftarrow{(\phi_*)^{\mathrm{t}}} & I'(M, L; \Lambda) \\
\downarrow & & \downarrow \\
C^{-\infty}(M'; \Lambda) & \xleftarrow{\phi^*} & C^{-\infty}(M; \Lambda) \,,
\end{array}
$$

where the vertical arrows are again inclusion maps. Comparing the second and third diagrams, we get

$$
(\phi_*)^{\mathrm{t}} = \phi^* : I'(M, L; \Lambda) \to I'(M', L'; \Lambda) \,. \tag{2.8.36}
$$

The analogous argument with the commutative diagrams

$$
\begin{array}{ccccccc}
I(M', L'; \Lambda) & \xleftarrow{\phi^*} & I(M, L; \Lambda) & \qquad & I'_{\mathrm{c}}(M', L'; \Lambda) & \xrightarrow{\phi_*} & I'_{\mathrm{c}}(M, L; \Lambda) \\
\uparrow & & \uparrow & & \downarrow & & \downarrow \\
C^\infty(M'; \Lambda) & \xleftarrow{\phi^*} & C^\infty(M; \Lambda) & & C^{-\infty}_{\mathrm{c}}(M'; \Lambda) & \xrightarrow{\phi_*} & C^{-\infty}_{\mathrm{c}}(M; \Lambda)
\end{array}
$$

shows that

$$
(\phi^*)^{\mathrm{t}} = \phi_* : I'_{\mathrm{c}}(M', L'; \Lambda) \to I'_{\mathrm{c}}(M, L; \Lambda) \,. \tag{2.8.37}
$$

Next, consider the commutative diagrams

$$
\begin{array}{ccccccc}
K_{\mathrm{c}}(M', L'; \Lambda) & \xrightarrow{\phi_*} & K_{\mathrm{c}}(M, L; \Lambda) & \qquad & K'(M', L'; \Lambda) & \xleftarrow{\phi^*} & K'(M, L; \Lambda) \\
\iota\downarrow & & \downarrow\iota & & R'\uparrow & & \uparrow R' \\
I_{\mathrm{c}}(M', L'; \Lambda) & \xrightarrow{\phi_*} & I_{\mathrm{c}}(M, L; \Lambda) & & I'(M', L'; \Lambda) & \xleftarrow{\phi^*} & I'(M, L; \Lambda) \,.
\end{array}
$$

As above, comparing the second one with the transposition of the first one, and using (2.8.36) and the surjectivity of $R' : I'(M, L; \Lambda) \to K'(M, L; \Lambda)$

(Sect. 2.7.4), we get

$$(\phi_*)^{\mathrm{t}} = \phi^* : K'(M, L; \Lambda) \to K'(M', L'; \Lambda) .$$

A similar argument with the commutative diagrams

$$\begin{CD} K(M', L'; \Lambda) @<{\phi^*}<< K(M, L; \Lambda) \\ @V{\iota}VV @VV{\iota}V \\ I(M', L'; \Lambda) @<{\phi^*}<< I(M, L; \Lambda) \end{CD} \qquad \begin{CD} K'_{\mathrm{c}}(M', L'; \Lambda) @>{\phi_*}>> K'_{\mathrm{c}}(M, L; \Lambda) \\ @A{R'}AA @AA{R'}A \\ I'_{\mathrm{c}}(M', L'; \Lambda) @>{\phi_*}>> I'_{\mathrm{c}}(M, L; \Lambda) \end{CD} ,$$

using (2.8.37), shows that

$$(\phi^*)^{\mathrm{t}} = \phi_* : K'_{\mathrm{c}}(M', L'; \Lambda) \to K'_{\mathrm{c}}(M, L; \Lambda) .$$

Now, consider the commutative diagrams

$$\begin{CD} J_{\mathrm{c}}(M', L'; \Lambda) @>{\phi_*}>> J_{\mathrm{c}}(M, L; \Lambda) \\ @A{R}AA @AA{R}A \\ I_{\mathrm{c}}(M', L'; \Lambda) @>{\phi_*}>> I_{\mathrm{c}}(M, L; \Lambda) \end{CD} \qquad \begin{CD} J'(M', L'; \Lambda) @<{\phi^*}<< J'(M, L; \Lambda) \\ @V{\iota'}VV @VV{\iota'}V \\ I'(M', L'; \Lambda) @<{\phi^*}<< I'(M, L; \Lambda) \end{CD} .$$

Again, comparing the second one with the transposition of the first one, and using (2.8.36) and the injectivity of $\iota' : J'(M, L; \Lambda) \to I'(M, L; \Lambda)$ (Sect. 2.7.4), we get

$$(\phi_*)^{\mathrm{t}} = \phi^* : J'(M, L; \Lambda) \to J'(M', L'; \Lambda) .$$

Finally, the same argument with the commutative diagrams

$$\begin{CD} J(M', L'; \Lambda) @<{\phi^*}<< J(M, L; \Lambda) \\ @A{R}AA @AA{R}A \\ I(M', L'; \Lambda) @<{\phi^*}<< I(M, L; \Lambda) \end{CD} \qquad \begin{CD} J'_{\mathrm{c}}(M', L'; \Lambda) @>{\phi_*}>> J'_{\mathrm{c}}(M, L; \Lambda) \\ @V{\iota'}VV @VV{\iota'}V \\ I'_{\mathrm{c}}(M', L'; \Lambda) @>{\phi_*}>> I'_{\mathrm{c}}(M, L; \Lambda) \end{CD} ,$$

using (2.8.37), gives

$$(\phi^*)^{\mathrm{t}} = \phi_* : J'_{\mathrm{c}}(M', L'; \Lambda) \to J'_{\mathrm{c}}(M, L; \Lambda) .$$

□

2.9 Witten's Perturbation of the de Rham Complex

2.9.1 *Witten's Complex*

The notation $z = \mu + i\lambda \in \mathbb{C}$ ($i = \sqrt{-1}$) will be used for a complex parameter. Any closed real $\eta \in C^\infty(M;\Lambda^1)$ induces the *Witten's operators* on $C^\infty(M;\Lambda)$, depending on the parameter $z \in \mathbb{C}$ [Wit82, Nov81, Nov82, Paj87, BF97],

$$\begin{cases} d_z = d + z\,\eta\wedge\,, \quad \delta_z = d_z^* = \delta - \bar{z}\,\eta\lrcorner\,, \\ D_z = d_z + \delta_z = D + \mu\hat{c}(\eta) + i\lambda c(\eta)\,, \\ \Delta_z = D_z^2 = d_z\delta_z + \delta_z d_z = \Delta + \mu \mathsf{H}_\eta - i\lambda \mathsf{J}_\eta + |z|^2|\eta|^2\,, \end{cases} \tag{2.9.1}$$

where $\mathsf{H}_\eta = \mathcal{L}_V + \mathcal{L}_V^*$ is of order zero and $\mathsf{J}_\eta = \mathcal{L}_V - \mathcal{L}_V^*$ is of order one. Here, d_z is an elliptic complex, giving rise to the *twisted cohomology* $H_z^\bullet(M)$, whose isomorphism class depends only on the real class $\xi := [\eta] \in H^1(M)$ and $z \in \mathbb{C}$. The more explicit notation $d_{z\eta}$, $\delta_{z\eta}$, $D_{z\eta}$ and $\Delta_{z\eta}$ may be used if needed.

Suppose the manifold M is closed, and let $n = \dim M$. Then Δ_z has a discrete spectrum, and the perturbed operators satisfy (2.1.37). We get the *twisted Betti numbers*, $\beta_z^k = \beta_z^k(M,\xi) = \dim H_z^k(M)$ ($k = 0,\dots,n$), whose alternate sum is the Euler characteristic, $\sum_k(-1)^k\beta_z^k = \chi(M)$ [Far04, Proposition 1.40]. Every β_z^k is independent of z outside a discrete subset of $\mathbb{C}$, where β_z^k jumps (Mityagin and Novikov [Nov02, Theorem 1]). This ground value of β_z^k, denoted by $\beta_{\mathrm{No}}^k = \beta_{\mathrm{No}}^k(M,\xi)$, is called the kth *Novikov Betti number*. Moreover $\beta_z^k = \beta_{\mathrm{No}}^k$ for $|\mu| \gg 0$ [Far95, Theorem 2.8], [BF97, Lemma 1.3] [ÁLKL25, Eq. (2.9)].

Since η is real, we have $\overline{d_z\alpha} = d_{\bar{z}}\bar{\alpha}$ for all $\alpha \in C^\infty(M;\Lambda)$. So conjugation induces a $\mathbb{C}$-antilinear isomorphism $H_z^k(M) \cong H_{\bar{z}}^k(M)$, yielding $\beta_z^k = \beta_{\bar{z}}^k$.

For $\alpha \in C^\infty(M;\Lambda^r)$ and $\beta \in C^\infty(M;\Lambda)$, we have

$$d(\alpha\wedge\beta) = d_z\alpha\wedge\beta + (-1)^r\alpha\wedge d_{-z}\beta\,. \tag{2.9.2}$$

It follows that the mappings $(\alpha,\beta)\mapsto\alpha\wedge\beta$ and $(\alpha,\beta)\mapsto\alpha\wedge\bar{\beta}$ induce maps,

$$H_z^r(M)\times H_{-z}^s(M)\to H^{r+s}(M)\,,\quad H_z^r(M)\times H_{-\bar{z}}^s(M)\to H^{r+s}(M)\,, \tag{2.9.3}$$

the first one is bilinear and the second one is sesquilinear. By density and continuity, the formula (2.9.2) has an extension to the product (2.8.1) of smooth differential forms and currents.

2.9.2 Interpretation as Coefficients in a Flat Line Bundle

If $\eta = dF$ for some real function $F \in C^\infty(M)$, we get the original operators introduced by Witten [Wit82], which satisfy

$$\begin{cases} d_z = e^{-zF}\,d\,e^{zF} = e^{-i\lambda F}\,d_\mu\,e^{i\lambda F}\,, & \delta_z = e^{\bar{z}F}\,\delta\,e^{-\bar{z}F} = e^{-i\lambda F}\,\delta_\mu\,e^{i\lambda F}\,,\\ D_z = e^{-i\lambda F}\,D_\mu\,e^{i\lambda F}\,, & \Delta_z = e^{-i\lambda F}\,\Delta_\mu\,e^{i\lambda F}\,. \end{cases} \tag{2.9.4}$$

Thus we have an isomorphism of differential complexes,

$$e^{zF} : (C^\infty(M;\Lambda), d_z) \xrightarrow{\cong} (C^\infty(M;\Lambda), d)\,,$$

which induces an isomorphism $H_z^\bullet(M) \cong H^\bullet(M)$.

Let $\mathcal{L}$ be the trivial line bundle $M \times \mathbb{C}$ with the flat structure that corresponds to the trivial flat structure by the multiplication isomorphism $e^F : \mathcal{L} \to M \times \mathbb{C}$, $(p,u) \mapsto (p, e^{F(p)}u)$. Its flat covariant derivative is determined by the condition $d^{\mathcal{L}}1 = dF$. Every power $\mathcal{L}^z$ is similarly defined by the function zF. We have $d_z \equiv d^{\mathcal{L}^z}$ on $C^\infty(M;\Lambda) \equiv C^\infty(M;\Lambda \otimes \mathcal{L}^z)$. Moreover $\delta_z \equiv \delta^{\mathcal{L}^z}$ and $\Delta_z \equiv \Delta^{\mathcal{L}^z}$ using the standard Hermitian structure on $\mathcal{L}^z$.

For arbitrary η, take the minimal regular covering $\pi : \widetilde{M} \to M$ so that the lift $\tilde{\eta}$ of η is exact, say $\tilde{\eta} = dF$ for some real function $F \in C^\infty(\widetilde{M})$. Thus $d_{\widetilde{M},z} = e^{-zF}\,d_{\widetilde{M}}\,e^{zF}$ on $C^\infty(\widetilde{M};\Lambda)$ corresponds to $d_{M,z}$ on $C^\infty(M;\Lambda)$ via the injection $\pi^* : C^\infty(M;\Lambda) \to C^\infty(\widetilde{M};\Lambda)$. Let $\Gamma = \mathrm{Aut}(\pi)$ be the group of deck transformations of $\widetilde{M}$. The action of every $\gamma \in \Gamma$ will be denoted by T_γ or by $\tilde{p} \mapsto \gamma \cdot \tilde{p}$. Since dF is Γ-invariant, there is a monomorphism $\Gamma \to \mathbb{R}$, $\gamma \mapsto c_\gamma$, so that $F(\gamma \cdot \tilde{p}) = F(\tilde{p}) + c_\gamma$ for all $\tilde{p} \in \widetilde{M}$; its image is the group of periods of the cohomology class $[\eta]$.

Let $\widetilde{\mathcal{L}}$ be the flat line bundle over $\widetilde{M}$ defined with F as above. The flat structure of $\widetilde{\mathcal{L}}$ is invariant by the first factor action of Γ on $\widetilde{\mathcal{L}}$, given by $\gamma \cdot_1 (\tilde{p}, u) = (\gamma \cdot \tilde{p}, u)$. Thus the corresponding quotient Hermitian line bundle $\mathcal{L} \equiv M \times \mathbb{C}$ has an induced flat structure determined by the condition $d^{\mathcal{L}}1 = z\eta$. We have $d_z \equiv d^{\mathcal{L}^z}$, $\delta_z \equiv \delta^{\mathcal{L}^z}$ and $\Delta_z \equiv \Delta^{\mathcal{L}^z}$ on $C^\infty(M;\Lambda) \equiv C^\infty(M;\Lambda \otimes \mathcal{L}^z)$.

Using the monomorphism $\Gamma \to \mathbb{R}^\times$, $\gamma \mapsto a_\gamma := e^{c_\gamma}$, we can also define the diagonal action of Γ on $\widetilde{M} \times \mathbb{C}$, $\gamma \cdot (\tilde{p}, u) = (\gamma \cdot \tilde{p}, a_\gamma u)$, which preserves the vector bundle and trivial flat structures. Moreover the isomorphism $e^F : \widetilde{\mathcal{L}} \to \widetilde{M} \times \mathbb{C}$ is equivariant with respect to the first factor and diagonal actions of Γ. Hence $\mathcal{L}$ can be also described as the quotient of the trivial flat line bundle $\widetilde{M} \times \mathbb{C}$ by the diagonal action of Γ. Let $\tilde{\omega} \in C^\infty(\widetilde{M};\widetilde{\mathcal{L}})$ be defined by $\tilde{\omega}(\tilde{p}) = (\tilde{p}, e^{F(\tilde{p})})$, which corresponds to $1 \in C^\infty(\widetilde{M}) \equiv C^\infty(\widetilde{M};\widetilde{\mathcal{L}})$ by the isomorphism $e^F : \widetilde{\mathcal{L}} \to \widetilde{M} \times \mathbb{C}$. This section is Γ-invariant and satisfies $d^{\widetilde{\mathcal{L}}}\tilde{\omega} = \tilde{\eta} \otimes \tilde{\omega}$ in $C^\infty(\widetilde{M};\Lambda \otimes \widetilde{\mathcal{L}})$. So it induces a non-vanishing section ω of $\mathcal{L}$ satisfying $d^{\mathcal{L}}\omega = \eta \otimes \omega$ in $C^\infty(M;\Lambda \otimes \mathcal{L})$.

Furthermore

$$\begin{cases} C^{\pm\infty}(M;\Lambda\otimes\mathcal{L}^z)\equiv C^{\pm\infty}(M;\Lambda)\otimes\mathbb{R}\omega^z\equiv C^{\pm\infty}(M;\Lambda)\,, \\ d^{\mathcal{L}^z}\equiv d_z\otimes 1\equiv d_z\,,\quad H^\bullet_z(M)\equiv H^\bullet(M,\mathcal{L}^z)\,, \\ \delta\equiv\delta_z\otimes 1\equiv\delta_z\,,\quad D\equiv D_z\otimes 1\equiv D_z\,,\quad \Delta\equiv\Delta_z\otimes 1\equiv\Delta_z\,, \end{cases} \tag{2.9.5}$$

writing $d=d^{\mathcal{L}^z}$, $\delta=\delta^{\mathcal{L}^z}$, $D=D^{\mathcal{L}^z}$, $\Delta=\Delta^{\mathcal{L}^z}$. Since $(\mathcal{L}^z)^*\equiv\mathcal{L}^{-z}$, this gives an interpretation of (2.9.2) and (2.9.3).

2.9.3 *Witten's Perturbation vs Pull-Back and Push-Forward Homomorphisms*

For a smooth map $\phi : M'\to M$, let $\eta'=\phi^*\eta$. The homomorphism $\phi^* : C^\infty(M;\Lambda)\to C^\infty(M';\Lambda)$ satisfies $\phi^*d_{z\eta}=d_{z\eta'}\phi^*$. If ϕ is a smooth submersion, then $\phi_* : C^\infty_{\mathrm{c/cv}}(M';\Lambda)\to C^\infty_{\mathrm{c/.}}(M;\Lambda)$ satisfies $\phi_*d_{z\eta'}=d_{z\eta}\phi_*$ by [BT82, Proposition I.6.14 and I.6.15 (a)].

2.9.4 *Perturbation of Pull-Back Homomorphisms*

Consider the notation of Sect. 2.9.2. For a smooth map $\phi : M\to M$, take a lift $\tilde\phi:\widetilde M\to\widetilde M$. Then $\tilde\phi^*_z := e^{-zF}\,\tilde\phi^*\,e^{zF}=e^{z(\tilde\phi^*F-F)}\,\tilde\phi^*$ is an endomorphism of $(C^\infty(\widetilde M;\Lambda),d_{\widetilde M,z})$ by (2.9.4). We have $T^*_\gamma(\tilde\phi^*F-F)=\tilde\phi^*F-F$ for all $\gamma\in\Gamma$, obtaining $T^*_\gamma\tilde\phi^*_z=\tilde\phi^*_zT^*_\gamma$. So $\tilde\phi^*_z$ induces an endomorphism ϕ^*_z of $(C^\infty(M;\Lambda),d_z)$, which depends on the choice of the lift $\tilde\phi$ of ϕ. In the case of a flow $\phi=\{\phi^t\}$ on M, there is a unique lift to a flow $\tilde\phi=\{\tilde\phi^t\}$ on $\widetilde M$, giving rise to a canonical definition of ϕ^{t*}_z, called the *perturbation* of ϕ^{t*} defined by η with parameter z.

2.9.5 *Witten's Operators on Oriented Manifolds*

In this subsection, assume M is oriented. If moreover M is closed, then the maps (2.9.3) and integration on M define nondegenerate pairings,

$$H^k_z(M)\times H^{n-k}_{-z}(M)\to\mathbb{C}\,,\quad H^k_z(M)\times H^{n-k}_{-\bar z}(M)\to\mathbb{C}\,, \tag{2.9.6}$$

the first one is bilinear and the second one is sesquilinear. Therefore $\beta^k_z=\beta^{n-k}_{-z}=\beta^{n-k}_{-\bar z}=\beta^k_{\bar z}$.

2.9.6 Witten's Operators vs Hodge Star Operator

Continuing with the condition of orientation, the equalities (2.8.5) yield

$$\begin{cases} \delta_z = (-1)^{nk+n+1} \star d_{-\bar{z}} \star = (-1)^{nk+n+1} \bar{\star} d_{-z} \bar{\star} \,, \\ d_z \star = (-1)^k \star \delta_{-\bar{z}} \,, \quad \delta_z \star = (-1)^{k+1} \star d_{-\bar{z}} \,, \quad \Delta_z \star = \star \Delta_{-\bar{z}} \,, \\ d_z \bar{\star} = (-1)^k \bar{\star} \delta_{-z} \,, \quad \delta_z \bar{\star} = (-1)^{k+1} \bar{\star} d_{-z} \,, \quad \Delta_z \bar{\star} = \bar{\star} \Delta_{-z} \,. \end{cases} \tag{2.9.7}$$

Then we get a linear isomorphism $\star : \ker \Delta_z \to \ker \Delta_{-\bar{z}}$ and an antilinear isomorphism $\bar{\star} : \ker \Delta_z \to \ker \Delta_{-z}$. If M is closed, they induce an explicit linear isomorphism $H_z^k(M) \cong H_{-\bar{z}}^{n-k}(M)$ and an antilinear isomorphism $H_z^k(M) \cong H_{-z}^{n-k}(M)$ by (2.1.37).

By (2.9.2) and the Stokes theorem, we get

$$d_z \equiv (-1)^{k+1} d_{-z}^{\mathrm{t}} \,, \tag{2.9.8}$$

as maps $C^{-\infty}(M; \Lambda^k) \to C^{-\infty}(M; \Lambda^{k+1})$ using (2.8.4). This identity also follows from (2.8.5), (2.9.7) and (2.8.6): for $\alpha \in C^\infty(M; \Lambda^k)$ and $\beta \in C^\infty(M; \Lambda^{n-k-1})$,

$$\begin{aligned} (d_z\alpha, \beta) &= (-1)^{(k+1)n+k+1} \langle d_z\alpha, \bar{\star}\beta \rangle = (-1)^{(k+1)n+k+1} \langle \alpha, \delta_z \bar{\star}\beta \rangle \\ &= (-1)^{(kn+1} \langle \alpha, \bar{\star} d_{-z}\beta \rangle = (-1)^{k+1} (\alpha, d_{-z}\beta) \,. \end{aligned}$$

This argument also applies to δ_z and Δ_z, giving

$$\begin{aligned} \delta_z &\equiv (-1)^k \delta_{-z}^{\mathrm{t}} : C^{-\infty}(M; \Lambda^k) \to C^{-\infty}(M; \Lambda^{k-1}) \,, \\ \Delta_z &\equiv \Delta_{-z}^{\mathrm{t}} : C^{-\infty}(M; \Lambda^k) \to C^{-\infty}(M; \Lambda^k) \,. \end{aligned} \tag{2.9.9}$$

2.9.7 Perturbed Operators with Two Parameters

We will also consider perturbed operators of the form

$$D_{z,z'} = d_z + \delta_{z'} \,, \quad \Delta_{z,z'} = D_{z,z'}^2 = d_z \delta_{z'} + \delta_{z'} d_z \,,$$

depending on two parameters $z, z' \in \mathbb{C}$. They are not symmetric if $z \neq z'$, but their leading symbol is symmetric.

2.9.8 Witten's Operators on Manifolds of Bounded Geometry

Consider now the notation of Sects. 2.8.1 and 2.9.1. Assume M is of bounded geometry and $\eta \in C_{\mathrm{ub}}^\infty(M; \Lambda^1)$ (Sect. 2.4.3). Then the differential complex d_z is uniformly bounded and uniformly elliptic for all $z \in \mathbb{C}$.

Using also the notation of Sect. 2.9.4, assume that $\phi : M \to M$ is of bounded geometry. Then $\tilde{\phi}^* F - F$ induces a function in $C^\infty_{\mathrm{ub}}(M)$. For $m \in \mathbb{N}_0 \cup \{\infty\}$, it follows from (2.4.6) that ϕ_z^* defines a continuous linear endomorphism of $C^m_{\mathrm{ub}}(M; \Lambda)$. If moreover $\phi : M \to M$ is uniformly metrically proper, then, by (2.4.7), ϕ_z^* also defines a continuous linear endomorphism of $H^m(M; \Lambda)$.

If ϕ is a diffeomorphism and both of $\phi^{\pm 1}$ are of bounded geometry, then ϕ_z^* defines a continuous linear endomorphism of $H^m(M; \Lambda)$ for all $m \in \mathbb{Z} \cup \{\pm\infty\}$. To show this, we can assume M is oriented with a standard argument using the covering of orientations. Then, by the version of the second equality of (2.1.30) for open manifolds and (2.8.3), ϕ_z^* on $H^{-m}(M; \Lambda)$ ($m \in \mathbb{N}_0 \cup \{\infty\}$) is the transpose of $(\phi^{-1})^*_{-z}$ on $H^m(M; \Lambda^{n-\bullet})$.

In the cases of $C^\infty_{\mathrm{ub}}(M; \Lambda)$ and $H^{\pm\infty}(M; \Lambda)$, all of the above endomorphisms are cochain maps with d_z.

Any solution of the symmetric hyperbolic equation

$$\partial_t \alpha_t = i D_z \alpha_t \;, \quad \alpha_0 = \alpha \;, \tag{2.9.10}$$

on an open subset of M and with t in any interval containing 0, satisfies the finite propagation speed property [Che73, Proof of Proposition 1.1] (see also [CGT82, Theorem 1.4], [Roe98, Proof of Proposition 7.20])

$$\operatorname{supp} \alpha_t \subset \operatorname{Pen}(\operatorname{supp} \alpha, |t|) \;. \tag{2.9.11}$$

In particular, given any $\alpha \in C^\infty(M; \Lambda)$, this is true for $\alpha_t = e^{itD_z}\alpha$.

For $\psi \in \mathcal{R}$ (Sect. 2.4.5), we may use the notation $k_z = k_{\psi,z} = K_{\psi(D_z)}$, where $\psi(D_z)$ is given by the spectral theorem. We may also use the notation $k_{u,z} = k_{\psi_u,z}$ for any family of functions $\psi_u \in \mathcal{R}$ depending on a parameter u.

For any $\psi \in \mathcal{S}$ (Sect. 2.9.12), we have [Roe88, Proof of Theorem 5.5]

$$\psi(D_z) = (2\pi)^{-1} \int_{-\infty}^{+\infty} e^{i\xi D_z} \hat{\psi}(\xi)\, d\xi \;. \tag{2.9.12}$$

According to Remark 2.4.14, it follows from (2.9.11) and (2.9.12) that, for all $r > 0$,

$$\operatorname{supp} \hat{\psi} \subset [-r, r] \Rightarrow \operatorname{supp} k_{\psi,z} \subset \{\, (p, q) \in M^2 \mid d(p, q) \le r \,\} \;. \tag{2.9.13}$$

For instance, for $\psi_u(x) = e^{-ux^2}$ ($u > 0$), we get the *perturbed heat kernel* $k_{u,z} = K_{e^{-u\Delta_z}}$. It satisfies the following estimate like the usual heat kernel [BE91]: for all $u_0 > 0$ and $m_1, m_2, m_3 \in \mathbb{N}_0$, there are $C_1, C_2 > 0$ so that, for all $0 < u \le u_0$,

$$|\partial_u^{m_1} \nabla_p^{m_2} \nabla_q^{m_3} k_{z,u}(p, q)| \le C_1 u^{-(n+m_2+m_3)/2-m_1} e^{-C_2 d^2(p,q)/u} \;. \tag{2.9.14}$$

In particular, $k_{z,u} \in C^\infty_{\mathrm{ub}}(M^2; \Lambda \boxtimes (\Lambda^* \otimes \Omega))$ for every $u > 0$.

To estimate more general kernels, consider the Fréchet algebra and $\mathbb{C}[z]$-module $\mathcal{A}$ which consists of the functions $\psi : \mathbb{R} \to \mathbb{C}$ that can be extended to entire functions on $\mathbb{C}$ such that, for every compact $K \subset \mathbb{R}$, the set $\{ x \mapsto \psi(x+iy) \mid y \in K \}$ is bounded in $\mathcal{S}$ [Roe87, Section 4]. It has the following properties: $\mathcal{A} \subset \mathcal{S}$; $\mathcal{A}$ contains all functions with compactly supported smooth Fourier transform, as well as the Gaussian $x \mapsto e^{-x^2}$; if $\psi \in \mathcal{A}$ and $u > 0$, then $\psi_u \in \mathcal{A}$, where $\psi_u(x) = \psi(ux)$; and, by the Paley-Wiener theorem, for every $\psi \in \mathcal{A}$ and $c > 0$, there is some $A_c > 0$ such that, for all $\xi \in \mathbb{R}$,

$$\left|\hat{\psi}(\xi)\right| \le A_c e^{-c|\xi|} . \tag{2.9.15}$$

Define the semi-norms $\|\cdot\|_{\mathcal{A},C,r}$ ($C > 0$ and $r \in \mathbb{N}_0$) on $\mathcal{A}$ by

$$\|\psi\|_{\mathcal{A},C,r} = \max_{j+k\le r} \int_{-\infty}^{+\infty} |\xi^j \partial_\xi^k \hat{\psi}(\xi)| \, e^{C|\xi|} \, d\xi .$$

Lemma 2.9.1 *If $\psi \in \mathcal{A}$ and $N > n/2$, then, for any $W > 0$, there is some $C_1' = C_1'(z, W) > 0$ such that, for all $p, q \in M$ and $m, m_1, m_2 \in \mathbb{N}_0$ with $m_1 + m_2 \le m$,*

$$|\nabla_p^{m_1} \nabla_q^{m_2} k_z(p,q)| \le C_1' e^{-Wd(p,q)} \|\psi\|_{\mathcal{A},W,N+m} .$$

Proof Using (2.9.11), (2.9.12) and the Sobolev embedding theorem, one can show that, for every $\epsilon > 0$, there is some $C_0 = C_0(z, \epsilon) > 0$ so that, for all $\psi \in \mathcal{A}$ and $p, q \in M$,

$$|k_z(p,q)| \le C_0 \int_{|\xi|>d(p,q)-\epsilon} \left|(1-\partial_\xi^2)^N \hat{\psi}(\xi)\right| d\xi .$$

Hence, for some fixed $\epsilon > 0$, we obtain that, for any $W > 0$, there is some $C_1 = C_1(z, W) > 0$ such that, for all $p, q \in M$,

$$\begin{aligned} |k_z(p,q)| &\le C_0 \int_{|\xi|>d(p,q)-\epsilon} e^{-W|\xi|} \left|(1-\partial_\xi^2)^N \hat{\psi}(\xi)\right| e^{W|\xi|} \, d\xi \\ &\le C_1 e^{-Wd(p,q)} \int_{-\infty}^{+\infty} \left|(1-\partial_\xi^2)^N \hat{\psi}(\xi)\right| e^{W|\xi|} \, d\xi \\ &= C_1 e^{-Wd(p,q)} \|\psi\|_{\mathcal{A},W,N} . \end{aligned}$$

By using $(1+x^2)^m \psi(x)$ ($m \in \mathbb{N}_0$) instead of $\psi(x)$, we also get

$$|(1+\Delta_{z,p})^{m_1} (1+\Delta_{-z,q})^{m_2} k_z(p,q)| \le C_1 e^{-Wd(p,q)} \|\psi\|_{\mathcal{A},W,N+m} ,$$

according to (2.8.4) and (2.9.9), yielding the estimate of the statement. □

2.9.9 Witten's Operators on Regular Coverings of Compact Manifolds

Let $\pi : \widetilde{M} \to M$, Γ, $\gamma \cdot \tilde{p}$, T_γ and $g_{\widetilde{M}}$ be like in Sect. 2.9.2. Recall that $\widetilde{M}$ is of bounded geometry with $g_{\widetilde{M}}$.

Let $|\cdot| : \Gamma \to \mathbb{N}_0$ denote the word length function defined by any finite set of generators $\gamma_1, \dots, \gamma_k$ of Γ; recall that $|\gamma|$ is the minimum length of the expressions of γ as products of elements $\gamma_i^{\pm 1}$. It is well known that there is some $c_1 \geq 1$ such that

$$c_1^{-1}\,|\gamma| \leq d_{\widetilde{M}}(\gamma \cdot \tilde{p}, \tilde{p}) \leq c_1\,|\gamma| \tag{2.9.16}$$

for all $\tilde{p} \in \widetilde{M}$ and $\gamma \in \Gamma$. Therefore, given any compact $\boldsymbol{K} \subset \widetilde{M}^2$, we have

$$c_1^{-1}\,|\gamma| - c_2 \leq d_{\widetilde{M}}(\gamma \cdot \tilde{p}, \tilde{q}) \leq c_1\,|\gamma| + c_2 \tag{2.9.17}$$

for all $\gamma \in \Gamma$ and $(\tilde{p}, \tilde{q}) \in \boldsymbol{K}$, where $c_2 = \max d_{\widetilde{M}}(\boldsymbol{K}) \geq 0$.

Let η be a closed real 1-form on M whose lift to $\widetilde{M}$ is exact; say $\tilde{\eta} = d_{\widetilde{M}}F$ for some $F \in C^\infty(\widetilde{M}, \mathbb{R})$. For $z \in \mathbb{C}$, let $D_z = D_{M,z}$, $\Delta_z = \Delta_{M,z}$, $\widetilde{D}_z = D_{\widetilde{M},z}$ and $\widetilde{\Delta}_z = \Delta_{\widetilde{M},z}$ (Sect. 2.9.1). For any $\psi \in \mathcal{R}$, let $k_z = K_{\psi(D_z)}$ and $\tilde{k}_z = K_{\psi(\widetilde{D}_z)}$ (Sect. 2.9.8). For every $\tilde{p} \in \widetilde{M}$, let $[\tilde{p}] = \pi(\tilde{p})$. We look for conditions on ψ to get

$$k_z([\tilde{p}], [\tilde{q}]) \equiv \sum_\gamma T_\gamma^* \tilde{k}_z(\gamma \cdot \tilde{p}, \tilde{q}) \tag{2.9.18}$$

for all $\tilde{p}, \tilde{q} \in \widetilde{M}$, using the identity

$$\Lambda_{\gamma \cdot \tilde{p}} \widetilde{M} \boxtimes (\Lambda_{\tilde{q}} \widetilde{M}^* \otimes \Omega_{\tilde{q}} \widetilde{M}) \equiv \Lambda_{[p]} M \boxtimes (\Lambda_{[q]} M^* \otimes \Omega_{[q]} M)\,.$$

In particular, (2.9.18) holds if $\hat{\psi} \in C_c^\infty(\mathbb{R})$, which can be proved as follows. In this case, $\tilde{k}_z$ is supported in a penumbra of the diagonal (Sect. 2.4.4). By (2.9.17), taking $\boldsymbol{K} = \boldsymbol{F}^2$ for some fundamental domain $\boldsymbol{F} \subset \widetilde{M}$, it follows that the right-hand side of (2.9.18) has a finite number of nonzero terms. So it defines a smooth section on M^2, which can be checked to be k_z using (2.2.14).

Examples where (2.9.18) fails are easy to construct. For instance, if Γ is non-amenable, it is well known that the spectrum of $\widetilde{\Delta}$ on functions has a gap of the form $(0, \epsilon)$ for some $\epsilon > 0$, and therefore (2.9.18) fails for $\psi(D)$ and $\psi(\widetilde{D})$ if ψ is even and supported in $(-\epsilon, \epsilon)$, with $\psi(0) \neq 0$.

Consider the Fréchet algebra and $\mathbb{C}[z]$-module $\mathcal{A}$ of Sect. 2.9.8.

Proposition 2.9.2 *If $\psi \in \mathcal{A}$, then* (2.9.18) *holds, where the series is convergent in the Fréchet space $C^\infty(\widetilde{M}^2; \Lambda\widetilde{M} \boxtimes (\Lambda\widetilde{M}^* \otimes \Omega\widetilde{M}))$.*

Proof First, let us prove that the series is uniformly convergent with all covariant derivatives on any fixed compact subset $\boldsymbol{K} \subset \widetilde{M}^2$.

By Lemma 2.9.1, for any $W > 0$ and $N > n/2$, there is some $C_1 = C_1(z, W) > 0$ such that, for all $\tilde{p}, \tilde{q} \in \widetilde{M}$, $m \in \mathbb{N}_0$ and $m_1 + m_2 \leq m$,

$$\left|\nabla_{\tilde{p}}^{m_1} \nabla_{\tilde{q}}^{m_2} \tilde{k}_z(\tilde{p}, \tilde{q})\right| \leq C_1 e^{-W d_{\widetilde{M}}(\tilde{p}, \tilde{q})} \|\psi\|_{\mathcal{A}, W, N+m} .$$

Then, by (2.9.17),

$$\left|\nabla_{\tilde{p}}^{m_1} \nabla_{\tilde{q}}^{m_2} \tilde{k}_z(\gamma \cdot \tilde{p}, \tilde{q})\right| \leq C_1' e^{-\frac{W}{c_1}|\gamma|} \|\psi\|_{\mathcal{A}, W, N+m} \tag{2.9.19}$$

for $\gamma \in \Gamma$, $(\tilde{p}, \tilde{q}) \in \boldsymbol{K}$ and $m_1 + m_2 \leq m$, where $C_1' = C_1 e^{W c_2}$. Since the growth of Γ is at most exponential, there is some $W_0 > 0$ such that

$$\sum_{\gamma \in \Gamma} e^{-W_0|\gamma|} < \infty . \tag{2.9.20}$$

Choosing $W > c_1 W_0$, it follows from (2.9.19) and (2.9.20) that there is some $C = C(z, \boldsymbol{K}, W, N) > 0$ such that

$$\left|\sum_{\gamma \in \Gamma} \nabla_{\tilde{p}}^{m_1} \nabla_{\tilde{q}}^{m_2} T_\gamma^* \tilde{k}_z(\gamma \cdot \tilde{p}, \tilde{q})\right| \leq C \|\psi\|_{\mathcal{A}, W, N+m} . \tag{2.9.21}$$

So the series in (2.9.18) is uniformly convergent on $\boldsymbol{K}$ with all covariant derivatives.

The identity (2.9.18) for any $\psi \in \mathcal{A}$ follows from (2.9.21), approximating ψ in $\mathcal{A}$ by a sequence of functions with compactly supported Fourier transform. □

Remark 2.9.3 Proposition 2.9.2 will be applied to an abelian covering. In that case, or, more generally, when Γ has polynomial growth, its proof can be slightly modified so that it works for any $\psi \in \mathcal{S}$. However not only this proposition, but also the estimate (2.9.19) will be used later, and we need $\psi \in \mathcal{A}$ to get the exponential factor of this estimate.

2.9.10 Local Index Formula for the Witten's Complex

Suppose M is of bounded geometry and consider the perturbed heat operator $e^{-t\Delta_z}$ $(t > 0)$ in $L^2(M; \Lambda)$, defined by the spectral theorem. By the ellipticity of Δ_z, the operator $e^{-t\Delta_z}$ is smoothing and let $k_{z,t} \in C^\infty_{\text{ub}}(M^2; \Lambda \boxtimes (\Lambda^* \otimes \Omega))$ denote its Schwartz kernel (the perturbed heat kernel). It has an asymptotic expansion as $t \downarrow 0$ in $C^\infty_{\text{ub}}(M^2; \Lambda \boxtimes (\Lambda^* \otimes \Omega))$ of the form

$$k_{z,t}(p, q) \sim h_t(p, q) \sum_{j=0}^{\infty} t^j \Theta_{z,j}(p, q) \cdot |\text{dvol}|(q) , \tag{2.9.22}$$

where |dvol| denotes the Riemannian density and

$$h_t(p,q) = \frac{1}{(4\pi t)^{n/2}} e^{-d(p,q)^2/4t}\,, \qquad \Theta_{z,j} \in C^\infty_{\mathrm{ub}}(M^2; \Lambda \boxtimes \Lambda^*)\,.$$

This expression can be formally differentiated to obtain also asymptotic expansions of the derivatives of $k_{z,t}(p,q)$ with respect to t, p and q. On the diagonal $\Delta \subset M^2$, the terms $\Theta_{z,j}$ can be locally described with algebraic expressions of the local coefficients of the metric and the form η, and their derivatives. When $z = 0$, we simply write k_t and Θ_j. (See e.g. [Gil95, Section 1.8.1] or [BGV04, Section 2.5].) We have $\Theta_{z,j}(p,p) = \Theta_{\mu,j}(p,p)$ by (2.9.4) since η is locally exact.

For even n, let $e(M,g) \in C^\infty(M; \Lambda \otimes o(M)) = C^\infty(M; \Omega)$ denote the Euler density of (M,g) (the representative of the Euler class given by the Chern-Weil theory).

Theorem 2.9.4 ([BZ92, Theorem 13.4]; See also [ÁLG21, Theorem 1.5]) *We have:*

(i) $\operatorname{str} \Theta_{z,j}(p,p) = 0$ *for* $j < n/2$; *and,*
(ii) *if* n *is even, then* $\operatorname{str} \Theta_{z,n/2}(p,p)\,|\mathrm{dvol}(p)| = e(M,g)(p)$.

Remark 2.9.5 In the given references, Theorem 2.9.4 was stated for compact manifolds, but its proof is a local computation, and therefore compactness is irrelevant. We have an additional proof of Theorem 2.9.4 (ii) using Getzler's rescaling, following [BGV04, Section 4.3]. In the case $n = 2$, this can be also checked directly. We omit the details of our alternative proof for brevity reasons.

2.9.11 Local Lefschetz Trace Formula for the Witten's Complex

Let $\phi : U \to V$ be a smooth map between open subsets of M with $U \subset V$, whose fixed point set is denoted by $\mathrm{Fix}(\phi)$. Recall that a fixed point p of ϕ is called *simple* if the eigenvalues of $\phi_* : T_pM \to T_pM$ are different from 1. This means that the graph of ϕ is transverse to Δ in M^2 at (p,p); in particular, p is isolated in $\mathrm{Fix}(\phi)$. In this case, let

$$\epsilon_p = \epsilon_p(\phi) = \operatorname{sign}\det(\mathrm{id} - \phi_* : T_pM \to T_pM) \in \{\pm 1\}\,. \tag{2.9.23}$$

Assume V is simply connected, and therefore $\eta = dF$ on V for some $F \in C^\infty(V)$. Consider the perturbed linear map $\phi_z^* = e^{z(\phi^*F - F)}\,\phi^* : \phi^*\Lambda V \to \Lambda U$ ($z \in \mathbb{C}$) (Sect. 2.9.4). Take any relatively compact open neighborhood W of p in U such that $\overline{W} \cap \mathrm{Fix}(\phi) = \{p\}$. Without loss of generality, we can assume that U is an open subset of a manifold of bounded geometry (or even of a closed manifold), where η and ϕ can be extended to a closed real 1-form and a smooth map.

Proposition 2.9.6 *For all* $z \in \mathbb{C}$,

$$\lim_{t\downarrow 0} \int_{q\in W} \operatorname{str}(\phi_z^* k_{z,t}(\phi(q), q) = \epsilon_p(\phi) .$$

Proof This follows like in the analytic proof of the Lefschetz trace formula [AB67] (see also [Roe98, Chapter 10] or [Gil95, Section 3.9]), using (2.9.22) and the expression

$$e^{z(F\phi(x)-F(x))} = 1 + O(|x|) ,$$

in terms of normal coordinates $x = (x^1, \dots, x^n)$ centered at p. □

2.9.12 A Tempered Distribution Associated to Some Closed 1-Forms

Assume M is closed, and let $\mathcal{S} = \mathcal{S}(\mathbb{R})$ (Sect. 2.1.11). Recall that the notation $z = \mu + i\lambda \in \mathbb{C}$ ($i = \sqrt{-1}$) is used for a complex parameter (Sect. 2.9.1). We would like to define a limit

$$Z = Z(M, g, \eta) = \lim_{\mu\to+\infty} Z_\mu \tag{2.9.24}$$

in $\mathcal{S}'$, where $Z_\mu = Z_\mu(M, g, \eta) \in \mathcal{S}'$ ($\mu \gg 0$) should be given by

$$\langle Z_\mu, f\rangle = -\frac{1}{2\pi}\int_0^\infty \int_{-\infty}^{+\infty} \operatorname{Str}\left(\eta\wedge\, \delta_z e^{-u\Delta_z}\right)\, \hat{f}(v)\, d\lambda\, du , \tag{2.9.25}$$

for all $f \in \mathcal{S}$, where Str denotes the supertrace. If $Z(M, g, -\eta)$ is defined, then $Z_\mu(M, g, \eta) \in \mathcal{S}'$ is defined for $\mu \ll 0$, and, in $\mathcal{S}'$,

$$-Z(M, g, -\eta) = \lim_{\mu\to-\infty} Z_\mu(M, g, \eta) .$$

Theorem 2.9.7 ([ÁLKL25, Theorems 1.1–1.4]) *Let M be a closed manifold of dimension n. For every real class $\xi \in H^1(M)$ and $\tau \gg 0$, there is some $\eta \in \xi$ and some Riemannian metric g on M such that (2.9.24) and (2.9.25) define the tempered distribution $Z = \tau\delta_0$, using the Dirac distribution δ_0 on $\mathbb{R}$. If n is even, this property holds for all $\tau \in \mathbb{R}$, and we can choose $\eta \in \xi$ so that $Z(M, g, \pm\eta)$ is defined and $\pm Z(M, g, \pm\eta) = \tau\delta_0$.*

Remark 2.9.8 If n is even, we can choose η and g in Theorem 2.9.7 so that $Z(M, g, \pm\eta) = 0$.

Chapter 3
Foliation Tools

3.1 Foliations

Standard references on foliations are [HH81, HH83, CLN85, God91, CC00, CC03], and for analysis on foliations see [Con82, MS88].

3.1.1 *Basic Concepts*

Recall that a (*smooth*) *foliation* $\mathcal{F}$ on a manifold M, with *codimension* n' and *dimension* n'' ($\operatorname{codim}\mathcal{F} = n'$, $\dim\mathcal{F} = n''$), can be described by a *foliated atlas* $\{U_k, x_k\}$ of M. The *foliated charts* or *foliated coordinates* (U_k, x_k) are of the form

$$x_k = (x'_k, x''_k) : U_k \to x_k(U_k) = \Sigma_k \times B''_k \,, \tag{3.1.1}$$

where B''_k is an open ball of $\mathbb{R}^{n''}$ and Σ_k is open in $\mathbb{R}^{n'}$, and the corresponding changes of coordinates are locally of the form

$$x_l x_k^{-1}(u, v) = (h_{lk}(u), g_{lk}(u, v)) \,. \tag{3.1.2}$$

We will use the notation

$$x_k = (x_k^1, \dots, x_k^n) = (x_k'^1, \dots, x_k'^{n'}, x_k''^{n'+1}, \dots, x''^{n_k}) \,.$$

It is also said that $(M, \mathcal{F})$ is a *foliated manifold*. The open sets U_k and the projections $x'_k : U_k \to \Sigma_k$ are said to be *distinguished*, the fibers of x'_k are called *plaques*, and the fibers of x''_k are called *local transversals* defined by (U_k, x_k), which can be identified with Σ_k via x'_k. Thus the sets Σ_k can be considered as local transversals of $\mathcal{F}$ with disjoint closures. The open subsets of all plaques form a

J. A. Álvarez López et al., *A Trace Formula for Foliated Flows*, Lecture Notes in Mathematics 2387, https://doi.org/10.1007/978-3-032-15413-2_3

base of a topology on M, called the *leaf topology*, becoming a smooth manifold of dimension n'' with the obvious charts induced by $\{U_k, x_k\}$, and its connected components are called *leaves*. The leaf through any point p may be denoted by L_p. The $\mathcal{F}$-*saturation* of a subset $S \subset M$, denoted by $\mathcal{F}(S)$, is the union of leaves that meet S.

Foliations on manifolds with boundary are similarly defined, assuming the boundary is either tangent or transverse to the leaves; we will only use the case where the boundary is tangent to the leaves (it is a union of leaves).

If a smooth map $\phi : M' \to M$ is transverse to (the leaves of) $\mathcal{F}$, then the connected components of the inverse images $\phi^{-1}(L)$ of the leaves L of $\mathcal{F}$ are the leaves of a smooth foliation $\phi^*\mathcal{F}$ on M' of codimension n', called *pull-back* of $\mathcal{F}$ by ϕ. In particular, for the inclusion map of any open subset, $\iota : U \hookrightarrow M$, the pull-back $\iota^*\mathcal{F}$ is the *restriction* $\mathcal{F}|_U$.

Any connected manifold M can be considered as a foliation with one leaf, also denoted by M. On the other hand, we can consider the foliation by points on M, denoted by M^δ (δ refers to the discreteness of the leaf topology). Given foliations $\mathcal{F}_a$ on manifolds M_a ($a = 1, 2$), the products of leaves of $\mathcal{F}_1$ and $\mathcal{F}_2$ are the leaves of the *product foliation* $\mathcal{F}_1 \times \mathcal{F}_2$, whose charts can be defined using products of charts of $\mathcal{F}_1$ and $\mathcal{F}_2$.

3.1.2 Holonomy

After considering a refinement if necessary, we can assume the foliated atlas $\{U_k, x_k\}$ is *regular* in the following sense: it is locally finite; for every k, there is a foliated chart $(\widetilde{U}_k, \tilde{x}_k)$ such that $\overline{U_k} \subset \widetilde{U}_k$ and $\tilde{x}_k$ extends x_k; and, if $U_{kl} := U_k \cap U_l \neq \emptyset$, then there is another foliated chart (U, x) such that $\overline{U_k} \cup \overline{U_l} \subset U$. In this case, (3.1.2) holds on the whole of U_{kl}, obtaining the *elementary holonomy transformations* $h_{kl} : x'_l(U_{kl}) \to x'_k(U_{kl})$, determined by the condition $h_{kl}x'_l = x'_k$ on U_{kl}. The collection $\{U_k, x'_k, h_{kl}\}$ is called a *defining cocycle*. The maps h_{kl} generate the *holonomy pseudogroup* $\mathcal{H}$ on $\Sigma := \bigsqcup_k \Sigma_k$, which is unique up to certain *equivalence* of pseudogroups [Hae80]. This Σ can be considered as a *complete transversal* of $\mathcal{F}$, in the sense that it meets all leaves. The notation $(\Sigma, \mathcal{H})$ may also be used. The $\mathcal{H}$-orbit of every $\bar{p} \in \Sigma$ is denoted by $\mathcal{H}(\bar{p})$. The maps x'_k induce a homeomorphism between the leaf space, $M/\mathcal{F}$, and the orbit space, $\Sigma/\mathcal{H}$.

The paths in the leaves are called *leafwise paths* when considered in M. Let $c : I := [0, 1] \to M$ be a leafwise path with $p := c(0) \in U_k$ and $q := c(1) \in U_l$. There is a partition of $I = [0, 1]$, $0 = t_0 < t_1 < \cdots < t_m = 1$, and a sequence of indices, $k = k_1, k_2, \ldots, k_m = l$, such that $c([t_{i-1}, t_i]) \subset U_{k_i}$ for $i = 1, \ldots, m$. The composition $h_c = h_{k_m k_{m-1}} \cdots h_{k_2 k_1}$, wherever defined, is a diffeomorphism with $x'_k(p) \in \operatorname{dom} h_c \subset \Sigma_k$ and $x'_l(q) = h_c x'_k(p) \in \operatorname{im} h_c \subset \Sigma_l$. The tangent map $h_{c*} : T_{x'_k(p)}\Sigma_k \to T_{x'_l(q)}\Sigma_l$ is called *infinitesimal holonomy* of c. The germ $\boldsymbol{h}_c$ of h_c at $x'_k(p)$, called *germinal holonomy* of c, depends only on $\mathcal{F}$ and the end-point homotopy class of c in $L = L_p$. In particular, taking $q = p$ and $l = k$, this defines the *holonomy homomorphism* onto the *holonomy group*, $\boldsymbol{h} = \boldsymbol{h}_L : \pi_1(L, p) \to \mathrm{Hol}(L, p)$. The isomorphism class of $\mathrm{Hol}(L, p)$ is independent of p; thus the notation $\mathrm{Hol}\, L$ may be used, like $\pi_1 L$. If $\mathrm{Hol}\, L$ is trivial, then L is said to be *without holonomy*. Residually many leaves have no holonomy [Hec77, EMT77]. If all leaves have no holonomy, then $\mathcal{F}$ is said to be *without holonomy*. The kernel of $\boldsymbol{h} : \pi_1 L \to \mathrm{Hol}\, L$ defines the *holonomy cover* $\widetilde{L} = \widetilde{L}^{\mathrm{hol}}$ of L. If D is a compact domain of a leaf L with smooth boundary, then $\mathcal{F}$ can be completely described in some neighborhood of D in M by the composition

$$\pi_1 D \to \pi_1 L \xrightarrow{h} \mathrm{Hol}\, L\ ,$$

where the first homomorphism is induced by $D \hookrightarrow L$ [Hae62, Section 2.7] (see also [HH81, Theorem 2.1.7], [CLN85, Theorem IV.2], [God91, Theorem II.2.29], [CC00, Theorem 2.3.9]). This description, called Reeb's local stability, involves the so-called *suspension foliation*, which allows the lifting of smooth paths from L to nearby leaves, continuously in the C^∞ topology.

3.1.3 Infinitesimal Transformations and Transverse Vector Fields

The vectors tangent to the leaves form the *tangent bundle* $T\mathcal{F} \subset TM$, obtaining also the *normal bundle* $N\mathcal{F} = TM/T\mathcal{F}$, the *cotangent bundle* $T^*\mathcal{F} = (T\mathcal{F})^*$ and the *conormal bundle* $N^*\mathcal{F} = (N\mathcal{F})^*$, the flat line bundles of *tangent/normal orientations*, $o(\mathcal{F}) = o(T\mathcal{F})$ and $o(N\mathcal{F})$, the *tangent/normal density bundles*, $\Omega^a\mathcal{F} = \Omega^a T\mathcal{F}$ ($a \in \mathbb{R}$) and $\Omega^a N\mathcal{F}$ (removing "a" from the notation when it is 1), and the *tangent/normal exterior bundles*, $\Lambda\mathcal{F} = \bigwedge T^*\mathcal{F} \otimes \mathbb{C}$ and $\Lambda N\mathcal{F} = \bigwedge N^*\mathcal{F} \otimes \mathbb{C}$. Again, we typically consider these density and exterior bundles with complex coefficients, without changing the notation; the few cases of real coefficients will be indicated. The terms *tangent/normal* vector fields, densities and differential forms are used for their smooth sections. Sometimes, "*leafwise*" is used instead of "tangent". Any $X \in TM$ (resp., $X \in \mathfrak{X}(M)$) canonically defines an element of $N\mathcal{F}$ (resp., $C^\infty(M; N\mathcal{F})$) denoted by $\overline{X}$. For any smooth local transversal Σ of $\mathcal{F}$ through a point $p \in M$, there is a canonical isomorphism $T_p\Sigma \cong N_p\mathcal{F}$.

A smooth vector bundle E over M, endowed with a flat $T\mathcal{F}$-partial connection, is said to be *$\mathcal{F}$-flat*. For instance, $N\mathcal{F}$ is $\mathcal{F}$-flat with the Bott $T\mathcal{F}$-partial connection $\nabla^{\mathcal{F}}$, given by $\nabla^{\mathcal{F}}_V \overline{X} = \overline{[V, X]}$ for $V \in \mathfrak{X}(\mathcal{F}) := C^\infty(M; T\mathcal{F})$ and $X \in \mathfrak{X}(M)$. For every leafwise path c from p to q, its infinitesimal holonomy can be considered as a homomorphism $h_{c*} : N_p\mathcal{F} \to N_q\mathcal{F}$, which is the $\nabla^{\mathcal{F}}$-parallel transport along c.

$\mathfrak{X}(\mathcal{F})$ is a Lie subalgebra and $C^\infty(M)$-submodule of $\mathfrak{X}(M)$, whose normalizer is denoted by $\mathfrak{X}(M, \mathcal{F})$, obtaining the quotient Lie algebra $\overline{\mathfrak{X}}(M, \mathcal{F}) =$

$\mathfrak{X}(M,\mathcal{F})/\mathfrak{X}(\mathcal{F})$. The elements of $\mathfrak{X}(M,\mathcal{F})$ (resp., $\overline{\mathfrak{X}}(M,\mathcal{F})$) are called *infinitesimal transformations* (resp., *transverse vector fields*). The projection of every $X \in \mathfrak{X}(M,\mathcal{F})$ to $\overline{\mathfrak{X}}(M,\mathcal{F})$ is also denoted by $\overline{X}$; in fact,

$$\overline{\mathfrak{X}}(M,\mathcal{F}) \equiv \{\, \overline{X} \in C^\infty(M;N\mathcal{F}) \mid \nabla^{\mathcal{F}}\overline{X} = 0 \,\} \subset C^\infty(M;N\mathcal{F}) \,.$$

Any $X \in \mathfrak{X}(M)$ is in $\mathfrak{X}(M,\mathcal{F})$ if and only if every restriction $X|_{U_k}$ can be projected by x'_k, defining an $\mathcal{H}$-invariant vector field on Σ, also denoted by $\overline{X}$. This induces a canonical isomorphism of $\overline{\mathfrak{X}}(M,\mathcal{F})$ to the Lie algebra $\mathfrak{X}(\Sigma,\mathcal{H})$ of $\mathcal{H}$-invariant tangent vector fields on Σ.

When M is not closed, we can consider the subsets of complete vector fields, $\mathfrak{X}_{\mathrm{com}}(\mathcal{F}) \subset \mathfrak{X}(\mathcal{F})$ and $\mathfrak{X}_{\mathrm{com}}(M,\mathcal{F}) \subset \mathfrak{X}(M,\mathcal{F})$. Let $\overline{\mathfrak{X}}_{\mathrm{com}}(M,\mathcal{F}) \subset \overline{\mathfrak{X}}(M,\mathcal{F})$ be the projection of $\mathfrak{X}_{\mathrm{com}}(M,\mathcal{F})$.

3.1.4 Holonomy Groupoid

On the space of leafwise paths in M, with the compact-open topology, two leafwise paths are declared to be equivalent if they have the same end points and the same germinal holonomy. This is an equivalence relation, and the corresponding quotient space, $\mathfrak{G} = \mathrm{Hol}(M,\mathcal{F})$, becomes a smooth manifold of dimension $n + n'$ in the following way. An open neighborhood $\mathfrak{U}$ of a class $[c]$ in $\mathfrak{G}$, with $c(0) \in U_k$ and $c(1) \in U_l$, is defined by the leafwise paths d such that $d(0) \in U_k$, $d(1) \in U_l$, $x'_k d(0) \in \operatorname{dom} h_c$, and h_d and h_c have the same germ at $x'_k d(0)$. Local coordinates on $\mathfrak{U}$ are given by $[d] \mapsto (d(0), x''_l d(1))$. Moreover, $\mathfrak{G}$ is a Lie groupoid, called the *holonomy groupoid*, where the space of units $\mathfrak{G}^{(0)} \equiv M$ is defined by the constant paths, the source and range projections $\boldsymbol{s},\boldsymbol{r} : \mathfrak{G} \to M$ are given by the first and last points of the paths, the operation is induced by the opposite of the usual path product, and the inversion is induced by the usual path inversion. Note that $\mathfrak{G}$ is Hausdorff if and only if $\mathcal{H}$ is *quasi-analytic* in the following sense: for any $h \in \mathcal{H}$ and open $O \subset \Sigma$ with $\overline{O} \subset \operatorname{dom} h$, if $h|_O = \mathrm{id}_O$, then h is the identity on some neighborhood of $\overline{O}$. Observe also that $\boldsymbol{s},\boldsymbol{r} : \mathfrak{G} \to M$ are smooth submersions, and $(\boldsymbol{r},\boldsymbol{s}) : \mathfrak{G} \to M^2$ is a smooth immersion. Let $\mathcal{R}_{\mathcal{F}} = \{\,(p,q) \in M^2 \mid L_p = L_q\,\} \subset M^2$, which is not a regular submanifold in general, and let $\Delta \subset M^2$ be the diagonal. We have $(\boldsymbol{r},\boldsymbol{s})(\mathfrak{G}) = \mathcal{R}_{\mathcal{F}}$ and $(\boldsymbol{r},\boldsymbol{s})(\mathfrak{G}^{(0)}) = \Delta$. For any leaf L and $p \in L$, we have $\mathrm{Hol}(L,p) = \boldsymbol{s}^{-1}(p) \cap \boldsymbol{r}^{-1}(p)$, the map $\boldsymbol{r} : \boldsymbol{s}^{-1}(p) \to L$ is the covering projection $\widetilde{L}^{\mathrm{hol}} \to L$, and $\boldsymbol{s} : \boldsymbol{r}^{-1}(p) \to L$ corresponds to $\boldsymbol{r} : \boldsymbol{s}^{-1}(p) \to L$ by the inversion of $\mathfrak{G}$. Thus $(\boldsymbol{r},\boldsymbol{s}) : \mathfrak{G} \to M^2$ is injective if and only if all leaves have trivial holonomy groups, but, even in this case, this map may not be a topological embedding. The fibers of $\boldsymbol{s}$ and $\boldsymbol{r}$ define smooth foliations of codimension n on $\mathfrak{G}$. We also have the smooth foliation $\boldsymbol{s}^*\mathcal{F} = \boldsymbol{r}^*\mathcal{F}$ of codimension n' with leaves $\boldsymbol{s}^{-1}(L) = \boldsymbol{r}^{-1}(L) = (\boldsymbol{r},\boldsymbol{s})^{-1}(L^2)$ for leaves L of $\mathcal{F}$, and every restriction $(\boldsymbol{r},\boldsymbol{s}) : (\boldsymbol{r},\boldsymbol{s})^{-1}(L^2) \to L^2$ is a smooth covering projection.

Let $\mathcal{F}_k = \mathcal{F}|_{U_k}$, $\mathfrak{G}_k = \operatorname{Hol}(U_k, \mathcal{F}_k)$ and $\mathcal{R}_k = \mathcal{R}_{\mathcal{F}_k}$. The set $\bigcup_k \mathfrak{G}_k$ (resp., $\bigcup_k \mathcal{R}_k$) is an open neighborhood of $\mathfrak{G}^{(0)}$ in $\mathfrak{G}$ (resp., of Δ in $\mathcal{R}_{\mathcal{F}}$). Furthermore, by the regularity of $\{U_k, x_k\}$, the map $(\boldsymbol{r}, \boldsymbol{s}) : \bigcup_k \mathfrak{G}_k \to M^2$ is a smooth embedding with image $\bigcup_k \mathcal{R}_k$; we will write $\bigcup_k \mathfrak{G}_k \equiv \bigcup_k \mathcal{R}_k$.

3.1.5 *The Convolution Algebra on $\mathfrak{G}$ and Its Global Action*

Consider the notation of Sect. 3.1.4. For the sake of simplicity, assume $\mathfrak{G}$ is Hausdorff [Con79]. The extension of the following concepts to the case where $\mathfrak{G}$ is not Hausdorff can be made like in [Con82].

Given a vector bundle E over M, let $S = \boldsymbol{r}^*E \otimes \boldsymbol{s}^*(E^* \otimes \Omega\mathcal{F})$, which is a vector bundle over $\mathfrak{G}$. Let $C^\infty_{\text{cs}}(\mathfrak{G}; S) \subset C^\infty(\mathfrak{G}; S)$ denote the subspace of sections $k \in C^\infty(\mathfrak{G}; S)$ such that $\operatorname{supp} k \cap \boldsymbol{s}^{-1}(K)$ is compact for all compact $K \subset M$; in particular, $C^\infty_{\text{cs}}(\mathfrak{G}; S) = C^\infty_{\text{c}}(\mathfrak{G}; S)$ if M is compact. Similarly, define $C^\infty_{\text{cr}}(\mathfrak{G}; S)$ by using $\boldsymbol{r}$ instead of $\boldsymbol{s}$. Both $C^\infty_{\text{cs}}(\mathfrak{G}; S)$ and $C^\infty_{\text{cr}}(\mathfrak{G}; S)$ are associative algebras with the *convolution* product defined by

$$(k_1 * k_2)(\gamma) = \int_{\boldsymbol{s}(\epsilon)=\boldsymbol{s}(\gamma)} k_1(\gamma\epsilon^{-1})\, k_2(\epsilon) = \int_{\boldsymbol{r}(\delta)=\boldsymbol{r}(\gamma)} k_1(\delta)\, k_2(\delta^{-1}\gamma)\,,$$

and $C^\infty_{\text{csr}}(\mathfrak{G}; S) := C^\infty_{\text{cs}}(\mathfrak{G}; S) \cap C^\infty_{\text{cr}}(\mathfrak{G}; S)$ and $C^\infty_{\text{c}}(\mathfrak{G}; S)$ are subalgebras.

The *global action* of $C^\infty_{\text{cr}}(\mathfrak{G}; S)$ on $C^\infty(M; E)$ is the left action defined by

$$(k \cdot u)(p) = \int_{\boldsymbol{r}(\gamma)=p} k(\gamma)\, u(\boldsymbol{s}(\gamma))\,.$$

In this way, $C^\infty_{\text{cr}}(\mathfrak{G}; S)$ can be considered as an algebra of operators on $C^\infty(M; E)$. Moreover $C^\infty_{\text{csr}}(\mathfrak{G}; S)$ preserves $C^\infty_{\text{c}}(M; E)$, obtaining an algebra of operators on $C^\infty_{\text{c}}(M; E)$. It can be said that these operators are defined by a leafwise version of a smooth Schwartz kernel (cf. Sect. 2.1.5).

Let $S' = \boldsymbol{r}^*(E^* \otimes \Omega\mathcal{F}) \otimes \boldsymbol{s}^*E$. The mapping $k \mapsto k^{\text{t}}$, $k^{\text{t}}(\gamma) = k(\gamma^{-1})$, defines anti-homomorphisms $C^\infty_{\text{cs/cr}}(\mathfrak{G}; S) \to C^\infty_{\text{cr/cs}}(\mathfrak{G}; S')$ and $C^\infty_{\text{csr}}(\mathfrak{G}; S) \to C^\infty_{\text{csr}}(\mathfrak{G}; S')$, obtaining a leafwise version of the transposition of operators (cf. Sect. 2.1.5). Similarly, using $E = \Omega^{1/2}\mathcal{F}$, or if E has a Hermitian structure and we fix a non-vanishing leafwise density, we get a leafwise version of taking adjoint operators. Moreover, in this case, $C^\infty_{\text{c}}(\mathfrak{G}; S)$ is $*$-algebra.

3.1.6 *Leafwise Metric*

A Euclidean structure $g_{\mathcal{F}}$ on $T\mathcal{F}$ is called a *leafwise* (*Riemannian*) *metric* of $\mathcal{F}$. The corresponding *leafwise distance* is the map $d_{\mathcal{F}} : M^2 \to [0, \infty]$ given by the distance function of the leaves on $\mathcal{R}_{\mathcal{F}}$, taking $d_{\mathcal{F}}(M^2 \setminus \mathcal{R}_{\mathcal{F}}) = \infty$. For $p \in M$, $S \subset M$ and $r > 0$, the *open* and *closed leafwise balls*, $B_{\mathcal{F}}(p, r)$ and $\overline{B}_{\mathcal{F}}(p, r)$, and the *open* and *closed leafwise penumbras*, $\operatorname{Pen}_{\mathcal{F}}(S, r)$ and $\overline{\operatorname{Pen}}_{\mathcal{F}}(S, r)$, are defined with $d_{\mathcal{F}}$ like in the case of Riemannian metrics (Sect. 2.4). The Levi-Civita connection on the leaves defines a $T\mathcal{F}$-partial connection on $T\mathcal{F}$, also denoted by $\nabla^{\mathcal{F}}$.

Equip the foliation $\boldsymbol{r}^*\mathcal{F}$ on $\mathfrak{G}$ with the leafwise Riemannian metric so that the foliated immersion $(\boldsymbol{r}, \boldsymbol{s}) : (\mathfrak{G}, \boldsymbol{r}^*\mathcal{F}) \to (M^2, \mathcal{F}^2)$ is isometric on the leaves. Let $d_{\boldsymbol{r}} : \mathfrak{G} \to [0, \infty]$ denote the leafwise distance for the foliation on $\mathfrak{G}$ defined by the fibers of $\boldsymbol{r}$, and consider the corresponding open and closed leafwise penumbras, $\operatorname{Pen}_{\boldsymbol{r}}(\mathfrak{G}^{(0)}, r)$ and $\overline{\operatorname{Pen}}_{\boldsymbol{r}}(\mathfrak{G}^{(0)}, r)$. Note that we get the same penumbras by using $\boldsymbol{s}$ instead of $\boldsymbol{r}$; indeed, they are given by the conditions $d_{\mathcal{F}}^{\text{hol}} < r$ and $d_{\mathcal{F}}^{\text{hol}} \le r$, resp., where $d_{\mathcal{F}}^{\text{hol}} : \mathfrak{G} \to [0, \infty)$ is defined by

$$d_{\mathcal{F}}^{\text{hol}}(\gamma) = \inf_c \operatorname{length}(c) ,$$

with c running in the piecewise smooth representatives of γ.

For example, if M is endowed with a Riemannian metric, its restriction to the leaves defines a leafwise Riemannian metric. In this case, $d_{\mathcal{F}} \ge d_M$ (the distance function of M), and the leafwise metric of $\boldsymbol{r}^*\mathcal{F}$ is given by the Riemannian metric on $\mathfrak{G}$ so that the immersion $(\boldsymbol{r}, \boldsymbol{s}) : \mathfrak{G} \to M^2$ is isometric.

By the smooth lifting of leafwise paths to nearby leaves, it easily follows that $d_{\mathcal{F}}^{\text{hol}} : \mathfrak{G} \to [0, \infty)$ and $d_{\mathcal{F}} : \mathcal{R}_{\mathcal{F}} \to [0, \infty)$ are upper semicontinuous. Moreover $d_{\mathcal{F}}^{\text{hol}} \equiv d_{\mathcal{F}}$ on $\bigcup_k \mathfrak{G}_k \equiv \bigcup_k \mathcal{R}_k$. Using the convexity radius (see e.g. [Pet98, Section 6.3.2]), it follows that, after refining $\{U_k, x_k\}$ if necessary, we can assume $d_{\mathcal{F}}$ is continuous on $\bigcup_k \mathcal{R}_k$.

Lemma 3.1.1 *The following properties hold for any compact $K \subset M^2$:*

(i) *If $K \subset \mathcal{R}_{\mathcal{F}}$, then $d_{\mathcal{F}}|_K$ reaches a finite maximum at some point.*
(ii) *If $K \cap \Delta = \emptyset$, then $\inf d_{\mathcal{F}}(K) > 0$. If moreover $\inf d_{\mathcal{F}}(K)$ is small enough, then it is the minimum of $d_{\mathcal{F}}|_K$.*

Proof Using that K is compact, $\Delta = \{d_{\mathcal{F}} = 0\}$, $\mathcal{R}_{\mathcal{F}} = \{d_{\mathcal{F}} < \infty\}$, and $\bigcup_k \mathcal{R}_k$ is a neighborhood of Δ in $\mathcal{R}_{\mathcal{F}}$ containing $K \cap \{d_{\mathcal{F}} \le r\}$ for some $r < 0$, we get (i) by the upper semicontinuity of $d_{\mathcal{F}}$, and (ii) by the continuity of $d_{\mathcal{F}}$ on $\bigcup_k \mathcal{R}_k$. □

Remark 3.1.2 The obvious version of Lemma 3.1.1 for $d_{\mathcal{F}}^{\text{hol}}$ and compact subsets of $\mathfrak{G}$ can be proved with analogous arguments.

From now on, suppose the leaves with $g_{\mathcal{F}}$ are complete Riemannian manifolds. Then their exponential maps define a smooth map $\exp_{\mathcal{F}} : T\mathcal{F} \to M$.

With the notation of Sect. 3.1.5, let $C_{\text{p}}^{\infty}(\mathfrak{G}; S) \subset C^{\infty}(\mathfrak{G}; S)$ denote the subspace of sections supported in leafwise penumbras of $\mathfrak{G}^{(0)}$. This is a subalgebra of $C_{\text{csr}}^{\infty}(\mathfrak{G}; S)$, and the leafwise transposition restricts to an anti-homomorphism $C_{\text{p}}^{\infty}(\mathfrak{G}; S) \to C_{\text{p}}^{\infty}(\mathfrak{G}; S')$ [ÁLKL20, Section 4.6].

3.1.7 Foliated Maps and Foliated Flows

A *foliated map* $\phi : (M_1, \mathcal{F}_1) \to (M_2, \mathcal{F}_2)$ is a map $\phi : M_1 \to M_2$ that maps leaves of $\mathcal{F}_1$ to leaves of $\mathcal{F}_2$. In this case, assuming that ϕ is smooth, its tangent map defines homomorphisms $\phi_* : T\mathcal{F}_1 \to T\mathcal{F}_2$ and $\phi_* : N\mathcal{F}_1 \to N\mathcal{F}_2$, where the second one is compatible with the corresponding flat partial connections. We also get an induced Lie groupoid homomorphism $\mathrm{Hol}(\phi) : \mathrm{Hol}(M_1, \mathcal{F}_1) \to \mathrm{Hol}(M_2, \mathcal{F}_2)$, defined by $\mathrm{Hol}(\phi)([c]) = [\phi c]$. The set of smooth foliated maps $(M_1, \mathcal{F}_1) \to (M_2, \mathcal{F}_2)$ is denoted by $C^\infty(M_1, \mathcal{F}_1; M_2, \mathcal{F}_2)$. A smooth family $\phi = \{\phi^t \mid t \in T\}$ of foliated maps $(M_1, \mathcal{F}_1) \to (M_2, \mathcal{F}_2)$ can be considered as the smooth foliated map $\phi : (M_1 \times T, \mathcal{F}_1 \times T^\delta) \to (M_2, \mathcal{F}_2)$.

For example, if a smooth map $\psi : M' \to M$ is transverse to a foliation $\mathcal{F}$ on M, then it is a foliated map $(M', \psi^*\mathcal{F}) \to (M, \mathcal{F})$. Moreover $\psi_* : N\psi^*\mathcal{F} \to N\mathcal{F}$ restricts to isomorphisms between the fibers; i.e., it induces an isomorphism $\psi_* : N\psi^*\mathcal{F} \xrightarrow{\cong} \psi^*N\mathcal{F}$ of $\psi^*\mathcal{F}$-flat vector bundles over M'.

Let $\mathrm{Diffeo}(M, \mathcal{F})$ be the group of foliated diffeomorphisms (or transformations) of $(M, \mathcal{F})$. A smooth flow $\phi = \{\phi^t\}$ on M is called *foliated* if $\phi^t \in \mathrm{Diffeo}(M, \mathcal{F})$ for all $t \in \mathbb{R}$. More generally, a local flow $\phi : \Omega \to M$, defined on some open neighborhood Ω of $M \times \{0\}$ in $M \times \mathbb{R}$, is called *foliated* if it is a foliated map $(\Omega, (\mathcal{F} \times \mathbb{R}^\delta)|_\Omega) \to (M, \mathcal{F})$. Then $\mathfrak{X}(M, \mathcal{F})$ consists of the smooth vector fields whose local flow is foliated, and $\mathfrak{X}_{\mathrm{com}}(M, \mathcal{F})$ consists of the complete smooth vector fields whose flow is foliated.

Let $X \in \mathfrak{X}_{\mathrm{com}}(M, \mathcal{F})$, with foliated flow $\phi = \{\phi^t\}$, and let $\bar\phi$ be the local flow on Σ generated by $\overline{X} \in \mathfrak{X}(\Sigma, \mathcal{H})$ (Sects. 3.1.2 and 3.1.3). The following properties hold [ÁLKL20, Section 4.8]: via $x'_k : U_k \to \Sigma_k$, the local flow defined by ϕ on every U_k corresponds to the restriction of $\bar\phi$ to Σ_k; and $\bar\phi$ is $\mathcal{H}$-equivariant in an obvious sense.

Take another vector field $Y \in \mathfrak{X}_{\mathrm{com}}(M, \mathcal{F})$ with foliated flow $\psi = \{\psi^t\}$.

Lemma 3.1.3 *We have $\overline{Y} = \overline{X}$ if and only if $\phi^t(L) = \psi^t(L)$ for all $t \in \mathbb{R}$ and every leaf L.*

Proof The condition $\overline{Y} = \overline{X}$ is equivalent to $\bar\phi = \bar\psi$, which means that the local flows defined by ϕ and ψ on every U_k correspond to the same local flow on Σ_k via π'_k. In turn, this is equivalent to the existence of some open $\Omega \subset M \times \mathbb{R}$, containing $M \times \{0\}$, such that $\phi(p, t)$ and $\psi(p, t)$ are in the same leaf for all $(p, t) \in \Omega$. But this is equivalent to $\phi^t(L) = \psi^t(L)$ for all leaf L and $t \in \mathbb{R}$ because ϕ and ψ are foliated flows. □

A smooth homotopy $H : M_1 \times I \to M_2$ ($I = [0, 1]$) between foliated maps $\phi, \psi : (M_1, \mathcal{F}_1) \to (M_2, \mathcal{F}_2)$ is said to be *leafwise* (or *integrable*) if it is a foliated map $(M_1 \times I, \mathcal{F}_1 \times I) \to (M_2, \mathcal{F}_2)$. When there is such a leafwise homotopy, it is said that ϕ and ψ are *leafwisely homotopic*.

A smooth *leafwise homotopy* between foliated flows on $(M, \mathcal{F})$, $\phi = \{\phi^t\}$ and $\psi = \{\psi^t\}$, is a smooth family $H = \{H^t\}$, where every $H^t : M \times I \to M$ is a leafwise homotopy between ϕ^t and ψ^t; in other words, it can be considered as a leafwise homotopy $H : M \times \mathbb{R} \times I \to M$ between the corresponding foliated maps $\phi, \psi : (M \times \mathbb{R}, \mathcal{F} \times \mathbb{R}^\delta) \to (M, \mathcal{F})$. If moreover every $H(\cdot, \cdot, s) : M \times \mathbb{R} \to M$ is a flow, then H is called a smooth *flow leafwise homotopy*.

Proposition 3.1.4 *Let $X, Y \in \mathfrak{X}_{com}(M, \mathcal{F})$, with foliated flows $\phi = \{\phi^t\}$ and $\psi = \{\psi^t\}$, such that $V := Y - X \in \mathfrak{X}_c(\mathcal{F})$. Then there is a flow leafwise homotopy $H : M \times \mathbb{R} \times I \to M$ between ϕ and ψ such that $H(p, t, s) = \phi^t(p)$ for all $p \in M$ with $\phi^t(p) = \psi^t(p)$.*

Proof Since $X \in \mathfrak{X}_{\mathrm{com}}(M, \mathcal{F})$ and $V \in \mathfrak{X}_{\mathrm{c}}(\mathcal{F})$, we have $Z_s := X + sV \in \mathfrak{X}_{\mathrm{com}}(M, \mathcal{F})$ $(s \in I)$. Let $\xi_s : M \times \mathbb{R} \to M$ denote the flow of every Z_s. Since $\overline{Z_s} = \overline{X}$ for all s, it follows from Lemma 3.1.3 that the statement holds with $H : M \times \mathbb{R} \times I \to M$ defined by $H(\cdot, \cdot, s) = \xi_s$. □

3.1.8 Differential Operators on Foliated Manifolds

Like in Sect. 2.1.7, using $\mathfrak{X}(\mathcal{F})$ instead of $\mathfrak{X}(M)$, we get the filtered subalgebra and $C^\infty(M)$-submodule of *leafwise differential operators*, $\mathrm{Diff}(\mathcal{F}) \subset \mathrm{Diff}(M)$, and a *leafwise principal symbol* exact sequence for every order m,

$$0 \to \mathrm{Diff}^{m-1}(\mathcal{F}) \hookrightarrow \mathrm{Diff}^m(\mathcal{F}) \xrightarrow{{}^{\mathcal{F}}\sigma_m} P^{(m)}(T^*\mathcal{F}) \to 0\ .$$

Moreover these concepts can be extended to vector bundles E and F over M like in Sect. 2.1.7, obtaining the filtered $C^\infty(M)$-submodule $\mathrm{Diff}(\mathcal{F}; E, F)$ (or $\mathrm{Diff}(\mathcal{F}; E)$ if $E = F$) of $\mathrm{Diff}(M; E, F)$, and the *leafwise principal symbol* ${}^{\mathcal{F}}\sigma_m : \mathrm{Diff}^m(\mathcal{F}; E, F) \to P^{(m)}(T^*\mathcal{F}; F \otimes E^*)$. The diagram

$$\begin{array}{ccc}
\mathrm{Diff}^m(\mathcal{F}; E, F) & \xrightarrow{{}^{\mathcal{F}}\sigma_m} & P^{(m)}(T^*\mathcal{F}; F \otimes E^*) \\
\downarrow & & \downarrow \\
\mathrm{Diff}^m(M; E, F) & \xrightarrow{\sigma_m} & P^{(m)}(T^*M; F \otimes E^*)
\end{array}$$

is commutative, where the left-hand side vertical arrow denotes the inclusion homomorphism, and the right-hand side vertical arrow is induced by the restriction homomorphism $T^*M \to T^*\mathcal{F}$. The condition of being a leafwise differential operator is preserved by compositions and by taking transposes, and by taking formal adjoints in the case of Hermitian vector bundles; in particular, $\mathrm{Diff}(\mathcal{F}; E)$ is a filtered subalgebra of $\mathrm{Diff}(M; E)$. It is said that $A \in \mathrm{Diff}^m(\mathcal{F}; E, F)$ is *leafwisely elliptic* if the symbol $\sigma_m(A)(p, \xi)$ is an isomorphism for all $p \in M$ and $0 \neq \xi \in T_p^*\mathcal{F}$. In this way, the concepts of *leafwise differential complex* and its *leafwise ellipticity* can be defined like in Sect. 2.1.14.

A smooth family of leafwise differential operators, $A = \{ A_t \mid t \in T \} \subset \mathrm{Diff}^m(\mathcal{F}; E, F)$, can be canonically considered as a leafwise differential operator $A \in \mathrm{Diff}^m(\mathcal{F} \times T^\delta; \mathrm{pr}_1^* E, \mathrm{pr}_1^* F)$, where $\mathrm{pr}_1 : M \times T \to M$ is the first-factor projection.

On the other hand, considering the canonical injection $N^*\mathcal{F} \subset T^*M$, it is said that $A \in \mathrm{Diff}^m(M; E, F)$ is *transversely elliptic* if the symbol $\sigma_m(A)(p, \xi)$ is an isomorphism for all $p \in M$ and $0 \neq \xi \in N_p^*\mathcal{F}$. The concept of *transverse ellipticity* has an obvious extension to differential complexes like in Sect. 2.1.14.

We can use $\mathrm{Diff}(\mathcal{F}; E)$ to define variants of the section spaces recalled in Sect. 2.1.4. For instance, for $m \in \mathbb{N}_0$, we have the LCHS

$$C_{\mathcal{F}}^{0,m}(M; E) = \{ u \in C(M; E) \mid \mathrm{Diff}^m(\mathcal{F}; E) \cdot u \subset C(M; E) \} ,$$

with the topology defined like in (2.1.1). Let also $C_{\mathcal{F}}^{0,\infty}(M; E) = \bigcap_m C^{0,m}(M; E)$. If $\mathcal{F}$ is described by a submersion $\varpi : M \to M'$, then the subscript ϖ may be used instead of $\mathcal{F}$, which agrees with the notation already used in Remark 2.5.11.

3.1.9 Transverse Structures

Recall that $(\Sigma, \mathcal{H})$ denotes the holonomy pseudogroup of $\mathcal{F}$. An *(invariant) transverse structure* of $\mathcal{F}$ is an $\mathcal{H}$-invariant structure on Σ. It can be also considered as a $\nabla^{\mathcal{F}}$-parallel structure on $N\mathcal{F}$. For our purposes, it is enough to consider structures on Σ (resp., on $N\mathcal{F}$) defined by smooth sections of bundles associated with $T\Sigma$ (resp., $N\mathcal{F}$) satisfying some conditions. For instance, we will use the concepts of a *transverse orientation*, a *transverse Riemannian metric* and a *transverse parallelism*. The existence of these transverse structures defines the classes of *transversely orientable*, *(transversely) Riemannian*, and *transversely parallelizable (TP) foliations*.

A transverse orientation of $\mathcal{F}$ can be simply described as an orientation of $N\mathcal{F}$, which is necessarily $\nabla^{\mathcal{F}}$-parallel. It can be determined by a non-vanishing real form $\omega \in C^\infty(M; \Lambda^{n'} N\mathcal{F})$; i.e., some real $\omega \in C^\infty(M; \Lambda^{n'})$ defining $\mathcal{F}$ in the sense that $T\mathcal{F} = \{ Y \in TM \mid \iota_Y \omega = 0 \}$. By Frobenius theorem, the integrability of $T\mathcal{F}$ means that $d\omega = \eta \wedge \omega$ for some real $\eta \in C^\infty(M; \Lambda^1)$, which is unique modulo $C^\infty(M; \Lambda^1 N\mathcal{F})$. All other pairs of differential forms ω' and η' satisfying these conditions are of the form $\omega' = e^f \omega$ and $\eta' = \eta + df + h\omega$ for any real functions $f, h \in C^\infty(M)$. We have $d\omega = 0$ just when ω defines an *invariant transverse volume form*. Any invariant transverse volume form ω defines an invariant transverse density $|\omega| \in C^\infty(M; \Omega N\mathcal{F})$, which can be considered as an invariant transverse measure.

Remark 3.1.5 Even when $\mathcal{F}$ is not transversely oriented, it is defined by some real $\omega \in C^\infty(M; \Lambda^{n'} N\mathcal{F} \otimes o(N\mathcal{F})) \equiv C^\infty(M; \Omega N\mathcal{F})$, and we have $d\omega = \omega \wedge \eta$ for some real 1-form η, as above.

A transverse parallelism can be described as a global frame of $N\mathcal{F}$ consisting of transverse vector fields $\overline{X_1}, \dots, \overline{X_{n'}}$. If its linear span is a Lie subalgebra $\mathfrak{g} \subset \overline{\mathfrak{X}}(M, \mathcal{F})$, it is called a *transverse Lie structure*, giving rise to the concept of ($\mathfrak{g}$-)*Lie foliation*. If moreover $\overline{X_1}, \dots, \overline{X_{n'}} \in \overline{\mathfrak{X}}_{\mathrm{com}}(M, \mathcal{F})$, then the TP or Lie foliation $\mathcal{F}$ is said to be *complete*.

Let G be the simply connected Lie group with Lie algebra $\mathfrak{g}$ as above. Then $\mathcal{F}$ is a $\mathfrak{g}$-Lie foliation just when $\mathcal{H}$ is equivalent to some pseudogroup generated by restrictions of some left translations on some open $T \subset G$, which is complete just when we can take $T = G$.

Similarly, a transverse Riemannian metric can be described as a $\nabla^{\mathcal{F}}$-parallel Euclidean structure on $N\mathcal{F}$. It is always induced by a Riemannian metric on M such that every $x_k' : U_k \to \Sigma_k$ is a Riemannian submersion, which is called a *bundle-like metric*. Thus $\mathcal{F}$ is Riemannian if and only if it can be endowed with a bundle-like metric on M.

It is said that $\mathcal{F}$ is *transitive at* a point $p \in M$ when the evaluation map $\mathrm{ev}_p : \mathfrak{X}(M, \mathcal{F}) \to T_pM$ is surjective, or, equivalently, the evaluation map $\overline{\mathrm{ev}}_p : \overline{\mathfrak{X}}(M, \mathcal{F}) \to N_p\mathcal{F}$ is surjective. The transitive point set is open and saturated. If $\mathcal{F}$ is transitive at every point, then it is called *transitive*. If $\mathrm{ev}_p(\mathfrak{X}_{\mathrm{com}}(M, \mathcal{F}))$ spans T_pM for all $p \in M$, then $\mathcal{F}$ is called *transversely complete* (*TC*). Since $\mathrm{ev}_p : \mathfrak{X}_{\mathrm{com}}(\mathcal{F}) \to T_p\mathcal{F}$ is surjective [Mol88, Section 4.5], $\mathcal{F}$ is TC if and only if $\overline{\mathrm{ev}}_p(\overline{\mathfrak{X}}_{\mathrm{com}}(M, \mathcal{F}))$ spans $N_p\mathcal{F}$ for all $p \in M$.

All TP foliations are transitive, and all transitive foliations are Riemannian. On the other hand, Molino's theory describes Riemannian foliations in terms of TP foliations [Mol88]. A Riemannian foliation is called *complete* if, using Molino's theory, the corresponding TP foliation is TC. Furthermore Molino's theory describes TC foliations in terms of complete Lie foliations with dense leaves. In turn, complete Lie foliations have the following description due to Fedida [Fed71, Fed73] (see also [Mol88, Theorem 4.1 and Lemma 4.5]). Assume M is connected and $\mathcal{F}$ a complete $\mathfrak{g}$-Lie foliation. Let G be the simply connected Lie group with Lie algebra $\mathfrak{g}$. Then there is a regular covering $\pi : \widetilde{M} \to M$ (the *holonomy covering*), a fiber bundle $D : \widetilde{M} \to G$ (the *developing map*) and a monomorphism $h : \Gamma := \mathrm{Aut}(\pi) \equiv \pi_1 M / \pi_1 \widetilde{M} \to G$ (the *holonomy homomorphism*) such that the leaves of $\widetilde{\mathcal{F}} := \pi^*\mathcal{F}$ are the fibers of D, and D is h-equivariant with respect to the left action of G on itself by left translations. As a consequence, π restricts to diffeomorphisms between the leaves of $\widetilde{\mathcal{F}}$ and $\mathcal{F}$. The subgroup $\mathrm{Hol}\,\mathcal{F} = \mathrm{im}\, h \subset G$, isomorphic to Γ, is called the *global holonomy group*.

The Molino's description also gives a precise equivalence between the holonomy pseudogroup $\mathcal{H}$ and the pseudogroup on G generated by the action of $\mathrm{Hol}\,\mathcal{F}$ by left translations. Thus the leaves are dense if and only if $\mathrm{Hol}\,\mathcal{F}$ is dense in G, which means $\mathfrak{g} = \overline{\mathfrak{X}}(M, \mathcal{F})$.

The $\widetilde{\mathcal{F}}$-leaf through every $\tilde{p} \in \widetilde{M}$ will be denoted by $\widetilde{L}_{\tilde{p}}$. Since D induces an identity $\widetilde{M}/\widetilde{\mathcal{F}} \equiv G$, the π-lift and D-projection of vector fields define identities

$$\overline{\mathfrak{X}}(M, \mathcal{F}) \equiv \overline{\mathfrak{X}}(\widetilde{M}, \widetilde{\mathcal{F}}, \Gamma) \equiv \mathfrak{X}(G, \mathrm{Hol}\,\mathcal{F})\,. \tag{3.1.3}$$

(Given an action, the group is added to the notation of a space of vector fields to indicate the subspace of invariant elements.) These identities give a precise realization of $\mathfrak{g} \subset \overline{\mathfrak{X}}(M, \mathcal{F})$ as the Lie algebra of left invariant vector fields on G.

If a smooth map $\psi : M' \to M$ is transverse to $\mathcal{F}$, since $\psi_* : N\psi^*\mathcal{F} \to N\mathcal{F}$ restricts to isomorphisms between the fibers and is compatible with the corresponding flat partial connections (Sect. 3.1.7), it follows that any transverse structure of $\mathcal{F}$ canonically induces a transverse structure of $\psi^*\mathcal{F}$ of the same type.

3.1.10 Foliations of Codimension One

In this section, assume $\mathcal{F}$ is of codimension one ($n' = 1$ and $n'' = n - 1$). Then the notation $(x, y) = (x, y^1, \ldots, y^{n-1})$ is used for the foliated coordinates instead of (x', x'').

Suppose also that $\mathcal{F}$ is transversely oriented. Thus there are real forms $\omega, \eta \in C^\infty(M; \Lambda^1)$ such that ω defines $\mathcal{F}$ and its transverse orientation, and $d\omega = \eta \wedge \omega$ (Sect. 3.1.9). There is some $X \in \mathfrak{X}(M)$ with $\omega(X) = 1$; in fact, $\overline{X} \in C^\infty(M; N\mathcal{F})$ and ω determine each other. Now $\mathcal{F}$ is Riemannian just when ω can be chosen so that $d\omega = 0$; i.e., $X \in \mathfrak{X}(M, \mathcal{F})$. Actually, $\mathcal{F}$ is an $\mathbb{R}$-Lie foliation in this case because $\mathbb{R} \cdot \overline{X}$ is a Lie subalgebra of $\overline{\mathfrak{X}}(M, \mathcal{F})$.

3.1.11 Complete $\mathbb{R}$-Lie Foliations

$\mathcal{F}$ is a complete $\mathbb{R}$-Lie foliation when there is some $Z \in \mathfrak{X}_{\text{com}}(M, \mathcal{F})$ so that $\overline{Z}$ has no zeros. This means that the orbits of the foliated flow $\phi : M \times \mathbb{R} \to M$ of Z are transverse to $\mathcal{F}$. Its Fedida's description is given by some $\pi : \widetilde{M} \to M$, $D : \widetilde{M} \to \mathbb{R}$ and $h : \Gamma \to \mathbb{R}$ (Sect. 3.1.9). Let $\widetilde{Z} \in \mathfrak{X}_{\text{com}}(\widetilde{M}, \widetilde{\mathcal{F}})$ and $\tilde{\phi} : \widetilde{M} \times \mathbb{R} \to \widetilde{M}$ be the lifts of Z and ϕ. Then $\widetilde{Z}$ is Γ-invariant and D-projectable. Without loss of generality, we can assume $D_*\widetilde{Z} = \partial_x \in \mathfrak{X}(\mathbb{R})$, where x denotes the standard global coordinate of $\mathbb{R}$. Thus $\tilde{\phi}$ is Γ-equivariant and induces via D the flow $\bar{\phi} = \{\bar{\phi}^t\}$ on $\mathbb{R}$ defined by $\bar{\phi}^t(x) = t + x$. Since $\bar{\phi}^t$ preserves every $\operatorname{Hol}\mathcal{F}$-orbit in $\mathbb{R}$ if and only if $t \in \operatorname{Hol}\mathcal{F}$, it follows that ϕ^t preserves every leaf of $\mathcal{F}$ if and only if $t \in \operatorname{Hol}\mathcal{F}$.

3.1.12 Foliations Almost Without Holonomy

Assume M is compact. It is said that $\mathcal{F}$ is *almost without holonomy* when all non-compact leaves have no holonomy. The structure of such a foliation was described by Hector [Hec72, Hec78]. In the case where $\mathcal{F}$ has a finite number of leaves with holonomy and is transversely oriented, the description of $\mathcal{F}$ is as follows. Let M^0 be the finite union of compact leaves with holonomy. Let $M^1 = M \setminus M^0$, whose connected components are denoted by M^1_l ($l = 1, \ldots, k$), and let $\mathcal{F}^1_l =$

$\mathcal{F}|_{M_l^1}$. Then, for every l, there is a connected compact manifold M_l, possibly with boundary, endowed with a smooth transversely oriented foliation $\mathcal{F}_l$ tangent to the boundary, such that, equipping $\boldsymbol{M} := \bigsqcup_l M_l$ with the combination $\boldsymbol{\mathcal{F}}$ of the foliations $\mathcal{F}_l$, there is foliated smooth local embedding $\boldsymbol{\pi} : (\boldsymbol{M}, \boldsymbol{\mathcal{F}}) \to (M, \mathcal{F})$, preserving the transverse orientations, so that:

- $\boldsymbol{\pi} : \mathring{M}_l \to M_l^1$ is a diffeomorphism for all l (we may write $\mathring{M}_l \equiv M_l^1$);
- $\boldsymbol{\pi} : \partial \boldsymbol{M} \to M^0$ is a twofold covering map; and
- every $\mathcal{F}_l$ is one of the following models:
  - (0) $\mathcal{F}_l$ is given by a trivial bundle over $[0, 1]$,
  - (1) $\mathring{\mathcal{F}}_l := \mathcal{F}_l|_{\mathring{M}_l}$ is given by a fiber bundle over S^1, or
  - (2) all leaves of $\mathring{\mathcal{F}}_l$ are dense in $\mathring{M}_l$.

Thus M is obtained by gluing the manifolds M_l along corresponding pairs of boundary components. Equivalently, $\boldsymbol{M}$ can be described by cutting M along M^0 like in Sect. 2.6. Since $\mathcal{F}$ is transversely oriented, the restriction of $\boldsymbol{\pi} : \partial \boldsymbol{M} \to M^0$ to every connected component of $\partial \boldsymbol{M}$ is a diffeomorphism to its image. Thus $\partial \boldsymbol{M} \equiv M^0 \sqcup M^0$. The restriction of $\boldsymbol{\mathcal{F}}$ to the interior $\mathring{\boldsymbol{M}}$ is denoted by $\mathring{\boldsymbol{\mathcal{F}}}$. Thus $\boldsymbol{\pi}$ restricts to a foliated diffeomorphism $(\mathring{\boldsymbol{M}}, \mathring{\boldsymbol{\mathcal{F}}}) \xrightarrow{\approx} (M^1, \mathcal{F}^1)$.

Remark 3.1.6 In the above description, we have the following:

(i) If $\mathcal{F}_l$ is a model (2), then $\mathring{\mathcal{F}}_l$ becomes a complete $\mathbb{R}$-Lie foliation after a possible change of the differentiable structure of $\mathring{M}_l$, keeping the same differentiable structure on the leaves [Hec78, Theorem 2].

(ii) The description holds as well if M^0 is any finite union of compact leaves, including all leaves with holonomy. In particular, if $\mathcal{F}_l$ is a model (1) with $\partial M_l = \emptyset$, then $M_l = M$ can be cut into models (0) by adding compact leaves to M^0. Conversely, if all foliations $\mathcal{F}_l$ are models (0), then $\mathcal{F}$ is a model (1) with $\partial M = \emptyset$.

3.2 Differential Forms on Foliated Manifolds

3.2.1 The Leafwise Complex

Let $d_{\mathcal{F}} \in \operatorname{Diff}^1(\mathcal{F}; \Lambda\mathcal{F})$ be given by $(d_{\mathcal{F}}\alpha)|_L = d_L(\alpha|_L)$ for every leaf L and $\alpha \in C^\infty(M; \Lambda\mathcal{F})$. Then $(C^\infty(M; \Lambda\mathcal{F}), d_{\mathcal{F}})$ is a differential complex, called the *leafwise* or *tangential (de Rham) complex*. The elements of $C^\infty(M; \Lambda\mathcal{F})$ are called *leafwise forms*; the leafwise forms in $\ker d_{\mathcal{F}}$ (resp., $\operatorname{im} d_{\mathcal{F}}$) are called *leafwise-closed forms* (resp., *leafwise-exact forms*). The leafwise complex gives rise to the *leafwise* or *tangential cohomology* $H^\bullet(\mathcal{F})$. The leafwise complex is not elliptic if $n' > 0$, and therefore it makes sense to consider also its reduced cohomology $\bar{H}^\bullet(\mathcal{F})$ (Sect. 2.1.13). The more precise notation $H^\bullet C^\infty(\mathcal{F}) = H^\bullet(\mathcal{F})$ and $\bar{H}^\bullet C^\infty(\mathcal{F}) =$

$\bar{H}^\bullet(\mathcal{F})$ may also be used. Recall that we typically take complex coefficients without any comment; the case of real coefficients will be indicated. Compactly supported versions may be also considered when M is not compact.

We can also take coefficients in any complex $\mathcal{F}$-flat vector bundle E over M, obtaining the differential complex $C^\infty(M;\Lambda\mathcal{F}\otimes E)$ with $d_\mathcal{F}\in\operatorname{Diff}^1(\mathcal{F};\Lambda\mathcal{F}\otimes E)$, and the corresponding cohomology, $H^\bullet(\mathcal{F};E)$, and reduced cohomology, $\bar{H}^\bullet(\mathcal{F};E)$. For example, we can consider the vector bundle E defined by the $\mathrm{GL}(n')$-principal bundle of (real) normal frames and any unitary representation of $\mathrm{GL}(n')$, with the $\mathcal{F}$-flat structure induced by the $\mathcal{F}$-flat structure of $N\mathcal{F}$. A particular case is $\Lambda N\mathcal{F}$, which gives rise to the differential complex $(C^\infty(M;\Lambda\mathcal{F}\otimes\Lambda N\mathcal{F}),d_\mathcal{F})$. Note that

$$\Lambda\mathcal{F}\equiv\Lambda\mathcal{F}\otimes\Lambda^0N\mathcal{F}\subset\Lambda\mathcal{F}\otimes\Lambda N\mathcal{F}\,, \tag{3.2.1}$$

and therefore $C^\infty(M;\Lambda\mathcal{F})$ becomes a subcomplex of $C^\infty(M;\Lambda\mathcal{F}\otimes\Lambda N\mathcal{F})$ with $d_\mathcal{F}$.

3.2.2 *Bigrading of Differential Forms*

Consider any splitting

$$TM=T\mathcal{F}\oplus\boldsymbol{H}\cong T\mathcal{F}\oplus N\mathcal{F}\,, \tag{3.2.2}$$

for some vector subbundle $\boldsymbol{H}\subset TM$. Recall that $\Lambda\boldsymbol{H}=\bigwedge\boldsymbol{H}^*\otimes\mathbb{C}$. The splitting (3.2.2) induces a decomposition

$$\Lambda M\equiv\Lambda\mathcal{F}\otimes\Lambda\boldsymbol{H}\cong\Lambda\mathcal{F}\otimes\Lambda N\mathcal{F}\,, \tag{3.2.3}$$

giving rise to the bigrading of ΛM defined by

$$\Lambda^{u,v}M\equiv\Lambda^v\mathcal{F}\otimes\Lambda^u\boldsymbol{H}\cong\Lambda^v\mathcal{F}\otimes\Lambda^uN\mathcal{F}\,, \tag{3.2.4}$$

and the corresponding bigrading of $C^\infty(M;\Lambda)$ with bihomogeneous components

$$C^\infty(M;\Lambda^{u,v})\equiv C^\infty(M;\Lambda^v\mathcal{F}\otimes\Lambda^uN\mathcal{F})\,.$$

In particular, $\Lambda^{0,v}M\equiv\Lambda^v\mathcal{F}$ and $\Lambda^{u,0}M\equiv\Lambda^u\boldsymbol{H}$, and then the identity of (3.2.4) becomes[1] $\Lambda^{0,v}M\otimes\Lambda^{u,0}M\equiv\Lambda^{u,v}M$, $\alpha\otimes\beta\equiv\alpha\wedge\beta$.

---

[1] This order in the wedge product, introduced in [ÁLKL20] and different from [ÁLK01], produces simpler sign expressions. However, the transverse degree is written first in the bigrading, like in the extension to foliations of the Leray-Serre spectral sequence.

This bigrading depends on $\boldsymbol{H}$, but the spaces $\Lambda^{\geq u,\cdot}M$ and $C^\infty(M;\Lambda^{\geq u,\cdot})$ are independent of $\boldsymbol{H}$ (see e.g. [ÁL89]). There are canonical identities

$$\Lambda^{\geq u,\cdot}M/\Lambda^{\geq u+1,\cdot}M \equiv \Lambda^{u,\cdot}M \equiv \Lambda\mathcal{F}\otimes\Lambda^u N\mathcal{F}\,, \tag{3.2.5}$$

where only $\Lambda^{u,\cdot}M$ depends on $\boldsymbol{H}$.

3.2.3 Bihomogeneous Components of the Derivative

The de Rham derivative on $C^\infty(M;\Lambda)$ decomposes into bihomogeneous components,

$$d = d_{0,1} + d_{1,0} + d_{2,-1}\,, \tag{3.2.6}$$

where the double subscript denotes the corresponding bidegree. By comparing bidegrees in the anti-derivation formula of d, we also get that every $d_{i,1-i}$ ($i \in \{0,1,2\}$) satisfies the same anti-derivation formula. Thus $d_{2,-1}$ is of order 0. The other components, $d_{0,1}$ and $d_{1,0}$, are of order 1. Moreover, $d_{2,-1} = 0$ if and only if $\boldsymbol{H}$ is completely integrable. By comparing bi-degrees in $d^2 = 0$, we get [ÁL89]

$$d_{0,1}^2 = d_{0,1}d_{1,0} + d_{1,0}d_{0,1} = 0\,. \tag{3.2.7}$$

So $(C^\infty(M;\Lambda), d_{0,1})$ is a differential complex of order one. In fact, via (3.2.3),

$$d_{0,1} \equiv d_{\mathcal{F}}\,. \tag{3.2.8}$$

Moreover

$$d_{0,1} = d : C^\infty(M;\Lambda^{n',\bullet}) \to C^\infty(M;\Lambda^{n',\bullet+1})\,. \tag{3.2.9}$$

3.2.4 Basic Complex

It is said that $\alpha \in C^\infty(M;\Lambda)$ is a *basic form* if $\iota_X\alpha = \iota_X d\alpha = 0$ for all $X \in \mathfrak{X}(\mathcal{F})$. This means that α is an $\mathcal{F}$-parallel section of $\Lambda N\mathcal{F} \equiv \Lambda^{\bullet,0}M$; i.e., $\alpha \in C^\infty(M;\Lambda^{\bullet,0}) \cap \ker d_{0,1}$. The basic forms form a subcomplex of the de Rham complex, called the *basic complex*. It is isomorphic to the complex of $\mathcal{H}$-invariant forms on Σ via the distinguished projections $x'_k : U_k \to \Sigma_k$ (Sect. 3.1.1).

3.2.5 Bihomogeneous Components of the Coderivative

Given a leafwise metric $g_{\mathcal{F}}$, the coderivative on the leaves defines an operator $\delta_{\mathcal{F}} \in \operatorname{Diff}^1(M;\Lambda\mathcal{F})$, like in the case of $d_{\mathcal{F}}$.

Fix a Riemannian metric g on M. Using $\boldsymbol{H} = T\mathcal{F}^{\perp}$ and taking formal adjoints in (3.2.6) and (3.2.7), we get a decomposition of the coderivative on $C^{\infty}(M; \Lambda)$,

$$\delta = \delta_{0,-1} + \delta_{-1,0} + \delta_{-2,1} , \tag{3.2.10}$$

and the bihomogeneous components $\delta_{-i,i-1} = d^*_{i,1-i}$ satisfy the analog of (3.2.7).

The metric g induces a leafwise metric $g_{\mathcal{F}}$. It also induces an Euclidean structure on $N\mathcal{F}$, which in turn induces a Hermitian structure on $\Lambda N\mathcal{F}$. Thus the adjoint $\delta_{\mathcal{F}} = d^*_{\mathcal{F}}$ is also defined on $C^{\infty}(M; \Lambda\mathcal{F} \otimes \Lambda N\mathcal{F})$. The analogue of (3.2.8),

$$\delta_{0,-1} \equiv \delta_{\mathcal{F}} \tag{3.2.11}$$

via (3.2.3), holds if and only if g is bundle-like [ÁLKL20, Lemma 4.12]. Thus, in this case, $\delta = \delta_{0,-1} \equiv \delta_{\mathcal{F}}$ on $C^{\infty}(M; \Lambda^{0,\bullet}) \equiv C^{\infty}(M; \Lambda\mathcal{F})$ via (3.2.1) and (3.2.4).

The following operators will be also used:

$$\begin{cases} D_0 = d_{0,1} + \delta_{0,-1} , \quad D_{\perp} = d_{1,0} + \delta_{-1,0} , \\ \Delta_0 = D_0^2 = d_{0,1}\delta_{0,-1} + \delta_{0,-1}d_{0,1} . \end{cases} \tag{3.2.12}$$

3.2.6 *Bigrading vs Orientations*

Recall that a transverse orientation of $\mathcal{F}$ can be described by a non-vanishing real form $\omega \in C^{\infty}(M; \Lambda^{n'}N\mathcal{F}) \equiv C^{\infty}(M; \Lambda^{n',0})$. According to Sect. 3.1.9, there is a real 1-form η satisfying $d\omega = \eta \wedge \omega$. We write $\eta = \eta_0 + \eta_1$, where $\eta_0 \in C^{\infty}(M; \Lambda^{0,1})$ is determined by ω, and $\eta_1 \in C^{\infty}(M; \Lambda^{1,0})$ can be chosen arbitrarily.

On the other hand, an orientation of $T\mathcal{F}$ is called a (*leafwise* or *tangential*) *orientation* of $\mathcal{F}$, which can be described by a non-vanishing real form $\chi \in C^{\infty}(M; \Lambda^{n''}\mathcal{F}) \equiv C^{\infty}(M; \Lambda^{0,n''})$. It is said $\mathcal{F}$ is *oriented* if it is endowed with an orientation. Given transverse and tangential orientations of $\mathcal{F}$, described by forms ω and χ as above, we consider the induced orientation of M defined by the non-vanishing real form $\chi \wedge \omega \in C^{\infty}(M; \Lambda^{n',n''}) = C^{\infty}(M; \Lambda^{n})$.

Suppose that M is a Riemannian manifold and take $\boldsymbol{H} = T\mathcal{F}^{\perp}$. Then, using (3.2.3), the induced Hodge star operators, $\star$ on ΛM, $\star_{\mathcal{F}}$ on $\Lambda\mathcal{F}$ and $\star_{\perp}$ on $\Lambda\boldsymbol{H}$, satisfy[2] [ÁLT91, Lemma 4.8], [ÁLK01, Lemma 3.2], [ÁLKL20, Eq. (42)]

$$\star \equiv (-1)^{u(n''-v)} \star_{\mathcal{F}} \otimes \star_{\perp} : \Lambda^{u,v}M \to \Lambda^{n'-u,n''-v}M . \tag{3.2.13}$$

[2] The sign of this expression, used in [ÁLKL20, Eq. (42)], is different from the sign used in [ÁLK01, Lemma 3.2] by the different choices of induced orientation of M.

If $\omega = \star_\perp 1$ and $\chi = \star_\mathcal{F} 1$, then $\chi \wedge \omega = \star 1$. We have

$$\delta_{-i,i-1} = (-1)^{nk+n+1} \star d_{i,1-i} \star \tag{3.2.14}$$

on $C^\infty(M; \Lambda^k)$, and

$$\delta_\mathcal{F} = (-1)^{n''v+n''+1} \star_\mathcal{F} d_\mathcal{F} \star_\mathcal{F} \tag{3.2.15}$$

on $C^\infty(M; \Lambda^v\mathcal{F})$. Using (3.2.13)–(3.2.15), we easily get

$$\delta_{0,-1} \equiv \delta_\mathcal{F} + \eta_0 \lrcorner$$

on $C^\infty(M; \Lambda^{0,v}) \equiv C^\infty(M; \Lambda^v\mathcal{F})$.

3.2.7 *Leafwise Euler Form*

If $\mathcal{F}$ is oriented, then $\Omega\mathcal{F} \equiv \Lambda^{0,n''}M \equiv \Lambda^{n''}\mathcal{F}$. If moreover $\mathcal{F}$ is equipped with a leafwise Riemannian metric $g_\mathcal{F}$ and n'' is even, then the *leafwise Euler form* $e(\mathcal{F}, g_\mathcal{F}) \in C^\infty(M; \Lambda^{n''}\mathcal{F}) \equiv C^\infty(M; \Omega\mathcal{F})$ is defined by the Euler form of the leaves (Sect. 2.9.10). When $\mathcal{F}$ is not oriented, $e(\mathcal{F}, g_\mathcal{F})$ is defined as an element of $C^\infty(M; \Lambda^{n''}\mathcal{F} \otimes o(\mathcal{F})) \equiv C^\infty(M; \Omega\mathcal{F})$.

3.2.8 *Leafwise Currents*

We may also consider the continuous extension of $d_\mathcal{F}$ to $C^{-\infty}(M; \Lambda\mathcal{F})$ (Sect. 2.1.7), defining another topological complex whose cohomology and reduced cohomology are denoted by $H^\bullet C^{-\infty}(\mathcal{F})$ and $\bar{H}^\bullet C^{-\infty}(\mathcal{F})$ (see Sect. 3.2.1). The elements of $C^{-\infty}(M; \Lambda\mathcal{F})$ are called *leafwise currents*. In general, $C^\infty(M; \Lambda\mathcal{F}) \hookrightarrow C^{-\infty}(M; \Lambda\mathcal{F})$ does not induce an isomorphism in cohomology or reduced cohomology (consider a foliation by points).

Like in (2.8.1), the exterior product has continuous extensions,

$$C^{\pm\infty}(M; \Lambda\mathcal{F}) \otimes C^{\mp\infty}(M; \Lambda\mathcal{F}) \to C^{-\infty}(M; \Lambda\mathcal{F}) ,$$

with a corresponding extension of the property of $d_\mathcal{F}$ to be a derivation. Given a leafwise metric $g_\mathcal{F}$ on M, we can also consider the continuous extension $\delta_\mathcal{F}$ to $C^{-\infty}(M; \Lambda\mathcal{F})$

The concept of leafwise currents with coefficients in any $\mathcal{F}$-flat vector bundle E can be also considered, and the obvious notation is used for the corresponding topological complex and its cohomology and reduced cohomology. In particular, E can be any vector bundle associated with $N\mathcal{F}$.

3.2.9 Bigrading of Currents

Consider also the bigrading of $C^{-\infty}(M;\Lambda)$ induced by the bigrading of ΛM, and the continuous extensions to $C^{-\infty}(M;\Lambda)$ of the operators $d_{i,1-i}$, which satisfy the extensions to currents of (3.2.8) and (3.2.9). Given a metric g on M, we can also consider the continuous extensions of the operators $\delta_{-i,i-1}$ to $C^{-\infty}(M;\Lambda)$.

Recall that $d_{0,1}^{\mathrm{t}}$ denotes the transpose of $d_{0,1}$ (Sect. 2.1.5). If M is oriented, then (2.8.3), (2.8.4) and (2.9.8) for $z=0$ give

$$(\Lambda^{u,v}M)^*\otimes\Omega M\equiv\Lambda^{n'-u,n''-v}M\;, \tag{3.2.16}$$

$$C^{-\infty}(M;\Lambda^{u,v})\equiv C^{\infty}_{\mathrm{c}}(M;\Lambda^{n'-u,n''-v})'\;, \tag{3.2.17}$$

$$d_{0,1}\equiv(-1)^{u+v+1}\,d_{0,1}^{\mathrm{t}}:C^{-\infty}(M;\Lambda^{u,v})\to C^{-\infty}(M;\Lambda^{u,v+1})\;. \tag{3.2.18}$$

When M is not oriented, these identities hold after adding the tensor product with $o(M)$ to the exterior bundles in the right-hand sides, or working locally, or passing to the double cover of orientations. By (3.2.9), if $u=n'$, then (3.2.18) agrees with (2.9.8) for $z=0$ on $C^{-\infty}(M;\Lambda^{n',v})$. By (3.2.3), (3.2.8) and (3.2.9), if $u=0$, then (3.2.16)–(3.2.18) become

$$(\Lambda^{v}\mathcal{F})^*\otimes\Omega M\equiv\Lambda^{n',n''-v}M\equiv\Lambda^{n''-v}\mathcal{F}\otimes\Lambda^{n'}N\mathcal{F}\equiv\Lambda^{n''-v}\mathcal{F}\otimes\Omega N\mathcal{F}\;, \tag{3.2.19}$$

$$C^{-\infty}(M;\Lambda^{v}\mathcal{F})\equiv C^{\infty}_{\mathrm{c}}(M;\Lambda^{n',n''-v})'\equiv C^{\infty}_{\mathrm{c}}(M;\Lambda^{n''-v}\mathcal{F}\otimes\Omega N\mathcal{F})'\;, \tag{3.2.20}$$

$$d_{\mathcal{F}}\equiv(-1)^{v+1}\,d^{\mathrm{t}}\equiv(-1)^{v+1}\,d_{\mathcal{F}}^{\mathrm{t}}:C^{-\infty}(M;\Lambda^{v}\mathcal{F})\to C^{-\infty}(M;\Lambda^{v+1}\mathcal{F})\;. \tag{3.2.21}$$

3.2.10 Pull-Back of Leafwise Forms

Let $\phi\in C^{\infty}(M',\mathcal{F}';M,\mathcal{F})$. Like in (2.8.8) and (2.8.9), the homomorphisms $\phi_*: T\mathcal{F}'\to T\mathcal{F}$ and $\phi_*:N\mathcal{F}'\to N\mathcal{F}$ induce continuous homomorphisms,

$$\phi^*:(C^{\infty}(M;\Lambda\mathcal{F}\otimes\Lambda N\mathcal{F}),d_{\mathcal{F}})\to(C^{\infty}(M';\Lambda\mathcal{F}'\otimes\Lambda N\mathcal{F}'),d_{\mathcal{F}'})\;, \tag{3.2.22}$$

$$\phi^*:(C^{\infty}(M;\Lambda\mathcal{F}),d_{\mathcal{F}})\to(C^{\infty}(M';\Lambda\mathcal{F}'),d_{\mathcal{F}'})\;, \tag{3.2.23}$$

the second one is a restriction of the first one according to (3.2.1).

On the other hand, $\phi^*:C^{\infty}(M;\Lambda)\to C^{\infty}(M';\Lambda)$ has restrictions

$$\phi^*:C^{\infty}(M;\Lambda^{\geq u,\cdot})\to C^{\infty}(M';\Lambda^{\geq u,\cdot})\;,$$

which induce (3.2.22) using (3.2.5).

3.2.11 Bihomogeneous Components of Pull-Back Homomorphisms

For any smooth map $\phi : M' \to M$, the homomorphism $\phi^* : C^\infty(M;\Lambda) \to C^\infty(M';\Lambda)$ decomposes into bihomogeneous components,

$$\phi^* = \cdots + \phi^*_{-1,1} + \phi^*_{0,0} + \phi^*_{1,-1} + \cdots$$

If $\phi \in C^\infty(M',\mathcal{F}'; M,\mathcal{F})$, then $\phi^*_{i,-i} = 0$ for $i < 0$. Moreover, via (3.2.3),

$$\phi^*_{0,0} \equiv \phi^* , \tag{3.2.24}$$

where the right-hand side is (3.2.22).

3.2.12 Bihomogeneous Components of the Lie Derivative

For any $X \in \mathfrak{X}(M)$, by comparing bidegrees in Cartan's formula, $\mathcal{L}_X = d\iota_X + \iota_X d$, we get a decomposition into bi-homogeneous components,

$$\mathcal{L}_X = \mathcal{L}_{X,-1,1} + \mathcal{L}_{X,0,0} + \mathcal{L}_{X,1,-1} + \mathcal{L}_{X,2,-2} .$$

For instance,

$$\mathcal{L}_{X,0,0} = d_{0,1}\iota_{VX} + \iota_{VX}d_{0,1} + d_{1,0}\iota_{HX} + \iota_{HX}d_{1,0} , \tag{3.2.25}$$

where $\boldsymbol{V} : TM \to T\mathcal{F}$ and $\boldsymbol{H} : TM \to \boldsymbol{H}$ denote the projections defined by (3.2.2). By comparing bidegrees in the derivation formula of $\mathcal{L}_X$, we also get that every $\mathcal{L}_{X,i,-i}$ ($i \in \{-1,0,1,2\}$) satisfies the same derivation formula. Thus $\mathcal{L}_{X,-1,1}$, $\mathcal{L}_{X,1,-1}$ and $\mathcal{L}_{X,2,-2}$ are of order zero. For the sake of simplicity, we will write $\Theta_X = \mathcal{L}_{X,0,0}$, which is of order 1.

If $X \in \mathfrak{X}(M,\mathcal{F})$, then $\mathcal{L}_{X,-1,1} = 0$, obtaining

$$\Theta_X d_{0,1} = d_{0,1}\Theta_X \tag{3.2.26}$$

by comparing bi-degrees in the formula $\mathcal{L}_X d = d\mathcal{L}_X$. On the other hand, by (3.2.25), if $X \in C^\infty(M;\boldsymbol{H})$, then, for all $f \in C^\infty(M)$,

$$\Theta_{fX} = f\Theta_X \tag{3.2.27}$$

on $C^\infty(M;\Lambda^{0,\bullet}) \equiv C^\infty(M;\Lambda\mathcal{F})$. If $d_{1,0}f = 0$, then (3.2.27) holds on $C^\infty(M;\Lambda)$ by (3.2.25) and the derivation formula of Θ_X.

3.2.13 *Local Descriptions*

Let (U, x) be a foliated chart of $\mathcal{F}$, with $x = (x', x'')$, like in (3.1.1). To emphasize the difference between the coordinates x' and x'', we use the following notation on U or $x(U)$. Let $x'^i = x^i$ and $\partial'_i = \partial_i$ for $i \le n'$, and $x''^i = x^i$ and $\partial''_i = \partial_i$ for $i > n'$. Thus, when using x'^i or ∂'_i, it will be understood that i runs in $\{1, \dots, n'\}$, and, when using x''^i or ∂''_i, it will be understood that i runs in $\{n' + 1, \dots, n\}$. For multi-indices of the form $J = \{j_1, \dots, j_r\}$ with $1 \le j_1 < \cdots < j_r \le n$, let $dx^J = dx^{j_1} \wedge \cdots \wedge dx^{j_r}$ be denoted by dx'^J or dx''^J if J only contains indices in $\{1, \dots, n'\}$ or $\{n' + 1, \dots, n\}$, respectively. Using functions $f_I \in C^\infty(U)$, $d_{\mathcal{F}}$ on U can be described by

$$d_{\mathcal{F}}(f_I \, dx''^I) = \partial''_j f_I \, dx''^j \wedge dx''^I \; . \tag{3.2.28}$$

Since the forms dx'^J are basic, (3.2.8) on U means that

$$d_{0,1}(f_{IJ} \, dx''^I \wedge dx'^J) = d_{\mathcal{F}}(f_{IJ} \, dx''^I) \wedge dx'^J \; , \tag{3.2.29}$$

using functions $f_{IJ} \in C^\infty(U)$.

Given a metric g on M, the local description

$$\delta_{0,-1}(f_{IJ} \, dx''^I \wedge dx'^J) = \delta_{\mathcal{F}}(f_{IJ} \, dx''^I) \wedge dx'^J \tag{3.2.30}$$

is satisfied just when g is bundle-like [ÁLK01, Lemma 3.4]; in fact this is a local expression of (3.2.11).

From (3.2.25), we also get that, on $C^\infty(U, \Lambda^{0,\bullet})$,

$$d_{1,0} = dx'^i \wedge \Theta_{\boldsymbol{H}\partial'_i} \; . \tag{3.2.31}$$

Since $d_{1,0}$ is an anti-derivation, it follows that

$$d_{1,0}(f_{IJ} \, dx''^I \wedge dx'^J) = (-1)^{|I|} \Theta_{\boldsymbol{H}\partial'_i}(f_{IJ} \, dx''^I) \wedge dx'^i \wedge dx'^J \; . \tag{3.2.32}$$

3.2.14 *Bigrading of Leafwise Forms*

Suppose $\mathcal{F}$ is a subfoliation of another smooth foliation $\mathcal{G}$ on M. Like in Sect. 3.2.2, for any choice of a complement $\boldsymbol{G}$ of $T\mathcal{F}$ in $T\mathcal{G}$, we have $\Lambda\mathcal{G} = \Lambda\mathcal{F} \otimes \Lambda\boldsymbol{G}$, obtaining a bigrading of $\Lambda\mathcal{G}$ defined by $\Lambda^{u,v}\mathcal{G} = \Lambda^v\mathcal{F} \otimes \Lambda^u\boldsymbol{G}$, and a corresponding bigrading of $C^\infty(M; \Lambda\mathcal{G})$. The decomposition (3.2.6) has an obvious version for $d_{\mathcal{G}}$ satisfying analogous properties.

3.2.15 Push-Forward and Pull-Back of Leafwise Currents

With the notation of Sect. 2.8.5, assume $\phi : M' \to M$ is a smooth submersion and $\mathcal{V}$ oriented. Using any complement $\mathcal{H}$ of $\mathcal{V}$ in TM', we get a corresponding bigrading of $\Lambda M'$ with $\phi^*\Lambda M \otimes \Omega_{\text{fiber}}M' \equiv \Lambda^{\bullet,\text{top}}M'$. Suppose M is equipped with a smooth foliation $\mathcal{F}$, and let $\mathcal{F}' = \phi^*\mathcal{F}$. Choose complements, $\boldsymbol{H}$ of $T\mathcal{F}$ in TM and $\boldsymbol{H}'$ of $T\mathcal{F}'$ in TM'. The tangent map ϕ_* defines an identity $\boldsymbol{H}' \equiv \phi^*\boldsymbol{H}$. Consider the bigradings of ΛM and $\Lambda M'$ induced by $(\mathcal{F}, \boldsymbol{H})$ and $(\mathcal{F}', \boldsymbol{H}')$. Then the maps (2.8.11)–(2.8.14) have restrictions compatible with $d_{0,1}$,

$$\phi_* : C^{\pm\infty}_{\text{c/cv}}(M'; \Lambda^{u,\bullet}) \to C^{\pm\infty}_{\text{c/\cdot}}(M; \Lambda^{u,\bullet-p}) \quad (p = \dim \mathcal{V}) ,$$

$$\phi^* : C^{-\infty}(M; \Lambda^{u,\bullet}) \to C^{-\infty}(M'; \Lambda^{u,\bullet}) .$$

For $u = 0$, by (3.2.1), they are continuous homomorphisms,

$$\phi_* : (C^{\pm\infty}_{\text{c/cv}}(M'; \Lambda\mathcal{F}'), d_{\mathcal{F}'}) \to (C^{\pm\infty}_{\text{c/\cdot}}(M; \Lambda\mathcal{F}), d_{\mathcal{F}}) , \tag{3.2.33}$$

$$\phi^* : (C^{-\infty}(M; \Lambda\mathcal{F}), d_{\mathcal{F}}) \to (C^{-\infty}(M'; \Lambda\mathcal{F}'), d_{\mathcal{F}'}) . \tag{3.2.34}$$

Like in (2.8.13)–(2.8.16), the maps (3.2.33) and (3.2.34) can be also defined as the compositions

$$C^{\pm\infty}_{\text{c/cv}}(M'; \Lambda\mathcal{F}') \xrightarrow{\pi_{\text{top}}} C^{\pm\infty}_{\text{c/cv}}(M'; \phi^*\Lambda\mathcal{F}) \xrightarrow{\phi_*} C^{\pm\infty}_{\text{c/\cdot}}(M; \Lambda\mathcal{F}) , \tag{3.2.35}$$

$$C^{-\infty}(M; \Lambda\mathcal{F}) \xrightarrow{\phi^*} C^{-\infty}(M'; \phi^*\Lambda\mathcal{F}) \xrightarrow{\phi^*} C^{-\infty}(M'; \Lambda\mathcal{F}') . \tag{3.2.36}$$

We can directly extend the definition of (3.2.33) to the case where M' is a manifold with boundary, assuming $\mathcal{F}'$ is tangent or transverse to the boundary. It is a cochain map when $\mathcal{F}'$ is tangent to the boundary. If $\mathcal{F}'$ is transverse to the boundary and $\phi|_{\partial M'} : \partial M' \to M$ is a submersion, the Stokes' formula gives

$$\phi_* d_{\mathcal{F}'} - d_{\mathcal{F}}\phi_* = (\phi|_{\partial M'})_*\iota^* : C^{\infty}_{\text{c}}(M'; \Lambda\mathcal{F}') \to C^{\infty}_{\text{c}}(M; \Lambda\mathcal{F}) , \tag{3.2.37}$$

where $\iota : \partial M' \hookrightarrow M'$.

3.2.16 Leafwise Homotopy Operators

With the notation of Sect. 2.8.6, suppose M and M' are equipped with respective smooth foliations $\mathcal{F}$ and $\mathcal{F}'$, H is a leafwise homotopy, and consider $H_t^* : C^{\infty}(M; \Lambda\mathcal{F}) \to C^{\infty}(M'; \Lambda\mathcal{F}')$ ($t \in I$). Then we similarly get a continuous linear map $\mathsf{h} : C^{\infty}(M; \Lambda\mathcal{F}) \to C^{\infty}(M'; \Lambda\mathcal{F}')$, called a *leafwise homotopy operator*, which is homogeneous of degree -1 and satisfies $H_1^* - H_0^* = \mathsf{h}d_{\mathcal{F}} + d_{\mathcal{F}'}\mathsf{h}$. By using (3.2.23), (3.2.37) and (3.2.33), h can be given as the composition

$$C^\infty(M; \Lambda\mathcal{F}) \xrightarrow{H^*} C^\infty(M' \times I; \Lambda(\mathcal{F}' \times I)) \xrightarrow{f_I} C^\infty(M'; \Lambda\mathcal{F}') . \tag{3.2.38}$$

So H_0 and H_1 induce the same homomorphisms $H^\bullet C^\infty(\mathcal{F}) \to H^\bullet C^\infty(\mathcal{F}')$ and $\bar{H}^\bullet C^\infty(\mathcal{F}) \to \bar{H}^\bullet C^\infty(\mathcal{F}')$.

Suppose H is transverse to $\mathcal{F}$ and $H^*\mathcal{F} = \mathcal{F}' \times I$. Let $\mathrm{pr}_1 : M' \times I \to M'$ denote the first-factor projection. Consider the bigradings defined by $\mathcal{F}$, $\mathcal{F}'$, and complements $\boldsymbol{H}$ and $\boldsymbol{H}'$ of their tangent bundles. So H_* defines a homomorphism $\mathrm{pr}_1^* \boldsymbol{H}' \to \boldsymbol{H}$ whose restrictions to the fibers are isomorphisms. Then (3.2.38) is the bihomogeneous component of bidegree $(0, -1)$ of (2.8.17).

If moreover H is a submersion, then (3.2.33) and (3.2.34) give a continuous extension of the maps of (3.2.38),

$$C^{-\infty}(M; \Lambda\mathcal{F}) \xrightarrow{H^*} C^{-\infty}(M' \times I; \Lambda(\mathcal{F}' \times I)) \xrightarrow{f_I} C^{-\infty}(M'; \Lambda\mathcal{F}') .$$

Their composition, $\mathsf{h} : C^{-\infty}(M; \Lambda\mathcal{F}) \to C^{-\infty}(M'; \Lambda\mathcal{F}')$, satisfies $H_1^* - H_0^* = \mathsf{h}d_{\mathcal{F}} + d_{\mathcal{F}'}\mathsf{h}$. Thus H_0 and H_1 also induce the same homomorphisms $H^\bullet C^{-\infty}(\mathcal{F}) \to H^\bullet C^{-\infty}(\mathcal{F}')$ and $\bar{H}^\bullet C^{-\infty}(\mathcal{F}) \to \bar{H}^\bullet C^{-\infty}(\mathcal{F}')$.

3.3 Witten's Perturbation on Foliated Manifolds

The operators acting on differential forms on foliated manifolds (Sect. 3.2) are extended now by taking Witten's perturbations (Sect. 2.9).

3.3.1 Perturbation vs Bigrading

Using the notation of Sects. 2.9.1 and 3.2, write[3] $\eta = \eta_0 + \eta_1$ with $\eta_0 \in C^\infty(M; \Lambda^{0,1}) \equiv C^\infty(M; \Lambda^1\mathcal{F})$ and $\eta_1 \in C^\infty(M; \Lambda^{1,0})$. The condition $d\eta = 0$ means

$$d_{0,1}\eta_0 = d_{1,0}\eta_1 = d_{1,0}\eta_0 + d_{0,1}\eta_1 = 0 . \tag{3.3.1}$$

Like in (3.2.6) and (3.2.10), we get

$$d_z = d_{z,0,1} + d_{z,1,0} + d_{2,-1} , \quad \delta_z = \delta_{z,0,-1} + \delta_{z,-1,0} + \delta_{-2,1} ,$$

[3] In [ÁLKL20, Section 11], we took $\eta \in C^\infty(M; \Lambda^{0,1})$. However a general η is needed, and therefore additional work is required in Sects. 3.3 and 3.5.

where

$$d_{z,0,1} = d_{0,1} + z\,\eta_0\wedge\,, \qquad d_{z,1,0} = d_{1,0} + z\,\eta_1\wedge\,,$$
$$\delta_{z,0,-1} = \delta_{0,1} - \bar z\,\eta_0\lrcorner\,, \qquad \delta_{z,-1,0} = \delta_{-1,0} - \bar z\,\eta_1\lrcorner\,.$$

We will also use the perturbed versions of the operators (3.2.12), denoted by $D_{z,0}$, $D_{z,\perp}$ and $\Delta_{z,0}$, defined with the operators $d_{z,i,1-i}$ and $\delta_{z,i,i-1}$.

There is an obvious analog of (3.2.7) for the operators $d_{z,i,1-i}$, giving rise to analogous relations for the operators $\delta_{z,i,i-1}$. In particular, $d_{z,0,1}$ and $\delta_{z,0,-1}$ define leafwise differential complexes. By (2.9.7), the expressions (3.2.14) and (3.2.15) have direct extensions to this setting as well.

Concerning uniform leafwise/transverse ellipticity, symmetry and being non-negative, the perturbations $d_{z,0,1}$, $\delta_{z,0,-1}$, $D_{z,0}$, $D_{z,\perp}$ and $\Delta_{z,0}$ satisfy the same properties as $d_{0,1}$, $\delta_{0,-1}$, D_0, $D_\perp$ and Δ_0.

By (3.2.31), on a foliated chart (U,x), we get

$$d_{1,0}\eta_0 = dx'^i \wedge \Theta_{\boldsymbol{H}\partial_i}\eta_0 = -\Theta_{\boldsymbol{H}\partial_i}\eta_0 \wedge dx'^i\,.$$

But, writing $\eta_1 = h_i\,dx'^i$, by (3.3.1),

$$d_{1,0}\eta_0 = -d_{0,1}\eta_1 = -\partial''_j h_i\,dx''^j \wedge dx'^i\,.$$

So

$$\Theta_{\boldsymbol{H}\partial_i}\eta_0 = \partial''_j h_i\,dx''^j\,.$$

Then, since $\Theta_{\boldsymbol{H}\partial_i}$ is a derivation, on $C^\infty(M;\Lambda\mathcal{F})$,

$$[\Theta_{\boldsymbol{H}\partial_i}, \eta_0\wedge] = (\Theta_{\boldsymbol{H}\partial_i}\eta_0)\wedge = \partial''_j h_i\,dx''^j\wedge = (d_{\mathcal{F}}h_i)\wedge = [d_{\mathcal{F}}, h_i]\,. \tag{3.3.2}$$

Thus (3.2.26) has the following change in this setting:

$$[\Theta_{\boldsymbol{H}\partial_i}, d_{z,0,1}] = z[d_{\mathcal{F}}, h_i]\,.$$

3.3.2 *Perturbation of the Leafwise Complex*

Consider also the perturbed leafwise complex, $d_{\mathcal{F},z} = d_{\mathcal{F}} + z\eta_0\wedge$ on $C^\infty(M;\Lambda\mathcal{F})$, or on $C^\infty(M;\Lambda\mathcal{F}\otimes\Lambda N\mathcal{F})$, as well as its formal adjoint $\delta_{\mathcal{F},z} = \delta_{\mathcal{F}} - \bar z\,\eta_0\lrcorner$, and the induced perturbations, $D_{\mathcal{F},z}$ of $D_{\mathcal{F}}$ and $\Delta_{\mathcal{F},z}$ of $\Delta_{\mathcal{F}}$. They satisfy the obvious versions of (3.2.8) and (3.2.29). If g is bundle-like, they also satisfy the obvious versions of (3.2.11) and (3.2.30).

3.3.3 Perturbation with Two Parameters

For $z, z' \in \mathbb{C}$, the operators $D_{0,z,z'}$ and $\Delta_{0,z,z'}$ are defined like $D_{z,z'}$ and $\Delta_{z,z'}$ (Sect. 2.9.7), by using $d_{z,0,1}$ and $\delta_{z',0,-1}$ instead of d_z and $\delta_{z'}$. In other words, $D_{0,z,z'}$ is the component of $D_{z,z'}$ that preserves the transverse degree, and $\Delta_{z,z'} = D^2_{z,z'}$. They are uniformly leafwise elliptic, with a symmetric leading symbol.

The operators $D_{\mathcal{F},z,z'}$ and $\Delta_{\mathcal{F},z,z'}$ on $C^\infty(M; \Lambda\mathcal{F})$, or on $C^\infty(M; \Lambda\mathcal{F}\otimes\Lambda N\mathcal{F})$, are defined like $D_{z,z'}$ and $\Delta_{z,z'}$, by using $d_{\mathcal{F},z}$ and $\delta_{\mathcal{F},z'}$ instead of d_z and $\delta_{z'}$. They are also uniformly leafwise elliptic, with a symmetric leading symbol. If g is bundle-like, they also agree with $D_{z,z'}$ and $\Delta_{z,z'}$ via (3.2.3).

3.3.4 Perturbation vs Foliated Maps

With the notation of Sects. 2.9.2 and 2.9.4 for a smooth foliated map $\phi : (M, \mathcal{F}) \to (M, \mathcal{F})$, let $\widetilde{\mathcal{F}}$ and $\tilde{\eta}_j$ ($j = 1, 2$) be the lifts of $\mathcal{F}$ and η_j to $\widetilde{M}$. Thus $\tilde{\eta}_0 = d_{0,1}F \equiv d_{\widetilde{\mathcal{F}}}F$ and $\tilde{\eta}_1 = d_{1,0}F$. Any lift $\tilde{\phi}$ of ϕ to $\widetilde{M}$ is a foliated diffeomorphism of $(\widetilde{M}, \widetilde{\mathcal{F}})$. The endomorphism ϕ_z^{t*} of $(C^\infty(M; \Lambda), d_z)$ decomposes into the sum of bihomogeneous components $\phi^*_{z,i,-i}$, like in Sect. 3.2.11, whose lifts to $C^\infty(\widetilde{M}; \Lambda)$ are $e^{z(\tilde{\phi}^*F-F)}\tilde{\phi}^*_{i,-i}$. Then $\phi^*_{z,0,0}$ is an endomorphism of $(C^\infty(M; \Lambda), d_{z,0,1})$.

Similarly, the endomorphism ϕ^* of $(C^\infty(M; \Lambda\mathcal{F} \otimes \Lambda N\mathcal{F}), d_{\mathcal{F}})$ given by (3.2.22) has a perturbation ϕ^*_z, which is an endomorphism of $(C^\infty(M; \Lambda\mathcal{F} \otimes \Lambda N\mathcal{F}), d_{\mathcal{F},z})$. We have $\phi^*_{z,0,0} \equiv \phi^*_z$ like in (3.2.24). By restriction using (3.2.1), we get an endomorphism ϕ^*_z of $(C^\infty(M; \Lambda\mathcal{F}), d_{\mathcal{F},z})$, which is a perturbation of the endomorphism ϕ^* of $(C^\infty(M; \Lambda\mathcal{F}), d_{\mathcal{F}})$ given by (3.2.23).

3.4 Analysis on Riemannian Foliations of Bounded Geometry

In this section, $\mathcal{F}$ is a Riemannian foliation on a possibly open manifold M, equipped with a bundle-like metric g. We adopt the notation of Sect. 2.4 for the metric concepts of M.

3.4.1 Riemannian Foliations of Bounded Geometry

The vector subbundle $\boldsymbol{H} := T\mathcal{F}^\perp \subset TM$ is called *horizontal*, giving rise to the concepts of *horizontal* vectors, vector fields and frames. Consider the corresponding splitting (3.2.2), obtaining orthogonal projections $\boldsymbol{V} : TM \to T\mathcal{F}$ and $\boldsymbol{H} : TM \to \boldsymbol{H}$. The O'Neill tensors [O'N66] of the local Riemannian submersions defining $\mathcal{F}$

can be combined to produce $(1,2)$-tensors T and A on M, defined by

$$\mathsf{T}_E F = \boldsymbol{H}\nabla_{\boldsymbol{V}E}(\boldsymbol{V}F) + \boldsymbol{V}\nabla_{\boldsymbol{V}E}(\boldsymbol{H}F)\,,$$
$$\mathsf{A}_E F = \boldsymbol{H}\nabla_{\boldsymbol{H}E}(\boldsymbol{V}F) + \boldsymbol{V}\nabla_{\boldsymbol{H}E}(\boldsymbol{H}F)\,,$$

for $E, F \in \mathfrak{X}(M)$. By [O'N66, Theorem 4], if M is connected, given g and any $p \in M$, the foliation $\mathcal{F}$ is determined by T, A and $T_p\mathcal{F}$.

The *adapted* Riemannian connection $\mathring{\nabla}$ on M is defined by

$$\mathring{\nabla}_E F = \boldsymbol{V}\nabla_E(\boldsymbol{V}F) + \boldsymbol{H}\nabla_E(\boldsymbol{H}F)\,,$$

for $E, F \in \mathfrak{X}(M)$ [ÁLT91]. It satisfies the following properties [ÁLKL14, Section 3], [ÁLKL20, Section 5]: for $V \in \mathfrak{X}(\mathcal{F})$ and $X \in C^\infty(M; \boldsymbol{H})$,

$$\nabla_V - \mathring{\nabla}_V = \mathsf{T}_V\,, \quad \nabla_X - \mathring{\nabla}_X = \mathsf{A}_X\,, \tag{3.4.1}$$

$$\boldsymbol{V}([X, V]) = \mathring{\nabla}_X V - \mathsf{T}_V X\,. \tag{3.4.2}$$

Moreover, the leaves are $\mathring{\nabla}$-totally geodesic, the $\mathring{\nabla}$-geodesics in the leaves are the $\nabla^{\mathcal{F}}$-geodesics, and $\mathring{\nabla}$ and ∇ have the same geodesics orthogonal to the leaves.

Let $x' : U \to \Sigma$ be a distinguished submersion around any $p \in M$. Consider the Riemannian metric on Σ such that x' is a Riemannian submersion, and let $\check{\nabla}$ and $\check{\exp}$ denote the corresponding Levi-Civita connection and exponential map of Σ. For all horizontal $X, Y \in \mathfrak{X}(U, \mathcal{F}|_U)$, we have $\mathring{\nabla}_X Y \in \mathfrak{X}(U, \mathcal{F}|_U)$ and $\overline{\mathring{\nabla}_X Y} = \check{\nabla}_{\overline{X}}\overline{Y}$ [O'N66, Lemma 1 (3)].

Let $\mathring{\exp}$ denote the exponential map of the geodesic spray of $\mathring{\nabla}$ (see e.g. [Poo81, pp. 96–99]). The maps $\mathring{\exp}$ and $\check{\exp}$ restrict to diffeomorphisms of some open neighborhoods, V of 0 in T_pM and $\check{V}$ of 0 in $T_{x'(p)}\Sigma$, to some open neighborhoods, O of p in M and $\check{O}$ of $x'(p)$ in Σ. Moreover we can suppose $O \subset U$, $x'_*(V) \subset \check{V}$ and $x'(O) \subset \check{O}$, and we have $x'\mathring{\exp} = \check{\exp}\,x'_*$ on $V \cap T_p\mathcal{F}^\perp$. Let $\kappa = \kappa_p$ be the smooth map of some neighborhood W of 0 in T_pM to M defined by

$$\kappa_p(X) = \mathring{\exp}_q(\mathring{P}_{\boldsymbol{H}X}\boldsymbol{V}X)\,,$$

where $q = \mathring{\exp}_p(\boldsymbol{H}X)$, and $\mathring{P}_{\boldsymbol{H}X} : T_pM \to T_qM$ denotes the $\mathring{\nabla}$-parallel transport along the $\mathring{\nabla}$-geodesic $t \mapsto \mathring{\exp}_p(t\boldsymbol{H}X)$, $0 \le t \le 1$, which is orthogonal to the leaves. Assume $W \subset V$ and $\kappa(W) \subset O$, and therefore $x'_*(W) \subset \check{V}$ and $x'\kappa(W) \subset \check{O}$. For $X, Y \in W$, we have $X - Y \in T_p\mathcal{F}$ if and only if $\kappa(X)$ and $\kappa(Y)$ belong to the same plaque in U [ÁLKL14, Proposition 6.1]. Moreover $x'\kappa(X) = \check{\exp}\,x'_*(X)$ for all $X \in W \cap T_p\mathcal{F}^\perp$, and κ defines a diffeomorphism of some neighborhood of 0 in T_pM to some neighborhood of p in M with $\kappa_* \equiv \mathrm{id} : T_0(TM) \equiv T_pM \to T_pM$ [ÁLKL14, Proposition 6.2 and Corollary 6.3]. Consider identities $T_p\mathcal{F}^\perp \equiv \mathbb{R}^{n'}$ and $T_p\mathcal{F} \equiv \mathbb{R}^{n''}$ given by the choice of horizontal and vertical orthonormal frames at p.

Then, for some open balls centered at the origin, B' in $\mathbb{R}^{n'}$ and B'' in $\mathbb{R}^{n''}$, we can assume κ is a diffeomorphism of $B' \times B''$ to some open neighborhood of p, obtaining foliated coordinates $x = (x', x'') := \kappa^{-1} : U := \kappa(B' \times B'') \to B' \times B''$, which are said to be *normal*. As usual, g_{ij} denotes the corresponding metric coefficients and $(g^{ij}) = (g_{ij})^{-1}$. It is said that $\mathcal{F}$ has *positive injectivity bi-radius* if there are normal foliated coordinates $x_p : U_p \to B' \times B''$ at every $p \in M$ such that the balls B' and B'' are independent of p. Then $\mathcal{F}$ is said to be of *bounded geometry* if it has positive injectivity bi-radius, and the functions $|\nabla^m R|$, $|\nabla^m \mathsf{T}|$ and $|\nabla^m \mathsf{A}|$ are uniformly bounded on M for every $m \in \mathbb{N}_0$ [ÁLKL14, Definition 8.1].

Example 3.4.1 Let H be a connected Lie group, $L \lhd H$ a normal connected Lie subgroup and $\Gamma \subset H$ a discrete subgroup. Then the projection of the translates of L to $\Gamma\backslash H$ are the leaves of a Riemannian foliation of bounded geometry with the bundle-like metric induced by any left invariant metric on H.

The following chart characterization of bounded geometry for Riemannian foliations is connected with another definition given by Sanguiao [San08, Definition 1.7].

Theorem 3.4.2 ([ÁLKL14, Theorem 8.4]) *With the above notation, $\mathcal{F}$ is of bounded geometry if and only if there is a normal foliated chart $x_p : U_p \to B' \times B''$ at every $p \in M$, such that the balls B' and B'' are independent of p, and the corresponding coefficients g_{ij} and g^{ij}, as family of smooth functions on $B' \times B''$ parametrized by i, j and p, lie in a bounded subset of the Fréchet space $C^\infty(B' \times B'')$.*

For the rest of Sect. 3.4, let us assume that $\mathcal{F}$ is of bounded geometry. Then M and the disjoint union of the leaves are of bounded geometry [ÁLKL14, Remark 8.2 and Proposition 8.6]. Consider the foliated charts $y_p : V_p \to B$ and foliated charts $x_p : U_p \to B' \times B''$ given by Theorems 2.4.1 and 3.4.2. Let r_0, r'_0 and r''_0 denote the radii of the balls B, B' and B''. For $0 < r \le r_0$, $0 < r' \le r'_0$ and $0 < r'' \le r''_0$, let R_r, $B'_{r'}$ and $B''_{r''}$ denote the balls in $\mathbb{R}^{n'}$ and $\mathbb{R}^{n''}$ centered at the origin with radii r, r' and r'', respectively. If r is small enough, then $V_{p,r} := x_p^{-1}(B_r) \subset U_p$ for all p [ÁLKL14, Proposition 8.6]. On the other hand, if $r' + r'' \le r_0$, then $U_{p,r',r''} := x_p^{-1}(B'_{r'} \times B''_{r''}) \subset V_p$ for all p by the triangle inequality. Then the following subsets are bounded in the corresponding Fréchet spaces [ÁLKL20, Proposition 5.6 and 5.7]:

$$\begin{cases} \{\, x_p y_p^{-1} \mid p \in M \,\} \subset C^\infty(B, \mathbb{R}^{n'} \times \mathbb{R}^{n''}) \,, \\ \{\, y_p x_p^{-1} \mid p \in M \,\} \subset C^\infty(B'_{r'} \times B''_{r''}, \mathbb{R}^n) \,. \end{cases} \tag{3.4.3}$$

Let E be the Hermitian vector bundle of bounded geometry associated to the principal $\mathrm{O}(n)$-bundle of orthonormal frames on M and a unitary representation of $\mathrm{O}(n)$ (Example 2.4.6). Since ∇ on TM is of bounded geometry, it follows from (3.4.1) that $\mathring{\nabla}$ is also of bounded geometry. Thus we get induced connections ∇ and $\mathring{\nabla}$ of bounded geometry on E (Example 2.4.10). By (3.4.1), we also get that

$\mathring{\nabla}$ can be used instead of ∇ to define equivalent versions of $\|\cdot\|_{C^m_{\rm ub}}$ and $\langle\cdot,\cdot\rangle_m$ in the spaces $C^m_{\rm ub}(M;E)$ and $H^m(M;E)$. Since the subsets (3.4.3) are bounded, if B' and B'' are small enough, then we can use the coordinates (U_p,x_p) instead of coordinates of (V_p,y_p) to define equivalent versions of $\|\cdot\|'_{C^m_{\rm ub}}$ and $\langle\cdot,\cdot\rangle'_m$. Similarly, given another bundle F like E, we can use the coordinates (U_p,x_p) instead of (V_p,y_p) to describe $\mathrm{Diff}^m_{\rm ub}(M;E,F)$ by requiring that the local coefficients form a bounded subset of the Fréchet space $C^\infty(B'\times B'';\mathbb{C}^{l'}\otimes\mathbb{C}^{l*})$, where l and l' are the ranks of E and F.

The condition of being leafwise differential operators of bounded geometry is preserved by compositions, and by taking transposes and formal adjoints. They form a filtered $C^\infty_{\rm ub}(M)$-submodule $\mathrm{Diff}_{\rm ub}(\mathcal{F};E,F)\subset\mathrm{Diff}(\mathcal{F};E,F)$. The notation $\mathrm{Diff}_{\rm ub}(\mathcal{F};E)$ is used if $E=F$; this is a filtered subalgebra of $\mathrm{Diff}(\mathcal{F};E)$. The concepts of *uniform leafwise ellipticity* for operators in $\mathrm{Diff}^m(\mathcal{F};E,F)$ can be defined like uniform ellipticity (Sect. 2.4.4), and can be extended to leafwise differential complexes of order m like in Sect. 2.1.14. The same applies to *uniform transverse ellipticity* for operators in $\mathrm{Diff}^m(M;E,F)$ and for differential complexes of order m. If $P\in\mathrm{Diff}^2_{\rm ub}(\mathcal{F};E)$ is uniformly leafwise elliptic and $Q\in\mathrm{Diff}^2_{\rm ub}(M;E)$ is uniformly transversely elliptic, and both P and Q are symmetric and non-negative, then $H^s(M;E)$ $(s\in\mathbb{R})$ can be described with the scalar product $\langle u,v\rangle_s=\langle((1+P)^s+(1+Q)^s)u,v\rangle$.

Let $\mathfrak{X}_{\rm ub}(\mathcal{F})$ and $\mathfrak{X}_{\rm ub}(M,\mathcal{F})$ denote the intersections of $\mathfrak{X}_{\rm ub}(M)$ with $\mathfrak{X}(\mathcal{F})$ and $\mathfrak{X}(M,\mathcal{F})$, respectively. Then $\mathrm{Diff}_{\rm ub}(\mathcal{F})$ can be also described like in Sect. 2.1.7, using $C^\infty_{\rm ub}(M)$ and $\mathfrak{X}_{\rm ub}(\mathcal{F})$ instead of $C^\infty(M)$ and $\mathfrak{X}(M)$, and $\mathrm{Diff}_{\rm ub}(\mathcal{F};E,F)$ can be also described as the $C^\infty_{\rm ub}(M)$-tensor product of $\mathrm{Diff}_{\rm ub}(\mathcal{F})$ and $C^\infty_{\rm ub}(M;E,F)$.

3.4.2 Operators of Bounded Geometry on Differential Forms

Since ∇ and $\mathring{\nabla}$ are of bounded geometry on TM, the induced connections ∇ and $\mathring{\nabla}$ on ΛM are of bounded geometry as well (Example 2.4.10). Using Examples 2.4.6 and 2.4.10, we get that $\boldsymbol{H}$ and $T\mathcal{F}$ are also of bounded geometry, and the restrictions of $\mathring{\nabla}$ to $\boldsymbol{H}$ and $T\mathcal{F}$ are of bounded geometry [ÁLKL20, Section 6]. Thus every $\Lambda^{u,v}M$ is of bounded geometry (Example 2.4.7), and $\mathring{\nabla}$ is of bounded geometry on $\Lambda^{u,v}M$ (Example 2.4.11). So this also applies to $\Lambda\mathcal{F}\equiv\Lambda^{0,\bullet}M$.

By using $\mathring{\nabla}$ instead of ∇ in the definitions of $\|\cdot\|_{C^m_{\rm ub}}$ and $\langle\cdot,\cdot\rangle_s$ $(m\in\mathbb{N}_0$ and $s\in\mathbb{R})$, it follows that the spaces $C^m_{\rm ub}(M;\Lambda)$ and $H^s(M;\Lambda)$ inherit the bigrading of ΛM, and therefore $C^\infty_{\rm ub}(M;\Lambda)$ and $H^{\pm\infty}(M;\Lambda)$ have an induced bigrading.

The following properties hold [ÁLK01, Section 3], [ÁLKL20, Section 6]: the canonical projections $\Lambda M\to\Lambda^{u,v}M$, the operators $\star$, $\star_{\mathcal{F}}$ or $\star_\perp$ (under appropriate orientability assumptions), and the operators of (3.2.6), (3.2.10) and (3.2.12) are of bounded geometry; the differential complexes $d_{0,1}$ and $\delta_{0,-1}$ are uniformly leafwise elliptic; the differential operators D_0 and Δ_0 are symmetric and uniformly leafwise elliptic; the differential operator $D_\perp$ is uniformly transversely elliptic; and there is an endomorphism of bounded geometry, K of ΛM, such that[4]

$$D_\perp\delta_{0,-1}+\delta_{0,-1}D_\perp=K\delta_{0,-1}+\delta_{0,-1}K\;. \tag{3.4.4}$$

Let us recall the definition of K and the proof of (3.4.4) because an extension will be needed, which is slightly more general than the extension considered in [ÁLKL20, Section 11]. Let $\Theta:\mathfrak{X}(\mathcal{F})\to C^\infty(M;\boldsymbol{H}^*\otimes T\mathcal{F})$ be the differential operator defined by $\Theta_X V=V([X,V])$ (the expression (3.4.2)), which induces a differential operator $\Theta:C^\infty(M;\Lambda\mathcal{F})\to C^\infty(M;\boldsymbol{H}^*\otimes\Lambda\mathcal{F})$. If $X\in\mathfrak{X}(M,\mathcal{F})\cap C^\infty(M;\boldsymbol{H})$, then Θ_X on $C^\infty(M;\Lambda\mathcal{F})$ agrees with Θ_X on $C^\infty(M;\Lambda^{0,\bullet})$ via (3.2.3) (Sect. 3.2.12). A homomorphism $\Xi:\Lambda\mathcal{F}\to\boldsymbol{H}^*\otimes\Lambda\mathcal{F}$ can be locally defined by

$$\Xi_X=(-1)^{(n''-v)v}[\Theta_X,\star_{\mathcal{F}}]\star_{\mathcal{F}}$$

on $C^\infty(M;\Lambda^v\mathcal{F})$, for any $X\in C^\infty(M;\boldsymbol{H})$, where $\star_{\mathcal{F}}$ is defined with any choice of local orientation of $\mathcal{F}$. Using (3.2.3), its tensor product with the identity on $\Lambda\boldsymbol{H}$ is a homomorphism $\Xi:\Lambda M\to\boldsymbol{H}^*\otimes\Lambda M$. Using the notation of Sect. 3.2.13 on any normal foliated chart (U,x), the local expression

$$K=dx'^i\wedge\Xi_{\boldsymbol{H}\partial'_i}$$

defines an endomorphism of ΛM. A computation using (3.2.15), (3.2.26), (3.2.30), and (3.2.32) gives

$$d_{1,0}\delta_{0,-1}+\delta_{0,-1}d_{1,0}=K\delta_{0,-1}+\delta_{0,-1}K\;, \tag{3.4.5}$$

yielding (3.4.4) by the analog of (3.2.7) for the operators $\delta_{-i,i-1}$.

3.4.3 Foliated Maps of Bounded Geometry

For $a=1,2$, let $\mathcal{F}_a$ be a Riemannian foliation of bounded geometry on a manifold M_a with a bundle-like metric. To refer to each $\mathcal{F}_a$, the subscript "a" is added to the notation used in Sect. 3.4.1: n'_a, n''_a, $y_{a,p}:V_{a,p}\to B_a$, $x_{a,p}:U_{a,p}\to B'_a\times B''_a$, $r_{a,0}$, $r'_{a,0}$ and $r''_{a,0}$, $V_{a,p,r}$ and $U_{a,p,r',r''}$. Like in the case of uniform spaces and differential operators, in the definition of bounded geometry for maps $M_1\to M_2$, we can replace the charts $(V_{1,p},y_{1,p})$ and $(V_{2,\phi(p)},y_{2,\phi(p)})$, and sets $B_1(p,r)$ with the charts $(U_{1,p},x_{1,p})$ and $(U_{2,\phi(p)},x_{2,\phi(p)})$, and sets $U_{1,p,r',r''}$. Let $C^\infty_{\text{ub}}(M_1,\mathcal{F}_1;M_2,\mathcal{F}_2)$ be the subset of $C^\infty(M_1,\mathcal{F}_1;M_2,\mathcal{F}_2)$ consisting of foliated maps of bounded geometry. For any $m\in\mathbb{N}_0$ and $\phi\in C^\infty_{\text{ub}}(M_1,\mathcal{F}_1;M_2,\mathcal{F}_2)$, using

[4] In [ÁLKL20, Eq. (55)], D_0 should be $\delta_{0,-1}$, like in [ÁLK01, Proposition 3.1].

the versions of $\|\cdot\|'_{C^m_{\mathrm{ub}}}$ and $\langle\cdot,\cdot\rangle'_m$ defined with the foliated charts (U_p, x_p) in the case where $m < \infty$ (Sect. 3.4.1), we get the following versions of (2.4.6) and (2.4.7) [ÁLKL20, Section 8]: (3.2.22) induces continuous homomorphisms,

$$\phi^* : C^m_{\mathrm{ub}}(M_2; \Lambda\mathcal{F}_2 \otimes \Lambda N\mathcal{F}_2) \to C^m_{\mathrm{ub}}(M_1; \Lambda\mathcal{F}_1 \otimes \Lambda N\mathcal{F}_1) , \tag{3.4.6}$$

and, if ϕ is uniformly metrically proper,

$$\phi^* : H^m(M_2; \Lambda\mathcal{F}_2 \otimes \Lambda N\mathcal{F}_2) \to H^m(M_1; \Lambda\mathcal{F}_1 \otimes \Lambda N\mathcal{F}_2) . \tag{3.4.7}$$

In particular, we get (3.4.7) if ϕ is a foliated diffeomorphism with $\phi^{\pm 1}$ of bounded geometry. In this case, it can be continuously extended to Sobolev spaces of order $-m$ using the version of the second equality of (2.1.30) for open manifolds, (3.2.3) and (3.2.16), like in Sect. 2.4.6.

3.4.4 Leafwise Functional Calculus

Consider the notation of Sects. 3.4.1 and 3.4.2. Like in (2.9.10) and (2.9.11), the hyperbolic equation

$$\partial_t \alpha_t = i D_0 \alpha_t , \quad \alpha_0 = \alpha , \tag{3.4.8}$$

has a unique solution on any open subset of M and for t in any interval containing zero, which satisfies [Che73, Theorem 1.3], [Roe87, Proposition 1.2]

$$\operatorname{supp} \alpha_t \subset \operatorname{Pen}_{\mathcal{F}}(\operatorname{supp} \alpha, |t|) . \tag{3.4.9}$$

The operators D_0 and Δ_0, with domain $C^\infty_{\mathrm{c}}(M; \Lambda)$, are essentially self-adjoint in $L^2(M; \Lambda)$ [Che73, Theorem 2.2], and their self-adjoint extensions are also denoted by D_0 and Δ_0. The functional calculus of D_0, given by the spectral theorem, assigns a (bounded) operator $\psi(D_0)$ to every (bounded) measurable function ψ on $\mathbb{R}$; in particular, we have a unitary operator e^{itD_0} and a bounded self-adjoint operator $e^{-t\Delta_0}$ on $L^2(M; \Lambda)$. The notation $\Pi_0 = e^{-\infty\Delta_0}$ is used for the orthogonal projection of $L^2(M; \Lambda)$ to $\ker D_0 = \ker \Delta_0$ in $L^2(M; \Lambda)$.

If $\alpha \in C^\infty_{\mathrm{c}}(M; \Lambda)$, the solution of (3.4.8) is given by $\alpha_t = e^{itD_0}\alpha$. For every $m \in \mathbb{N}_0$, there is some $C_m \geq 0$ such that [Tay81, Section IV.2], [Roe87, Proposition 1.4], [ÁLKL20, Proposition 7.1]

$$\|e^{itD_0}\alpha\|_m \leq e^{C_m|t|} \|\alpha\|_m , \tag{3.4.10}$$

for all $\alpha \in C^\infty_{\mathrm{c}}(M; \Lambda)$.

On the other hand, like in (2.9.12), for $\psi \in \mathcal{S}$, we get

$$\psi(D_0) = \frac{1}{2\pi}\int_{-\infty}^{+\infty} \hat{\psi}(\xi)e^{i\xi D_0}\,d\xi \; . \tag{3.4.11}$$

Taking $\psi \in \mathcal{A}$ (Sect. 2.9.8), it follows from (2.9.15), (3.4.10) and (3.4.11) that, for every $m \in \mathbb{Z} \cup \{\pm\infty\}$, the functional calculus $\psi \mapsto \psi(D_0)$ restricts to a continuous homomorphisms of $\mathbb{C}[z]$-modules and algebras [Roe87, Proposition 4.1], [ÁLKL20, Proposition 7.2],

$$\mathcal{A} \to \operatorname{End}(H^m(M; \Lambda)) \; , \quad \mathcal{A} \to \operatorname{End}(H^\infty(M; \Lambda)) \; . \tag{3.4.12}$$

By taking coefficients in $o(M)$ and transposition (see Sect. 3.2.9), $\psi \mapsto \psi(D_0)$ also induces continuous homomorphisms of $\mathbb{C}[z]$-modules and algebras,

$$\mathcal{A} \to \operatorname{End}(H^{-m}(M; \Lambda)) \; , \quad \mathcal{A} \to \operatorname{End}(H^{-\infty}(M; \Lambda)) \; . \tag{3.4.13}$$

3.4.5 Leafwise Hodge Decomposition

According to (3.4.12), the operator $e^{-t\Delta_0}$ ($t > 0$) restricts to a continuous endomorphism of $H^\infty(M; \Lambda)$. As pointed out in [San08], using the bounded geometry and uniform leafwise/transverse ellipticity of the operators considered in Sect. 3.4.2, and applying (3.4.4) and (3.4.12), the arguments of [ÁLK01] can be adapted to show the following, where Δ_0 is considered on $H^\infty(M; \Lambda)$ [San08], [ÁLKL20, Theorem 7.3 and Corollary 7.4]: there is a TVS-direct-sum decomposition,

$$H^\infty(M; \Lambda) = \ker \Delta_0 \oplus \overline{\operatorname{im} d_{0,1}} \oplus \overline{\operatorname{im} \delta_{0,-1}} \; , \tag{3.4.14}$$

whose terms are orthogonal in $L^2(M; \Lambda)$; the map

$$[0, \infty] \times H^\infty(M; \Lambda) \to H^\infty(M; \Lambda) \; , \quad (t, \alpha) \mapsto e^{-t\Delta_0}\alpha \; , \tag{3.4.15}$$

is well-defined and continuous; and $\Pi_0 : H^\infty(M; \Lambda) \to \ker \Delta_0$ induces a TVS-isomorphism

$$\bar{H}(H^\infty(M; \Lambda), d_{0,1}) \xrightarrow{\cong} \ker \Delta_0 \; , \tag{3.4.16}$$

whose inverse is induced by $\ker \Delta_0 \hookrightarrow H^\infty(M; \Lambda)$. The analogs of (3.4.14)–(3.4.16) with $H^{-\infty}(M; \Lambda)$ are also true.

By (3.2.3) and (3.2.8), we can consider $(H^\infty(M; \Lambda\mathcal{F}), d_{\mathcal{F}})$ as a topological subcomplex of $(H^\infty(M; \Lambda), d_{0,1})$, and the notation $H^\bullet H^\infty(\mathcal{F})$ and $\bar{H}^\bullet H^\infty(\mathcal{F})$ is used for its cohomology and reduced cohomology. By (3.2.30), $\delta_{\mathcal{F}}$ on $H^\infty(M; \Lambda\mathcal{F})$

is also given by $\delta_{0,-1}$. Thus we get the operators $D_{\mathcal{F}} = d_{\mathcal{F}} + \delta_{\mathcal{F}}$ and $\Delta_{\mathcal{F}} = D_{\mathcal{F}}^2 = \delta_{\mathcal{F}} d_{\mathcal{F}} + d_{\mathcal{F}} \delta_{\mathcal{F}}$ on $H^\infty(M; \Lambda\mathcal{F})$, which are essentially self-adjoint in $L^2(M; \Lambda\mathcal{F})$. Let $\Pi_{\mathcal{F}} = e^{-\infty\Delta_{\mathcal{F}}}$ be the orthogonal projection to $\ker D_{\mathcal{F}} = \ker \Delta_{\mathcal{F}}$ in $L^2(M; \Lambda\mathcal{F})$. Then (3.4.10)–(3.4.16) have obvious versions with $D_{\mathcal{F}}$, $\Delta_{\mathcal{F}}$, $\bar{H}^\bullet H^\infty(\mathcal{F})$ and $\Pi_{\mathcal{F}}$ [San08], [ÁLKL20, Section 7].

3.4.6 A Class of Smoothing Operators

Suppose $\mathcal{F}$ is of codimension one for the sake of simplicity. (The case of codimension > 1 can be treated like in [ÁLK08].) Assume also that M is endowed with a bundle-like metric g so that $\mathcal{F}$ is of bounded geometry. Let $\phi : M \times \mathbb{R} \to M$ be a foliated flow of $\mathbb{R}$-local bounded geometry, whose infinitesimal generator is $Z \in \mathfrak{X}_{\mathrm{ub}}(M, \mathcal{F})$ (Sect. 2.4.7). Assume $\inf_M |\overline{Z}| > 0$; in particular, the orbits of ϕ are transverse to the leaves. Given $f \in C_{\mathrm{c}}^\infty(\mathbb{R})$, consider the following operators on $H^{-\infty}(M; \Lambda\mathcal{F})$. For every $\psi \in \mathcal{A}$, the operator

$$P = \int_{\mathbb{R}} \phi^{t*} f(t)\, dt\, \psi(D_{\mathcal{F}}) \tag{3.4.17}$$

is defined by the version of (3.4.13) for $D_{\mathcal{F}}$ and the version of (3.4.7) for ϕ^{t*} on $H^{-\infty}(M; \Lambda\mathcal{F})$. The subscripts "$\psi$" or "$f$" may be added to the notation of P if needed, or the subscript "u" in the case of functions $\psi_u \in \mathcal{A}$ depending on a parameter u. For example, we may take $\psi_u(x) = e^{-ux^2}$ and the corresponding operators P_u $(u > 0)$ on $H^{-\infty}(M; \Lambda\mathcal{F})$. Let also

$$P_\infty = \int_{\mathbb{R}} \phi^{t*} f(t)\, dt\, \Pi_{\mathcal{F}}\ .$$

The following properties hold [ÁLKL20, Propositions 9.1, 9.4 and 9.6 and Corollaries 9.2, 9.3 and 9.5]: every $P_{\psi,f}$, given by (3.4.17), is smoothing, obtaining continuous bilinear maps[5]

$$\begin{cases} \mathcal{A} \times C_{\mathrm{c}}^\infty(\mathbb{R}) \to L(H^{-\infty}(M; \Lambda\mathcal{F}), H^\infty(M; \Lambda\mathcal{F}))\ , & (\psi, f) \mapsto P_{\psi,f}\ , \\ \mathcal{A} \times C_{\mathrm{c}}^\infty(\mathbb{R}) \to C_{\mathrm{ub}}^\infty(M^2; \Lambda\mathcal{F} \boxtimes (\Lambda\mathcal{F}^* \otimes \Omega M))\ , & (\psi, f) \mapsto K_{P_{\psi,f}}\ ; \end{cases} \tag{3.4.18}$$

[5] In [ÁLKL20, Propositions 9.1 and Corollary 9.2], only the continuous dependence on $\psi \in \mathcal{A}$ is indicated, but the additional continuous dependence on $f \in C_{\mathrm{c}}^\infty(\mathbb{R})$ is given by [ÁLKL20, Proposition 9.6], indicated in (3.4.20).

P_∞ is smoothing, with

$$\begin{cases} \lim_{u\to\infty} P_u = P_\infty & \text{in} \quad L(H^{-\infty}(M;\Lambda\mathcal{F}), H^\infty(M;\Lambda\mathcal{F}))\,, \\ \lim_{u\to\infty} K_{P_u} = K_{P_\infty} & \text{in} \quad C^\infty_{\mathrm{ub}}(M^2;\Lambda\mathcal{F}\boxtimes(\Lambda\mathcal{F}^*\otimes\Omega M))\,; \end{cases} \tag{3.4.19}$$

and, for any compact $I \subset \mathbb{R}$ containing $\operatorname{supp} f$ and $m \le m'$ in $\mathbb{N}_0$, there are some $C, C' > 0$ and $N \in \mathbb{N}_0$, depending on m, m' and I, such that

$$\|P_{\psi,f}\|_{m,m'} \le C'\|\psi\|_{\mathcal{A},C,N}\,\|f\|_{I,C^N}\,. \tag{3.4.20}$$

3.4.7 Description of Some Schwartz Kernels

In Sect. 3.4.6, $\overline{Z}$ defines the structure of a transversely complete $\mathbb{R}$-Lie foliation on $\mathcal{F}$, and therefore we can consider also the notation of Sect. 3.1.11. Then the lift $\tilde{g}$ of g to $\widetilde{M}$ is a bundle-like metric of $\widetilde{\mathcal{F}} = \pi^*\mathcal{F}$, and $\widetilde{Z} \in \mathfrak{X}_{\mathrm{ub}}(\widetilde{M},\widetilde{\mathcal{F}})$. Assume $D_*\widetilde{Z} = \partial_x \in \mathfrak{X}(\mathbb{R})$ and $\bar{\phi}^t(x) = t + x$; hence ϕ^t preserves every leaf of $\mathcal{F}$ if and only if $t \in \operatorname{Hol}\mathcal{F}$ (Sect. 3.1.11).

For any $\psi \in \mathcal{A}$ and $f \in C^\infty_{\mathrm{c}}(\mathbb{R})$, we have the smoothing operator P given by (3.4.17), and a similar smoothing operator $\widetilde{P}$ is defined by using $\tilde{\phi}$ and $\widetilde{\mathcal{F}}$ instead of ϕ and $\mathcal{F}$. We are going to describe their Schwartz kernels.

Let $\mathfrak{G} = \operatorname{Hol}(M,\mathcal{F})$ and $\widetilde{\mathfrak{G}} = \operatorname{Hol}(\widetilde{M},\widetilde{\mathcal{F}})$, whose source and range maps are denoted by $\boldsymbol{s}, \boldsymbol{r} : \mathfrak{G} \to M$ and $\tilde{\boldsymbol{s}}, \tilde{\boldsymbol{r}} : \widetilde{\mathfrak{G}} \to \widetilde{M}$ (Sect. 3.1.4). Since the leaves of $\mathcal{F}$ and $\widetilde{\mathcal{F}}$ have trivial holonomy groups, the smooth immersions $(\boldsymbol{r},\boldsymbol{s}) : \mathfrak{G} \to M^2$ and $(\tilde{\boldsymbol{r}},\tilde{\boldsymbol{s}}) : \widetilde{\mathfrak{G}} \to \widetilde{M}^2$ are injective, with images $\mathcal{R}_{\mathcal{F}}$ and $\mathcal{R}_{\widetilde{\mathcal{F}}}$. Via these injections, the restriction $\pi \times \pi : \mathcal{R}_{\widetilde{\mathcal{F}}} \to \mathcal{R}_{\mathcal{F}}$ corresponds to the Lie groupoid homomorphism $\pi_{\mathfrak{G}} := \operatorname{Hol}(\pi) : \widetilde{\mathfrak{G}} \to \mathfrak{G}$ (Sect. 3.1.7), which is a covering map with $\operatorname{Aut}(\pi_{\mathfrak{G}}) \equiv \Gamma$. In fact, since $\widetilde{\mathcal{F}}$ is defined by the fiber bundle D, we get that $\mathcal{R}_{\widetilde{\mathcal{F}}}$ is a regular submanifold of $\widetilde{M}^2$, and $(\tilde{\boldsymbol{r}},\tilde{\boldsymbol{s}}) : \widetilde{\mathfrak{G}} \to \mathcal{R}_{\widetilde{\mathcal{F}}}$ is a diffeomorphism. We may write $\mathfrak{G} \equiv \mathcal{R}_{\mathcal{F}}$ and $\widetilde{\mathfrak{G}} \equiv \mathcal{R}_{\widetilde{\mathcal{F}}}$.

Consider the C^∞ vector bundles, $S = \boldsymbol{r}^*\Lambda\mathcal{F} \otimes \boldsymbol{s}^*(\Lambda\mathcal{F} \otimes \Omega\mathcal{F})$ over $\mathfrak{G}$ and $\widetilde{S} = \tilde{\boldsymbol{r}}^*\Lambda\widetilde{\mathcal{F}} \otimes \tilde{\boldsymbol{s}}^*(\Lambda\widetilde{\mathcal{F}} \otimes \Omega\widetilde{\mathcal{F}})$ over $\widetilde{\mathfrak{G}}$. Note that $\widetilde{S} \equiv \pi_{\mathfrak{G}}^*S$, and any $k \in C^\infty(\mathfrak{G};S)$ lifts via $\pi_{\mathfrak{G}}$ to a section $\tilde{k} \in C^\infty(\widetilde{\mathfrak{G}};\widetilde{S})$. Since π restricts to diffeomorphisms of the leaves of $\widetilde{\mathcal{F}}$ to the leaves of $\mathcal{F}$, it follows that $\tilde{k} \in C^\infty_{\mathrm{p}}(\widetilde{\mathfrak{G}};\widetilde{S})$ if and only if $k \in C^\infty_{\mathrm{p}}(\mathfrak{G};S)$.

For any $\psi \in \mathcal{R}$, the collection of Schwartz kernels $k_L := K_{\psi(D_L)}$, for all leaves L of $\mathcal{F}$, defines a section $k = k_\psi$ of S. This also applies to the operators $\psi(D_{\widetilde{L}})$ on the leaves $\widetilde{L}$ of $\widetilde{\mathcal{F}}$, obtaining a section $\tilde{k} = \tilde{k}_\psi$ of $\widetilde{S}$.

If $\hat{\psi} \in C^\infty_{\mathrm{c}}(\mathbb{R})$, then $k_\psi \in C^\infty_{\mathrm{p}}(\mathfrak{G};S)$, and the global action of k_ψ on $C^\infty_{\mathrm{c}}(M;\Lambda\mathcal{F})$ (Sect. 3.1.5) agrees with the restriction of the operator $\psi(D_{\mathcal{F}})$ on $H^\infty(M;\Lambda\mathcal{F})$ defined by the version of (3.4.12) for $D_{\mathcal{F}}$ [ÁLKL20, Proposition 10.1]. Precisely, if $\operatorname{supp}\hat{\psi} \subset [-R,R]$ for some $R > 0$, then $\operatorname{supp} k_\psi \subset$

$\overline{\operatorname{Pen}}_{\mathcal{F}}(\mathfrak{G}^{(0)}, R)$ by (2.9.13), and therefore $\operatorname{supp}\psi(D_{\mathcal{F}})\alpha \subset \overline{\operatorname{Pen}}_{\mathcal{F}}(\operatorname{supp}\alpha, R)$ for all $\alpha \in H^\infty(M; \Lambda\mathcal{F})$ by Remark 2.4.14.

Let $\widetilde{\Lambda} = D^*dx \equiv dx$, which is an invariant transverse volume form of $\widetilde{\mathcal{F}}$ defining the same transverse orientation as $\widetilde{\overline{Z}}$. Since $\widetilde{\Lambda}$ is Γ-invariant by the h-equivariance of D, it defines a transverse volume form Λ of $\mathcal{F}$, which defines the same transverse orientation as $\overline{Z}$. These $\widetilde{\Lambda}$ and Λ define invariant transverse densities $|\widetilde{\Lambda}|$ and $|\Lambda|$ of $\widetilde{\mathcal{F}}$ and $\mathcal{F}$.

Let $\tilde{p}, \tilde{q} \in \widetilde{M}$ over $p, q \in M$, and write $t_{\tilde{p},\tilde{q}} = D(\tilde{q}) - D(\tilde{p})$. If $\psi \in \mathcal{A}$, then[6]

$$K_P(p,q) \equiv \sum_{\gamma\in\Gamma} T_\gamma^*\,\tilde{\phi}^{t_{\tilde{p},\tilde{q}}-h(\gamma)*}\tilde{k}\big(T_\gamma\tilde{\phi}^{t_{\tilde{p},\tilde{q}}-h(\gamma)}(\tilde{p}),\tilde{q}\big)\,f(t_{\tilde{p},\tilde{q}}-h(\gamma))\,|\Lambda|(q)\,, \tag{3.4.21}$$

defining a convergent series in $C^\infty_{\text{ub}}(\widetilde{M}^2; \widetilde{S})$ [ÁLKL20, Proposition 10.3]. Here, the identity $\widetilde{S}_{(\tilde{p},\tilde{q})} \equiv S_{(p,q)}$ is used, and the leafwise part of the density of $K_P(\cdot, q)$ at q is given by the density of $\tilde{k}(\cdot, \tilde{q})$ at $\tilde{q}$.

3.5 Witten's Operators on Riemannian Foliations of Bounded Geometry

Consider the notation of Sect. 3.3.1 with our assumption that $\mathcal{F}$ is Riemannian of bounded geometry. Suppose also that $\eta \in C^\infty_{\text{ub}}(M; \Lambda^1)$, and therefore $\eta_0 \in C^\infty_{\text{ub}}(M; \Lambda^{0,1}) \equiv C^\infty_{\text{ub}}(M; \Lambda^1\mathcal{F})$ and $\eta_1 \in C^\infty_{\text{ub}}(M; \Lambda^{1,0})$. Thus the operators $d_{z,i,1-i}$, $\delta_{z,i,i-1}$, $D_{z,0}$, $D_{z,\perp}$ and $\Delta_{z,0}$ are of bounded geometry. Arguing like in (3.4.4), we get

$$\begin{aligned}
&(d_{1,0}\,\eta_0\lrcorner + \eta_0\lrcorner\, d_{1,0})(f_{IJ}\,dx''^I \wedge dx'^J)\\
&\quad = (K\,\eta_0\lrcorner + \eta_0\lrcorner\,K)(f_{IJ}\,dx''^I \wedge dx'^J)\\
&\qquad + (-1)^{(n''+1)|I|+n''}\big(\star_{\mathcal{F}}[\Theta_{H\partial'_i}, \eta_0\wedge]\star_{\mathcal{F}}(f_{IJ}\,dx''^I)\big) \wedge dx'^i \wedge dx''^J\,.
\end{aligned}$$

Using (3.2.30) and (3.3.2), it follows that

$$(d_{1,0}\,\eta_0\lrcorner + \eta_0\lrcorner\,d_{1,0} - K\,\eta_0\lrcorner - \eta_0\lrcorner\,K)(f_{IJ}\,dx''^I \wedge dx'^J)$$

[6] There is an error in the statement of [ÁLKL20, Proposition 10.3]: it is written $f(t_{\tilde{p},\tilde{q}})$ instead of $f(t_{\tilde{p},\tilde{q}} - h(\gamma))$. However, its proof shows the expression given in (3.4.21).

$$
\begin{aligned}
&= (-1)^{(n''+1)|I|+n''}\big(\star_{\mathcal{F}}[d_{\mathcal{F}}, h_i]\star_{\mathcal{F}}(f_{IJ}\,dx''^I)\big)\wedge dx'^i\wedge dx''^J\\
&= -(-1)^{|I|}\big(\delta_{\mathcal{F}}(h_i f_{IJ}\,dx''^I)\big)\wedge dx'^i\wedge dx''^J\\
&\quad + (-1)^{|I|}\big(\delta_{\mathcal{F}}(f_{IJ}\,dx''^I)\big)\wedge\eta_1\wedge dx''^J\\
&= -\delta_{0,-1}\big(\eta_1\wedge f_{IJ}\,dx''^I\wedge dx''^J\big) - \eta_1\wedge\delta_{0,-1}\big(f_{IJ}\,dx''^I\wedge dx''^J\big)\,.
\end{aligned}
$$

This shows that

$$
d_{1,0}\,\eta_0\lrcorner + \eta_0\lrcorner\, d_{1,0} = K\,\eta_0\lrcorner + \eta_0\lrcorner\, K - \delta_{0,-1}\,\eta_1\wedge - \eta_1\wedge\delta_{0,-1}\,. \tag{3.5.1}
$$

Combining (3.4.5) and (3.5.1), and using that $\eta_0\lrcorner$ is an anti-derivation, we compute

$$
\begin{aligned}
&d_{z,1,0}\delta_{z,0,-1} + \delta_{z,0,-1}d_{z,1,0}\\
&\qquad = K\delta_{0,-1} + \delta_{0,-1}K - \bar{z}(K\,\eta_0\lrcorner + \eta_0\lrcorner\,K - \delta_{0,-1}\,\eta_1\wedge - \eta_1\wedge\delta_{0,-1})\\
&\qquad\quad + z(\eta_1\wedge\delta_{0,-1} + \delta_{0,-1}\,\eta_1\wedge) + |z|^2(\eta_1\wedge\eta_0\lrcorner + \eta_0\lrcorner\,\eta_1\wedge)\\
&\qquad = K\delta_{z,0,-1} + \delta_{z,0,-1}K + 2\Re z\,(\eta_1\wedge\delta_{0,-1} + \delta_{0,-1}\,\eta_1\wedge)\\
&\qquad\quad - 2\Re z\,\bar{z}(\eta_1\wedge\eta_0\lrcorner + \eta_0\lrcorner\,\eta_1\wedge)\\
&\qquad = K_z\delta_{z,0,-1} + \delta_{z,0,-1}K_z\,,
\end{aligned}
$$

where $K_z = K + 2\Re z\,\eta_1\wedge$ is an endomorphism of ΛM of bounded geometry. Using also the analog of (3.2.7) for the operators $d_{z,i,1-i}$, it follows that[7]

$$
D_{z,\perp}\delta_{z,0,-1} + \delta_{z,0,-1}D_{z,\perp} = K_z\delta_{z,0,-1} + \delta_{z,0,-1}K_z\,. \tag{3.5.2}
$$

Using this key equality, we get straightforward generalizations of all results in Sects. 3.4.4 and 3.4.5 for $d_{z,0,1}$, $\delta_{z,0,-1}$, $D_{z,0}$ and $\Delta_{z,0}$, which also have obvious versions for $d_{\mathcal{F},z}$, $\delta_{\mathcal{F},z}$, $D_{\mathcal{F},z}$ and $\Delta_{\mathcal{F},z}$. Let $\Pi_{0,z}$ and $\Pi_{\mathcal{F},z}$ denote the corresponding versions of Π_0 and $\Pi_{\mathcal{F}}$.

Let $\phi : (M,\mathcal{F}) \to (M,\mathcal{F})$ be a smooth foliated map of bounded geometry. Since $\eta \in C^\infty_{\mathrm{ub}}(M;\Lambda)$, we get versions of the continuity of (3.4.6) and (3.4.7) for ϕ_z^*, assuming ϕ is uniformly metrically proper for the second one (Sect. 2.4.6). In particular, this applies to any foliated flow of $\mathbb{R}$-local bounded geometry (Sect. 2.4.7), $\phi = \{\phi^t\}$ on $(M,\mathcal{F})$, using its unique lift $\tilde{\phi} = \{\tilde{\phi}^t\}$ to $\widetilde{M}$. Then the definitions and results of Sects. 3.4.6 and 3.4.7 have obvious twisted extensions

[7] The equality

$$
d_{-\bar{z},1,0}\delta_{z,0,-1} + \delta_{z,0,-1}d_{-\bar{z},1,0} = K\delta_{z,0,-1} + \delta_{z,0,-1}K
$$

is also true, but this does not fit the analog of (3.2.7).

using ϕ_z^{t*}, $D_{\mathcal{F},z}$ and $\Pi_{\mathcal{F},z}$. The subscript "z" may be added to the notation P, P_u, P_∞, k, $\tilde{k}$, k_u and $\tilde{k}_u$ in this setting.

Recall that $D_{0,z,z'}$ and $\Delta_{0,z,z'}$ are uniformly leafwise elliptic with a symmetric leading symbol (Sect. 3.3.3). Moreover, they are of bounded geometry. Then the obvious version of (3.4.8) with $D_{0,z,z'}$ has a unique solution, which satisfies the obvious analogs of (3.4.9) and (3.4.10). Thus $\psi(D_{0,z,z'})$ ($\psi \in \mathcal{S}$) can be defined by the analog of (3.4.11), obtaining corresponding analogs of (3.4.12) and (3.4.13). Then, using ϕ_z^{t*} and $D_{\mathcal{F},z,z'}$, we get obvious extensions of the definitions and results of Sects. 3.4.6 and 3.4.7, except for the statements involving Π_∞ and P_∞. The double subscript "z, z'" may be added to the notation P, P_u, k, $\tilde{k}$, k_u and $\tilde{k}_u$ in this setting. However, if $z \neq z'$ and $\eta \neq 0$, $D_{0,z,z'}$ and $\Delta_{0,z,z'}$ are not symmetric, and therefore the results of Sect. 3.4.5 cannot be generalized for these operators.

Chapter 4
Foliations with Simple Foliated Flows

4.1 Simple Foliated Flows

4.1.1 Simple Flows

Let $\phi : \Omega \to M$ be a smooth local flow, where Ω is an open neighborhood of $M \times \{0\}$ in $M \times \mathbb{R}$. Let $Z \in \mathfrak{X}(M)$ be the infinitesimal generator. For $p \in M$ and $t \in \mathbb{R}$, let

$$\Omega_p = \{ t \in \mathbb{R} \mid (p, t) \in \Omega \} , \quad \Omega^t = \{ q \in M \mid (q, t) \in \Omega \} ,$$

and let $\phi^t = \phi(\cdot, t) : \Omega^t \to M$. The fixed point set is

$$\operatorname{Fix}(\phi) = \{ p \in M \mid p \in \operatorname{Fix}(\phi^t) \ \forall t \in \Omega_p \text{ close enough to } 0 \} ,$$

which equals the zero set of Z. Recall that a fixed point p of ϕ is called *simple* (or *transverse*) if it is a simple fixed point of ϕ^t for all $t \neq 0$ close enough to 0 in Ω_p (see Sect. 2.9.11). In this case, the associated number $\epsilon_p(\phi^t)$, defined in (2.9.23), is independent of $t > 0$ close enough to 0 in Ω_p, and is denoted by $\epsilon_p = \epsilon_p(\phi)$. If the fixed points of ϕ are simple, then $\operatorname{Fix}(\phi)$ is a discrete subset of M. For a fixed point p, we can write $\phi^t_* = e^{tA}$ on T_pM for some endomorphism A of T_pM. Then p is simple just when A is an automorphism.

Now, assume Z is complete, and therefore we can take $\Omega = M \times \mathbb{R}$. On $M \setminus \operatorname{Fix}(\phi)$, let $N\phi$ denote the normal bundle of the foliation defined by the orbits of ϕ; i.e., $N_p\phi = T_pM/\mathbb{R}Z(p)$ for every $p \in M \setminus \operatorname{Fix}(\phi)$. Let $\mathcal{C} = \mathcal{C}(\phi)$ denote the set of closed orbits of ϕ (without including fixed points). For any $c \in \mathcal{C}$, let $\ell(c)$ denote its minimum positive period. For every subset $I \subset \mathbb{R}$, let

$$\mathcal{C}_I = \mathcal{C}_I(\phi) = \{ c \in \mathcal{C} \mid \ell(c) \in I \} .$$

J. A. Álvarez López et al., *A Trace Formula for Foliated Flows*, Lecture Notes in Mathematics 2387, https://doi.org/10.1007/978-3-032-15413-2_4

The nonzero periods of all closed orbits form the set

$$\mathcal{P} = \mathcal{P}(\phi) = \{\, k\ell(c) \mid c \in \mathcal{C},\ k \in \mathbb{Z}^\times \,\} .$$

For all $c \in \mathcal{C}$, $k \in \mathbb{Z}$ and $p \in c$, let $\phi_*^{k\ell(c)} : N_p\phi \to N_p\phi$ be the homomorphism induced by $\phi_*^{k\ell(c)} : T_pM \to T_pM$. Recall that c is called *simple* when the eigenvalues of $\phi_*^{k\ell(c)} : N_p\phi \to N_p\phi$ are different from 1 for some (and therefore for all) $p \in c$ and $k \in \mathbb{Z}^\times$; in this case, let

$$\epsilon_c(k) = \epsilon_c(k, \phi) = \operatorname{sign}\det\left(\operatorname{id} - \phi_*^{k\ell(c)} : N_p\phi \to N_p\phi\right) \in \{\pm 1\} .$$

For every simple closed orbit c, there are neighborhoods, V of c in M and I of $\ell(c)$ in $\mathbb{R}$, such that c is the only closed orbit in V whose first positive period is in I, and moreover that $V \cap \operatorname{Fix}(\phi) = \emptyset$.

The flow ϕ is called *simple* if all of its fixed points and closed orbits are simple. If moreover M is closed, then $\operatorname{Fix}(\phi)$ is finite, and $\mathcal{C}_I(\phi)$ are finite for all compact $I \subset \mathbb{R}$. Therefore $\mathcal{P}(\phi)$ is a discrete subset of $\mathbb{R}$.

4.1.2 Transversely Simple Foliated Flows

Let $\mathcal{F}$ be a transversely oriented smooth foliation of codimension one on a closed manifold M. We assume M is closed for the sake of simplicity, but the concepts and properties recalled here also have obvious versions when M is a manifold with boundary, where both $\mathcal{F}$ and ϕ are tangent to ∂M. Some generalizations to non-compact manifolds will be also indicated and needed.

Let $\phi = \{\phi^t\}$ be a foliated flow on M and let $Z \in \mathfrak{X}(M, \mathcal{F})$ be its infinitesimal generator (Sect. 3.1.7). Let M^0 be the union of leaves preserved by ϕ, and let $M^1 = M \setminus M^0$. The ϕ-invariant set M^0 is compact because it is the zero set of $\overline{Z} \in \overline{\mathfrak{X}}(M, \mathcal{F}) \subset C^\infty(M; N\mathcal{F})$. Therefore the ϕ-invariant set M^1 is open in M. Moreover ϕ is transverse to the leaves on M^1. So there is a canonical isomorphism $N\phi \cong T\mathcal{F}$ on M^1, and $\mathcal{F}$ is transitive at every point of M^1 (Sect. 3.1.9); in particular, the leaves in M^1 have no holonomy. Consider the notation of Sects. 3.1.2 and 3.1.3, 3.1.7, using the notation (x_k, y_k) instead of (x'_k, x''_k) because $\operatorname{codim}\mathcal{F} = 1$. Let $\bar\phi$ be the local flow on Σ generated by $\overline{Z} \in \mathfrak{X}(\Sigma, \mathcal{H})$. Via the homeomorphism $M/\mathcal{F} \to \Sigma/\mathcal{H}$ induced by the coordinates $x_k : U_k \to \Sigma_k$, the leaves preserved by ϕ correspond to the $\mathcal{H}$-orbits preserved by $\bar\phi$, which indeed form $\operatorname{Fix}(\bar\phi)$ because the $\mathcal{H}$-orbits are totally disconnected. $\overline{Z}$ is $\mathcal{H}$-invariant, and $\bar\phi$ is $\mathcal{H}$-equivariant in an obvious sense.

Since $\dim\Sigma = 1$, for all simple $\bar p \in \operatorname{Fix}(\bar\phi)$, there is some $\varkappa = \varkappa_{\bar p} \in \mathbb{R}^\times$ such that $\bar\phi_*^t \equiv e^{\varkappa t}$ on $T_{\bar p}\Sigma \equiv \mathbb{R}$. By the $\mathcal{H}$-equivariance of $\bar\phi$, we get $\varkappa_{\bar p} = \varkappa_{\bar q}$ for all $\bar q \in \mathcal{H}(\bar p) \subset \operatorname{Fix}(\bar\phi)$. Thus we can use the notation $\varkappa_L = \varkappa_{\bar p}$ if $\mathcal{H}(\bar p)$ corresponds to a leaf L.

The leaves preserved by ϕ that correspond to simple fixed points of $\bar{\phi}$ are said to be *transversely simple*. If all leaves preserved by ϕ are transversely simple, then ϕ (or Z) is called *transversely simple*; if moreover its closed orbits are simple, then ϕ (or Z) is said to be *weakly simple*. If ϕ is weakly simple, every closed orbit is contained either in M^0 or in M^1, and its (possibly non-simple) fixed points belong to M^0.

Suppose ϕ is transversely simple unless otherwise stated. Then M^0 is a finite union of compact leaves because every fixed point of $\bar{\phi}$ is isolated. For any point p in a preserved leaf L, there are foliated coordinates $(x, y) : U \to \Sigma \times B$, where $\Sigma \subset \mathbb{R}$ is an open interval containing 0, so that [ÁLKL22, Lemma 3.2]

$$x(p) = 0\,, \quad \overline{Z} = \varkappa_L x \partial_x\,, \quad \bar{\phi}^t(x) = e^{\varkappa_L t} x\,. \tag{4.1.1}$$

Hence the following properties hold [ÁLKL22, Propositions 3.4 and 3.5]:

(C) $\mathcal{F}$ is almost without holonomy with finitely many leaves with holonomy.
(D) The holonomy groups of the compact leaves are groups of germs at 0 of homotheties on $\mathbb{R}$, for some choice of $\{U_k, (x_k, y_k)\}$.

According to (C) and Remark 3.1.6 (ii), we can consider Hector's description with this choice of M^0 and M^1, even though there may be leaves without holonomy in M^0. With the notation of Sect. 3.1.12, since $\pi : (M_l, \mathcal{F}_l) \to (M, \mathcal{F})$ is a foliated local embedding and $\pi : \partial M_l \to M^0$ a local diffeomorphism, any $A \in \mathfrak{X}(M, M^0)$ has a lift $A_l \in \mathfrak{X}_{\mathrm{b}}(M_l)$. Moreover $A_l \in \mathfrak{X}(M_l, \mathcal{F}_l)$ if $A \in \mathfrak{X}(M, \mathcal{F})$. Thus any (foliated) flow $\zeta = \{\zeta^t\}$ on $(M, \mathcal{F})$ preserving M^0 can be lifted via π to a (foliated) flow $\zeta_l = \{\zeta_l^t\}$ on $(M_l, \mathcal{F}_l)$ preserving ∂M_l. If a foliated flow ζ on $(M, \mathcal{F})$ is weakly simple, then ζ_l is also weakly simple (on the foliated manifold with boundary $(M_l, \mathcal{F}_l)$). The restrictions $A_l|_{\mathring{M}_l} \equiv A|_{M_l^1}$ and $\zeta_l|_{\mathring{M}_l} \equiv \zeta|_{M_l^1}$ are also denoted by A_l and ζ_l. In particular, this notation applies to Z and ϕ, obtaining Z_l and $\phi_l = \{\phi_l^t\}$, which induces the structure of a complete $\mathbb{R}$-Lie foliation on $\mathcal{F}_l^1$. According to Sect. 3.1.12, we consider the transverse orientation of every $\mathcal{F}_l$ so that $\pi : (M_l, \mathcal{F}_l) \to (M, \mathcal{F})$ is compatible with the transverse orientations. However, we will consider the transverse orientation of every $\mathcal{F}_l^1$ defined by $\overline{Z}_l$. Now Hector's description has the following more specific cases [ÁLKL22, Section 3]:

(c) $\mathcal{F}$ is given by a fiber bundle $M \to S^1$ with connected fibers.
(d) $\mathcal{F}$ is an $\mathbb{R}$-Lie foliation with dense leaves.
(e) $M^0 \neq \emptyset$, $\operatorname{Hol} L \cong \mathbb{Z}$ for all leaves $L \subset M^0$, and the foliations $\mathcal{F}_l^1$ are given by fiber bundles $M_l^1 \to S^1$ with connected fibers.
(f) $M^0 \neq \emptyset$, $\operatorname{Hol} L$ is a finitely generated abelian group of rank > 1 for all leaves $L \subset M^0$, and all foliations $\mathcal{F}_l^1$ are minimal $\mathbb{R}$-Lie foliations.

The case (c) can be considered as a model (1) with empty boundary, avoiding the use of models (0), or it can be cut into models (0) by adding a finite number of leaves without holonomy to M^0. Except in this case, M^1 is just the transitive point set of $\mathcal{F}$, and $\mathfrak{X}(M, \mathcal{F})$ spans $\mathfrak{X}(M, M^0)$ as $C^\infty(M)$-module by (4.1.1).

For every leaf $L \subset M^0$, its holonomy homomorphism $\boldsymbol{h} = \boldsymbol{h}_L$ is induced by a homomorphism $\hat{h} = \hat{h}_L : \pi_1 L \to \mathrm{Diffeo}^+(\mathbb{R}, 0)$ whose image consists of homotheties; i.e., writing $\Gamma = \Gamma_L = \pi_1 L / \ker \hat{h}$, $\hat{h}$ induces a monomorphism $h = h_L : \Gamma \to \mathrm{Diffeo}^+(\mathbb{R}, 0)$, $\gamma \mapsto h_\gamma$, with $h_\gamma(x) = a_\gamma x$ for some monomorphism $\Gamma \to \mathbb{R}^+ \equiv (\mathbb{R}^+, \times)$, $\gamma \mapsto a_\gamma = a_{L,\gamma}$. The restriction of $\mathcal{F}$ to some neighborhood of L can be described as the suspension of h (Sect. 3.1.2); its definition for this case will be recalled in Sect. 4.2.1. On the other hand, every $\mathcal{F}_l^1$ has a Fedida's description, which will be better analyzed in Sect. 4.3.4.

Remark 4.1.1 The concepts recalled in this subsection do not need the compactness of M. Only the completeness of Z and compactness of M^0 are needed to extend the indicated notions and properties.

4.1.3 Existence of Simple Foliated Flows

For a transversely oriented foliation $\mathcal{F}$ of codimension one on a closed foliated manifold M, the following conditions are equivalent [ÁLKL22, Propositions 6.1 and 6.3 and Theorem 6.9]:

(g) It satisfies (C) and (D).
(h) There is a transversely simple foliated flow.
(i) There is a weakly simple foliated flow.
(j) There is a simple foliated flow.

Moreover the families of foliated flows ϕ satisfying (h), (i) or (j) induce the same family of local flows $\bar{\phi}$ on Σ. In the case (i), it can be also assumed that $M^0 \subset \mathrm{Fix}(\phi)$ and there are no closed orbits in some neighborhood of M^0, obtaining the same family of local flows $\bar{\phi}$ on Σ.

A more precise description of the foliations satisfying these equivalent conditions is given in [ÁLKL22, Theorem 6.9], but it will not be needed here.

4.2 Case of Suspension Foliations

4.2.1 Suspension Foliations Defined with Homotheties

For a pointed connected closed manifold (L, p), let $\hat{h} : \pi_1 L = \pi_1(L, p) \to \mathrm{Diffeo}^+(\mathbb{R}, 0)$ be a homomorphism whose image consists of homotheties, like in Sect. 4.1. Therefore, writing $\Gamma = \pi_1 L / \ker \hat{h}$, $\hat{h}$ induces a monomorphism $h : \Gamma \to \mathrm{Diffeo}^+(\mathbb{R}, 0)$, $\gamma \mapsto h_\gamma$, where $h_\gamma(x) = a_\gamma x$ for some monomorphism $\Gamma \to \mathbb{R}^+$, $\gamma \mapsto a_\gamma$; in particular, Γ is abelian, torsion free and finitely generated. Let $\pi = \pi_L : (\widetilde{L}, \tilde{p}) \to (L, p)$ be the pointed regular covering map with $\pi_1 \widetilde{L} = \pi_1(\widetilde{L}, \tilde{p}) \equiv \ker \hat{h}$, and therefore $\mathrm{Aut}(\pi) \equiv \Gamma$. Like in Sects. 2.9.2 and

2.9.9, the canonical left action of every $\gamma \in \Gamma$ on $\widetilde{L}$ is denoted by T_γ or $\tilde{y} \mapsto \gamma \cdot \tilde{y}$, and write $[\tilde{y}] = \pi(\tilde{y})$ for $\tilde{y} \in \widetilde{L}$. For the diagonal left Γ-action on $\widetilde{M} = \mathbb{R} \times \widetilde{L}$, $\gamma \cdot (x, \tilde{y}) = (a_\gamma x, \gamma \cdot \tilde{y})$, the orbit space $M = \Gamma \backslash \widetilde{M}$ is called a *suspension manifold*. The canonical projection $\pi_M : \widetilde{M} \to M$ is a Γ-cover with deck transformations $h_\gamma \times T_\gamma$ ($\gamma \in \Gamma$). Write $[x, \tilde{y}] = \pi_M(x, \tilde{y})$ for $(x, \tilde{y}) \in \widetilde{M}$.

Let $\tilde{\varpi} : \widetilde{M} \to \widetilde{L}$ denote the second-factor projection, and let $\widetilde{\mathcal{F}}$ be the foliation on $\widetilde{M}$ with leaves $\{x\} \times \widetilde{L}$ ($x \in \mathbb{R}$). Since $\tilde{\varpi}$ is Γ-equivariant, we get an induced fiber bundle map $\varpi : M \to L$, defined by $\varpi([x, \tilde{y}]) = \pi(\tilde{y})$. On the other hand, since $\widetilde{\mathcal{F}}$ and its canonical transverse orientation are Γ-invariant, we also get an induced transversely oriented foliation $\mathcal{F}$ on M, called a *suspension foliation*, which is transverse to the fibers of ϖ. The typical fiber of ϖ is $\mathbb{R}$ because the corresponding fibers of $\tilde{\varpi}$ and ϖ can be identified via π_M. Since 0 is fixed by the Γ-action on $\mathbb{R}$, the leaf $\{0\} \times \widetilde{L} \equiv \widetilde{L}$ of $\widetilde{\mathcal{F}}$ is Γ-invariant, and $\pi_M(\{0\} \times \widetilde{L}) \equiv L$ is a compact leaf of $\mathcal{F}$. The other leaves of $\widetilde{\mathcal{F}}$ are diffeomorphic via π_M to the corresponding leaves of $\mathcal{F}$ because the elements of $\Gamma \setminus \{e\}$ have no fixed points in $\mathbb{R}^\times$. Given any $\tilde{p} \in \widetilde{L}$ with $\pi(\tilde{p}) = p \in L$, the fiber $\varpi^{-1}(p) \equiv \tilde{\varpi}^{-1}(\tilde{p}) = \mathbb{R} \times \{\tilde{p}\} \equiv \mathbb{R}$ is a global transversal of $\mathcal{F}$ through $p \equiv [0, \tilde{p}]$. Note that the holonomy homomorphism $\boldsymbol{h} : \pi_1 L \to \operatorname{Hol} L$ is induced by h, and therefore $\widetilde{L}^{\mathrm{hol}} \equiv \widetilde{L}$ (Sect. 3.1.2). The standard orientation of $\mathbb{R}$ induces a transverse orientation of $\widetilde{\mathcal{F}}$, which is Γ-invariant because the image of h consists of orientation preserving homotheties, giving rise to a transverse orientation of $\mathcal{F}$. Let $\boldsymbol{H} \subset TM$ and $\widetilde{\boldsymbol{H}} \subset T\widetilde{M}$ be the linear subbundles of vectors tangent to the fibers of ϖ and $\tilde{\varpi}$, which induce bigradings of ΛM and $\Lambda \widetilde{M}$ satisfying $d_{2,-1} = 0$ (Sects. 3.2.2 and 3.2.3). For $\tilde{p} \in \widetilde{M}$ and $p \in M$, we will use the identities $\Lambda_{\tilde{p}} \widetilde{\mathcal{F}} \equiv \Lambda_{\tilde{\varpi}(\tilde{p})} \widetilde{L}$ and $\Lambda_p \mathcal{F} \equiv \Lambda_{\varpi(p)} L$ induced by $\tilde{\varpi}$ and ϖ.

4.2.2 *Transversely Simple Flows on Suspension Foliations*

Let $\phi = \{\phi^t\}$ be any transversely simple foliated flow on M and let $Z \in \mathfrak{X}_{\mathrm{com}}(M, \mathcal{F})$ be its infinitesimal generator. Let us recall the notation of Sect. 4.1.2 in this case (see also Sect. 3.1.12). Without loss of generality, we can assume $M^0 = \pi_M(\{0\} \times \widetilde{L}) \equiv L$ for the description around a compact leaf. By (4.1.1), we can suppose the lifts of ϕ and Z to $\widetilde{M}$, denoted by $\tilde{\phi}$ and $\widetilde{Z}$, are of the form

$$\tilde{\phi}^t(x, \tilde{y}) = (e^{\varkappa t} x, \tilde{\phi}^t_x(\tilde{y})) , \quad \widetilde{Z} = (\varkappa x \partial_x, \widetilde{Z}_x) , \tag{4.2.1}$$

for some $\varkappa \in \mathbb{R}^\times$, and smooth families, $\{\tilde{\phi}^t_x \mid x, t \in \mathbb{R}\} \subset \operatorname{Diffeo}(\widetilde{L})$ and $\{\widetilde{Z}_x \mid x \in \mathbb{R}\} \subset \mathfrak{X}(\widetilde{L})$, with $\tilde{\phi}^0_x = \mathrm{id}_{\widetilde{L}}$. In particular, $\widetilde{Z}_0$ and $\tilde{\phi}^t_0$ are the restrictions of $\widetilde{Z}$ and $\tilde{\phi}^t$ to $\widetilde{L} \equiv \{0\} \times \widetilde{L}$. Thus $\widetilde{Z}_0$ is Γ-invariant and $\tilde{\phi}_0 = \{\tilde{\phi}^t_0\}$ is Γ-equivariant, inducing the restrictions of Z and ϕ^t to L, denoted by Z_0 and $\phi_0 = \{\phi^t_0\}$; we may also use the notation $Z_L = Z_0$ and $\phi_L = \{\phi^t_L\} = \{\phi^t_0\}$. The Γ-equivariance of $\tilde{\phi}^t$

and the Γ-invariance of $\widetilde{Z}$ mean that, for all $\gamma \in \Gamma$ and $x, t \in \mathbb{R}$,

$$T_\gamma \tilde{\phi}^t_x = \tilde{\phi}^t_{a_\gamma x} T_\gamma \;, \quad T_{\gamma *} \widetilde{Z}_x = \widetilde{Z}_{a_\gamma x} \;. \tag{4.2.2}$$

The only preserved leaf of $\tilde{\phi}$, $\{0\}\times\widetilde{L} \equiv \widetilde{L}$, is transversely simple. Now $\widetilde{M}^0 = \{0\}\times\widetilde{L}$ and $\widetilde{M}^1 = \widetilde{M}\setminus\widetilde{M}^0 = \mathbb{R}^\times\times\widetilde{L}$, which has two connected components, $\widetilde{M}^1_\pm = \mathbb{R}^\pm\times\widetilde{L}$. In this case, $\widetilde{M}_\pm = (\mathbb{R}^\pm \cup \{0\}) \times \widetilde{L}$, with $\mathring{\widetilde{\mathcal{F}}}_\pm = \widetilde{M}^1_\pm$ and $\partial\widetilde{M}_\pm = \widetilde{M}^0 \equiv \widetilde{L}$. The connected components of $M^1 = M \setminus M^0$ are $M^1_\pm = \pi_M(\widetilde{M}^1_\pm)$, and we have $M_\pm = \pi_M(\widetilde{M}_\pm)$, with $\mathring{M}_\pm = M^1_\pm$ and $\partial M_\pm = M^0 \equiv L$. The restriction $\pi_M : \widetilde{M}_\pm \to M_\pm$ will be denoted by $\pi_{M_\pm}$. The foliations $\widetilde{\mathcal{F}}^1_\pm = \mathring{\widetilde{\mathcal{F}}}_\pm$ on $\widetilde{M}^1_\pm = \mathring{\widetilde{M}}_\pm$ and $\widetilde{\mathcal{F}}_\pm$ on $\widetilde{M}_\pm$ are restrictions of $\widetilde{\mathcal{F}}$, and the foliations $\mathcal{F}^1_\pm = \mathring{\mathcal{F}}_\pm$ on $M^1_\pm = \mathring{M}_\pm$ and $\mathcal{F}_\pm$ on $M_\pm$ are restrictions of $\mathcal{F}$. We have $\boldsymbol{M} = M_+ \sqcup M_-$ (resp., $\widetilde{\boldsymbol{M}} = \widetilde{M}_+ \sqcup \widetilde{M}_-$), equipped with the combination $\boldsymbol{\mathcal{F}}$ (resp., $\widetilde{\boldsymbol{\mathcal{F}}}$) of $\mathcal{F}_+$ and $\mathcal{F}_-$ (resp., $\widetilde{\mathcal{F}}_+$ and $\widetilde{\mathcal{F}}_-$). The restriction of $\boldsymbol{\mathcal{F}}$ to $\mathring{\boldsymbol{M}}$ is denoted by $\mathring{\boldsymbol{\mathcal{F}}}$. Now the map $\boldsymbol{\pi} : \boldsymbol{M} \to M$ (resp., $\tilde{\boldsymbol{\pi}} : \widetilde{\boldsymbol{M}} \to \widetilde{M}$) is the combination of the inclusion maps $M_\pm \hookrightarrow M$ (resp., $\widetilde{M}_\pm \hookrightarrow \widetilde{M}$). The combination of the maps $\pi_{M_\pm}$ is a covering projection $\pi_{\boldsymbol{M}} : \widetilde{\boldsymbol{M}} \to \boldsymbol{M}$. Moreover ϖ (resp., $\tilde{\varpi}$) restricts to global collar neighborhoods of the boundaries, $\varpi_\pm : M_\pm \to \partial M_\pm \equiv L$ (resp., $\tilde{\varpi}_\pm : \widetilde{M}_\pm \to \partial\widetilde{M}_\pm \equiv \widetilde{L}$), whose combination is a global collar neighborhood of the boundary, $\boldsymbol{\varpi} : \boldsymbol{M} \to \partial\boldsymbol{M} \equiv L \sqcup L$ (resp., $\tilde{\boldsymbol{\varpi}} : \widetilde{\boldsymbol{M}} \to \partial\widetilde{\boldsymbol{M}} \equiv \widetilde{L} \sqcup \widetilde{L}$). Like in Sect. 4.2.1, for $\tilde{p} \in \widetilde{M}_\pm$ and $p \in M_\pm$, we have canonical identities $\Lambda_{\tilde{p}}\widetilde{\mathcal{F}}_\pm \equiv \Lambda_{\tilde{\varpi}_\pm(\tilde{p})}\widetilde{L}$ and $\Lambda_p\mathcal{F}_\pm \equiv \Lambda_{\varpi_\pm(p)}L$.

Recall also that any $A \in \mathfrak{X}(M, \mathcal{F})$, with foliated flow $\zeta = \{\zeta^t\}$, induces vector fields $A_\pm \in \mathfrak{X}(M_\pm, \mathcal{F}_\pm) \subset \mathfrak{X}_{\mathrm{b}}(M_\pm)$, with foliated flows $\zeta_\pm = \{\zeta^t_\pm\}$, whose restrictions to $\mathring{M}_\pm \equiv M^1_\pm$ are denoted in the same way. In particular, we get $Z_\pm$ with flow $\phi_\pm = \{\phi^t_\pm\}$. The same kind of notation is used for vector fields and flows induced by elements of $\mathfrak{X}_{\mathrm{com}}(\widetilde{M}, \widetilde{\mathcal{F}})$. Then $\mathcal{F}^1_\pm \equiv \mathring{\mathcal{F}}_\pm$ on $M^1_\pm \equiv \mathring{M}_\pm$ is a transversely complete $\mathbb{R}$-Lie foliation with the structure defined by $Z_\pm \in \mathfrak{X}_{\mathrm{com}}(M^1_\pm, \mathcal{F}^1_\pm)$ (see Remark 4.1.1). In its Fedida's description (Sect. 3.1.9), $\widetilde{M}^1_\pm$ is the holonomy covering of $M^1_\pm$, whose group of deck transformations is also Γ, the developing map $D_\pm : \widetilde{M}^1_\pm \to \mathbb{R}$ is given by $D_\pm(x, y) = \varkappa^{-1}\ln|x| =: t$, the holonomy monomorphism $h_\pm : \Gamma \to \mathbb{R}$ is given by $h_\pm(\gamma) = \varkappa^{-1}\ln a_\gamma$, and therefore $\operatorname{Hol}\mathcal{F}_\pm = \{\varkappa^{-1}\ln a_\gamma \mid \gamma \in \Gamma\}$. Thus $\widetilde{Z}_\pm \in \mathfrak{X}_{\mathrm{com}}(\widetilde{M}^1_\pm, \widetilde{\mathcal{F}}^1_\pm)$ is $D_\pm$-projectable and $(D_\pm)_*\widetilde{Z}_\pm = \partial_t$. Furthermore $\phi^t_\pm$ preserves every leaf of $\mathcal{F}_\pm$ if and only if $t = h_\pm(\gamma) = \varkappa^{-1}\ln a_\gamma$ for some $\gamma \in \Gamma$ (Sect. 3.1.11).

Let $\tilde{\xi} = \{\tilde{\xi}^t\}$ be the weakly simple foliated flow on $(\widetilde{M}, \widetilde{\mathcal{F}})$, with infinitesimal generator is $\widetilde{Y} \in \mathfrak{X}_{\mathrm{com}}(\widetilde{M}, \widetilde{\mathcal{F}})$, given by

$$\tilde{\xi}^t(x, \tilde{y}) = (e^{\varkappa t}x, \tilde{y}) \;, \quad \widetilde{Y} = (\varkappa x\partial_x, 0) \;. \tag{4.2.3}$$

We have $\overline{\widetilde{Y}} = \overline{\widetilde{Z}} \equiv \varkappa x\partial_x$, $\operatorname{Fix}(\tilde{\xi}^t) = \widetilde{L}$, and the orbits of $\tilde{xi}^t$ on $\mathring{\widetilde{M}}_\pm$ are the fibers of the restriction $\tilde{\varpi} : \mathring{\widetilde{M}}_\pm \to \widetilde{L}$. Since $\tilde{\xi}^t$ is Γ-equivariant and $\widetilde{Y}$ is Γ-invariant, they project to M obtaining a weakly simple foliated flow ξ^t on $(M, \mathcal{F})$ and its infinitesimal generator $Y \in \mathfrak{X}(M, \mathcal{F})$. We have $\overline{Y} = \overline{Z} \equiv \varkappa x \partial_x$, $\mathrm{Fix}(\xi^t) = L$, and the orbits of $\mathring{M}_\pm$ are the fibers of the restriction $\varpi : \mathring{M}_\pm \to L$.

On the one hand, we consider the restriction of the transverse orientations of $\widetilde{\mathcal{F}}$ and $\mathcal{F}$ to $\widetilde{\mathcal{F}}_\pm$ and $\mathcal{F}_\pm$, and, on the other hand, we consider the transverse orientations of $\widetilde{\mathcal{F}}_\pm^1$ and $\mathcal{F}_\pm^1$ induced by $\widetilde{Z}_\pm$ and $Z_\pm$, which corresponds to the standard orientation of $\mathbb{R}$ by $D_\pm$ (Sects. 3.1.12 and 4.1.2). They agree on $\widetilde{M}_+^1$ and M_+^1 (resp., $\widetilde{M}_-^1$ and M_-^1) if and only if $\kappa > 0$ (resp., $\kappa < 0$).

4.2.3 A Defining Form of $\mathcal{F}$

For $k = \operatorname{rank}\Gamma$, fix generators $\gamma_1, \ldots, \gamma_k$ of Γ. Let c_i be a piecewise smooth loop in L based at p such that $[c_i] \in \pi_1(L, p)$ projects to γ_i, and let $a_i = a_{\gamma_i}$. By the universal coefficients and Hurewicz theorems, there are closed 1-forms $\beta_1, \ldots, \beta_k$ on L such that $\delta_{ij} = \langle[\beta_i], [c_j]\rangle = \int_0^1 c_j^*\beta_i$ and $\langle[\beta_i], \ker\hat{h}\rangle = 0$. Every $\pi^*\beta_i$ is exact on $\widetilde{L}$. Let $\eta = -\ln(a_1)\,\beta_1 - \cdots - \ln(a_k)\,\beta_k$ and $\tilde{\eta} = \pi^*\eta = dF$ for some $F \in C^\infty(\widetilde{L})$. Note that $h(\Gamma) \cong \Gamma$ is the group of periods of η. With some abuse of notation, write also $F \equiv \tilde{\varpi}^*F \in C^\infty(\widetilde{M})$, and

$$\eta \equiv \varpi^*\eta \in C^\infty(M; \Lambda^{0,1}) \equiv C^\infty(M; \Lambda^1\mathcal{F}) ,$$

$$\tilde{\eta} \equiv \tilde{\varpi}^*\tilde{\eta} = \pi_M^*\eta = dF \in C^\infty(\widetilde{M}; \Lambda^{0,1}) \equiv C^\infty(M; \Lambda^1\widetilde{\mathcal{F}}) ,$$

using (3.2.5). Thus $\eta = \eta_0$ in this case, with the notation of Sect. 3.3.1. It is easy to check that

$$T_\gamma^*F = F - \ln a_\gamma \tag{4.2.4}$$

for all $\gamma \in \Gamma$, yielding $T_\gamma^*e^F = a_\gamma^{-1}e^F$. It easily follows that the 1-form $\tilde{\omega} = |\varkappa|^{-1}e^F\,dx$ on $\widetilde{M}$ is Γ-invariant. (Recall that the Γ-action on $\widetilde{M}$ is given by $\gamma \cdot (x, \tilde{y}) = (a_\gamma x, \gamma \cdot \tilde{y})$.) Furthermore $T\widetilde{\mathcal{F}} = \ker\tilde{\omega}$ and $\tilde{\omega}$ defines the transverse orientation of $\widetilde{\mathcal{F}}$. Therefore $\tilde{\omega}$ induces a 1-form ω on M satisfying $T\mathcal{F} = \ker\omega$ and defining the transverse orientation of $\mathcal{F}$. On the other hand, it is easy to compute $d\tilde{\omega} = \tilde{\eta} \wedge \tilde{\omega}$, yielding $d\omega = \eta \wedge \omega$. The vector field $\widetilde{X} = (|\varkappa|e^{-F}\partial_x, 0) \equiv |\varkappa|e^{-F}\partial_x \in C^\infty(\widetilde{M}; \widetilde{\boldsymbol{H}})$ is determined by $\tilde{\omega}(\widetilde{X}) = 1$. Thus $\widetilde{X}$ is Γ-invariant and induces the vector field $X \in C^\infty(M; \boldsymbol{H})$ satisfying $\omega(X) = 1$. So $\widetilde{X}$ and X also define the transverse orientations of $\widetilde{\mathcal{F}}$ and $\mathcal{F}$. On the other hand, $\tilde{\omega}(\widetilde{Z}) = \operatorname{sign}(\varkappa)e^F x$ by (4.2.1), yielding $\operatorname{sign}\tilde{\omega}(\widetilde{Z}_\pm) = \pm\operatorname{sign}(\varkappa)$, and therefore

$$\operatorname{sign}\omega(Z_\pm) = \pm\operatorname{sign}(\varkappa) .$$

So the transverse orientation of $\mathcal{F}_\pm^1$ is also defined by the restrictions to $M_\pm^1$ of $\pm\operatorname{sign}(\varkappa)\overline{X}$ or $\pm\operatorname{sign}(\varkappa)\omega$.

4.2.4 A Defining Function of M^0

Let $\tilde\rho = e^F x$, which is a defining function of $\widetilde{M}^0 \equiv \widetilde{L}$ on the whole of $\widetilde{M}$. Moreover $\tilde\rho$ is Γ-invariant by (4.2.4), and therefore it induces a defining function ρ of $M^0 \equiv L$ on the whole of M. It is easy to compute

$$d\tilde\rho = e^F(x\tilde\eta + dx) = \tilde\rho\tilde\eta + |\varkappa|\tilde\omega ,$$

yielding

$$d\rho = \rho\eta + |\varkappa|\omega , \tag{4.2.5}$$

and therefore

$$d\rho \wedge \omega = \rho\eta \wedge \omega . \tag{4.2.6}$$

Since $\tilde\xi^{t*}\tilde\rho = e^{\varkappa t}\tilde\rho$ by (4.2.3), we also get

$$\xi^{t*}\rho = e^{\varkappa t}\rho . \tag{4.2.7}$$

The global tubular neighborhood $\tilde\varpi : \widetilde{M} \to \widetilde{L} \equiv \widetilde{M}^0$ can be trivialized with $\tilde\rho$, obtaining $\widetilde{M} \equiv \mathbb{R}_{\tilde\rho} \times \widetilde{L}_{\tilde\varpi}$ besides $\widetilde{M} = \mathbb{R}_x \times \widetilde{L}_{\tilde\varpi}$. Thus the global tubular neighborhood $\varpi : M \to L \equiv M^0$ can be trivialized with ρ, obtaining $M \equiv \mathbb{R}_\rho \times L_\varpi$. According to Sect. 2.6, we have corresponding vector fields $\partial_x, \partial_{\tilde\rho} \in \mathfrak{X}(\widetilde{M})$ and $\partial_\rho \in \mathfrak{X}(M)$, and operators $\partial_x, \partial_{\tilde\rho} \in \operatorname{Diff}^1(\widetilde{M}; \Lambda)$ and $\partial_\rho \in \operatorname{Diff}^1(M; \Lambda)$. We compute

$$|\varkappa|\partial_{\tilde\rho} = |\varkappa|\partial_{\tilde\rho}(x)\,\partial_x = |\varkappa|\partial_{\tilde\rho}(e^{-F}\tilde\rho)\,\partial_x = |\varkappa|e^{-F}\,\partial_x = \widetilde{X} \in \mathfrak{X}(\widetilde{M}) .$$

It easily follows that

$$|\varkappa|\partial_{\tilde\rho} = |\varkappa|e^{-F}\,\partial_x \in \operatorname{Diff}^1(M; \Lambda) .$$

But, with the notation of Sect. 3.2.12, it is easy to check that

$$\partial_x = \Theta_{\partial_x} \in \operatorname{Diff}^1(\widetilde{M}; \Lambda) .$$

Since $\widetilde{X} \in C^\infty(\widetilde{M}; \widetilde{\boldsymbol{H}})$ and $d_{1,0}F = 0$, we can apply (3.2.27) on $C^\infty(\widetilde{M}; \Lambda)$ to get

$$\Theta_{\widetilde{X}} = |\varkappa|e^{-F}\Theta_{\partial_x} = |\varkappa|e^{-F}\partial_x = |\varkappa|\partial_{\tilde\rho} \in \operatorname{Diff}^1(\widetilde{M}; \Lambda) .$$

So, by the derivation formula of $\Theta_{\widetilde{X}}$, (3.2.26) and (3.2.27), and since $\partial_x \in \mathfrak{X}(\widetilde{M}, \widetilde{\mathcal{F}}) \cap C^\infty(\widetilde{M}; \widetilde{\boldsymbol{H}})$ and $\mathcal{L}_{\widetilde{X}}\tilde\eta = 0$,

$$[d_{0,1}, \Theta_{\widetilde{X}}] = |\varkappa|[d_{0,1}, e^{-F}\Theta_{\partial_x}] = |\varkappa|[d_{0,1}, e^{-F}]\Theta_{\partial_x}$$
$$= -|\varkappa|e^{-F}\tilde{\eta}\wedge\Theta_{\partial_x} = -\tilde{\eta}\wedge\Theta_{\widetilde{X}} = -\Theta_{\widetilde{X}}\,\tilde{\eta}\wedge\,.$$

Hence $|\varkappa|\partial_\rho = X \in \mathfrak{X}(M)$, and

$$|\varkappa|\partial_\rho = \Theta_X \in \mathrm{Diff}^1(M;\Lambda)\,, \tag{4.2.8}$$

$$[\Theta_X, d_{0,1}] = \eta\wedge\Theta_X = \Theta_X\,\eta\wedge \in \mathrm{Diff}^1(M;\Lambda)\,. \tag{4.2.9}$$

Note also that $\Theta_{\widetilde{X}}\tilde{\omega} = 0$, and therefore $\Theta_X\omega = 0$. Moreover, $\Theta_X\eta = 0$.

For any $\epsilon > 0$, the restriction $\tilde{\varpi} : \widetilde{T}_\epsilon := \{|\tilde{\rho}| < \epsilon\} \to \widetilde{L}$ is a smaller tubular neighborhood of $\widetilde{L}$ in $\widetilde{M}$, which induces a smaller tubular neighborhood $\varpi : T_\epsilon := \{|\rho| < \epsilon\} \to L$ of L in M. Let $\widetilde{T}^1_\epsilon = \widetilde{T}_\epsilon \cap \widetilde{M}^1$, $\widetilde{T}^1_{\pm,\epsilon} = \widetilde{T}_\epsilon \cap \widetilde{M}^1_\pm$, $T^1_\epsilon = T_\epsilon \cap M^1$ and $T^1_{\pm,\epsilon} = T_\epsilon \cap M^1_\pm$.

4.2.5 A Boundary-Defining Function of $M_\pm$

Now consider the boundary-defining function $\tilde{\rho}_\pm = e^F|x| = \pm\tilde{\rho} = |\tilde{\rho}|$ on $\widetilde{M}_\pm$, which is Γ-invariant, and therefore it induces a boundary-defining function $\rho_\pm$ on $M_\pm$ satisfying

$$\rho_\pm = \pm\rho = |\rho|\,. \tag{4.2.10}$$

If there is no danger of confusion, with some abuse of notation, these boundary-defining functions may be simply denoted by $\tilde{\rho}$ and ρ. Furthermore, we have the boundary-defining function $\tau = \tau_\pm := |x|$ on $\widetilde{M}_\pm$.

The global collar neighborhood $\tilde{\varpi} : \widetilde{M}_\pm \to \partial\widetilde{M}_\pm \equiv \widetilde{L}$ can be trivialized with, either τ, or $\tilde{\rho}$, obtaining

$$\widetilde{M}_\pm \equiv [0,\infty)_\tau \times \widetilde{L}_{\tilde{\varpi}} \tag{4.2.11}$$
$$\equiv [0,\infty)_{\tilde{\rho}} \times \widetilde{L}_{\tilde{\varpi}}\,. \tag{4.2.12}$$

So the global collar neighborhood $\varpi : M_\pm \to \partial M \equiv L$ can be trivialized with ρ, obtaining

$$M_\pm \equiv [0,\infty)_\rho \times L_\varpi\,. \tag{4.2.13}$$

By (4.2.5), (4.2.8), (4.2.9), and (4.2.10),

$$d\rho = \rho\eta \pm |\varkappa|\omega\,, \tag{4.2.14}$$

$$[\partial_\rho, d_{0,1}] = \eta\wedge\partial_\rho \in \mathrm{Diff}^1(M_\pm;\Lambda)\,, \tag{4.2.15}$$

using the operator $\partial_\rho \in \mathrm{Diff}^1(M_\pm;\Lambda)$ introduced in Remarks 2.5.2 and 2.5.3. We have $\partial_\rho\omega = 0$, and therefore $\partial_\rho\eta = 0$.

Observe that $\tilde\rho^{-1}\tilde\omega = |\varkappa x|^{-1}dx$ is a basic form of $\widetilde{\mathring{\mathcal{F}}}_\pm$, and therefore $\rho^{-1}\omega$ is a basic form of $\mathring{\mathcal{F}}_\pm$. The transverse orientation of $\mathring{\mathcal{F}}_\pm = \mathcal{F}^1_\pm$ is also defined by the basic form

$$\omega_{\mathrm{b},\pm} = \operatorname{sign}(\varkappa)\,\rho^{-1}\omega = \pm\operatorname{sign}(\varkappa)\,\rho_\pm^{-1}\omega = \operatorname{sign}(\omega(Z_\pm))\,\rho_\pm^{-1}\omega\,, \tag{4.2.16}$$

whose lift to $\tilde{\mathring{M}}_\pm$ is $\tilde\omega_{\mathrm{b},\pm} = (\varkappa x)^{-1}dx$; in fact, $\tilde\omega_{\mathrm{b},\pm}(\widetilde{Z}_\pm) = 1$, and therefore $\omega_{\mathrm{b},\pm}(Z_\pm) = 1$. By (4.2.14) and (4.2.16),

$$\rho_\pm^{-1}\,d\rho_\pm = \eta + \varkappa\,\omega_{\mathrm{b},\pm}\,. \tag{4.2.17}$$

Let $\nu = \nu_\pm$ be the unique smooth trivialization of ${}_+N\partial M_\pm$ so that $d\rho(\nu) = 1$ (Sect. 2.5.1). By (4.2.14), ν is represented by the restriction of $\pm|\varkappa|^{-1}X$ to $L \equiv \partial M_\pm$.

The combination of ρ_+ and ρ_- is a boundary-defining function $\boldsymbol{\rho}$ on $\boldsymbol{M}$, and the combination of ν_+ and ν_- is the unique smooth trivialization $\boldsymbol{\nu}$ of ${}_+N\partial\boldsymbol{M}$ so that $d\boldsymbol{\rho}(\boldsymbol{\nu}) = 1$. Similarly, we define $\tilde{\boldsymbol{\rho}}$ on $\widetilde{\boldsymbol{M}}$ and $\tilde{\boldsymbol{\nu}}$ on $\partial\widetilde{\boldsymbol{M}}$.

For $\epsilon > 0$, the restriction $\tilde\varpi_\pm = \tilde\varpi : \widetilde{T}_{\pm,\epsilon} := \{\tilde\rho_\pm < \epsilon\} \to \widetilde{L}$ is a smaller collar neighborhood of the boundary in $\widetilde{M}_\pm$, which induces a smaller collar neighborhood $\varpi_\pm = \varpi : T_{\pm,\epsilon} := \{\rho_\pm < \epsilon\} \to L$ of the boundary in $M_\pm$. By combination, we get smaller collar neighborhoods of the boundaries, $\tilde{\boldsymbol{\varpi}} : \widetilde{\boldsymbol{T}}_\epsilon := \{\tilde{\boldsymbol{\rho}} < \epsilon\} \to \partial\widetilde{\boldsymbol{M}} = \widetilde{L} \sqcup \widetilde{L}$ and $\boldsymbol{\varpi} : \boldsymbol{T}_\epsilon := \{\boldsymbol{\rho} < \epsilon\} \to \partial\boldsymbol{M} = L \sqcup L$. We have $\widetilde{\mathring{T}}_\epsilon \equiv \widetilde{T}^1_\epsilon$, $\tilde{\mathring{T}}_{\pm,\epsilon} \equiv \widetilde{T}^1_{\pm,\epsilon}$, $\mathring{T}_\epsilon \equiv T^1_\epsilon$ and $\mathring{T}_{\pm,\epsilon} \equiv T^1_{\pm,\epsilon}$.

4.2.6 The Metric g_M

Take a Riemannian metric g_L on L, and let $g_{\widetilde{L}}$ be its lift to $\widetilde{L}$. Consider the leafwise metrics, $g_{\mathcal{F}} = \varpi^* g_L$ for $\mathcal{F}$ and $g_{\widetilde{\mathcal{F}}} = \tilde\varpi^* g_{\widetilde{L}}$ for $\widetilde{\mathcal{F}}$; their restrictions to $\mathcal{F}_\pm$ and $\widetilde{\mathcal{F}}_\pm$ may be denoted by $g_{\mathcal{F}_\pm}$ and $g_{\widetilde{\mathcal{F}}_\pm}$. Consider also the metric $g_M = \omega^2 + g_{\mathcal{F}}$ on M. The lift of g_M to $\widetilde{M}$ is

$$g_{\widetilde{M}} = \tilde\omega^2 + g_{\widetilde{\mathcal{F}}} = |\varkappa|^{-2}e^{2F}\,(dx)^2 + g_{\widetilde{\mathcal{F}}}\,.$$

With respect to g_M, the transverse volume form is ω, X is unitary and orthogonal to $\mathcal{F}$, and $T\mathcal{F}^\perp = \boldsymbol{H}$.

4.2.7 The b-Metrics $g_{b,\pm}$

Define also the metric $g_{\mathrm{b},\pm} = \rho_\pm^{-2}\omega^2 + g_{\mathcal{F}} = \omega_{\mathrm{b},\pm}^2 + g_{\mathcal{F}}$ on $\mathring{M}_\pm = M^1_\pm$, where the last equality uses (4.2.16). It is bundle-like for $\mathring{\mathcal{F}}_\pm = \mathcal{F}^1_\pm$, and its lift to $\mathring{\widetilde{M}}_\pm$ is

$$\tilde{g}_{\mathrm{b},\pm} = \tilde{\rho}_\pm^{-2}\,\tilde{\omega}^2 + g_{\widetilde{\mathcal{F}}} = \tilde{\omega}_{\mathrm{b},\pm}^2 + g_{\widetilde{\mathcal{F}}} = (\varkappa x)^{-2}\,(dx)^2 + g_{\widetilde{\mathcal{F}}}\,. \tag{4.2.18}$$

With respect to $g_{\mathrm{b},\pm}$, the transverse volume form is $\omega_{\mathrm{b},\pm}$, $Z_\pm$ is unitary and orthogonal to $\mathring{\mathcal{F}}_\pm$, and $T\mathring{\mathcal{F}}_\pm^\perp = \boldsymbol{H}|_{\mathring{M}_\pm}$. The metrics $\tilde{g}_{\mathrm{b},\pm}$ and $g_{\mathrm{b},\pm}$ on $\mathring{\widetilde{M}}_\pm$ and $\mathring{M}_\pm$ are restrictions of b-metrics on $\widetilde{M}_\pm$ and $M_\pm$, also denoted by $\tilde{g}_{\mathrm{b},\pm}$ and $g_{\mathrm{b},\pm}$. In the rest of this subsection, $\mathring{\widetilde{M}}_\pm$ and $\mathring{M}_\pm$ (resp., $\widetilde{M}_\pm$ and $M_\pm$) are assumed to be endowed with the metrics (resp., b-metrics) $\tilde{g}_{\mathrm{b},\pm}$ and $g_{\mathrm{b},\pm}$. By (4.2.17), if $\eta \neq 0$ or $\varkappa^2 \neq 1$, then the b-metrics $\tilde{g}_{\mathrm{b},\pm}$ and $g_{\mathrm{b},\pm}$ are not exact (Sect. 2.5).

Proposition 4.2.1 *$\mathring{\mathcal{F}}_\pm$ is of bounded geometry.*

Proof $\mathring{\widetilde{M}}_\pm$ is of bounded geometry because it is the Riemannian product of $(\mathbb{R}^\pm, (\varkappa x)^{-2}(dx)^2)$ and $(\widetilde{L}, g_{\widetilde{L}})$, which are of bounded geometry since L is compact and the change of coordinate $t = \varkappa^{-1}\ln|x|$ defines an isometry between $(\mathbb{R}^\pm, (\varkappa x)^{-2}(dx)^2)$ and $(\mathbb{R}, (dt)^2)$. Via this isometry, $\tilde{\omega}_{\mathrm{b},\pm} \equiv (\varkappa x)^{-1}\,dx$ on $\mathbb{R}^\pm$ is the pull-back of dt on $\mathbb{R}$. On the other hand, the leaves of $\mathring{\widetilde{\mathcal{F}}}_\pm$ are the fibers $\{x\} \times \widetilde{L}$ ($x \in \mathbb{R}^\pm$), and the O'Neill tensors of $\mathring{\widetilde{\mathcal{F}}}_\pm$ with $\tilde{g}_{\mathrm{b},\pm}$ vanish. Hence, on $\mathring{M}_\pm$ with $\mathring{\mathcal{F}}_\pm$ and $g_{\mathrm{b},\pm}$, all covariant derivatives of the curvature tensor are uniformly bounded, and the O'Neill tensors vanish.

Finally, the injectivity bi-radius of $\mathring{\widetilde{\mathcal{F}}}_\pm$ with $\tilde{g}_{\mathrm{b},\pm}$ is positive because a normal foliated chart centered at any $\tilde{p} = (x_0, \tilde{q})$ is given by

$$\tilde{\chi}_{\tilde{p},\pm} = (t_{x_0}, \tilde{y}_{\tilde{q}}) : \widetilde{U}_{\tilde{p},\pm} = \mathbb{R}^\pm \times B_{\widetilde{L}}(\tilde{q}, r) \to \mathbb{R} \times B\,,$$

where $t_{x_0} = \varkappa^{-1}(\ln|x| - \ln|x_0|)$, $r = \mathrm{inj}_L \le \mathrm{inj}_{\widetilde{L}}$, B is the open ball in $\mathbb{R}^{n-1}$ of radius r and center 0, and $\tilde{y}_{\tilde{q}} : B_{\widetilde{L}}(\tilde{q}, r) \to B$ is a normal chart of $\widetilde{L}$. Let $q = \pi(\tilde{q})$, $p = \pi_M(\tilde{p})$ and $U_{p,\pm} = \pi_M(\widetilde{U}_{\tilde{p},\pm})$. Then $\pi : B_{\widetilde{L}}(\tilde{q}, r) \to B_L(q, r)$ is a diffeomorphism, obtaining a normal chart $y_q : B_L(q, r) \to B$ of L that corresponds to $\tilde{y}_{\tilde{q}}$ via π. So $\pi_M : \widetilde{U}_{\tilde{p},\pm} \to U_{p,\pm}$ is also a diffeomorphism, and $\tilde{\chi}_{\tilde{p},\pm}$ induces via π_M a normal foliated chart of $\mathring{\mathcal{F}}_\pm$ with $g_{\mathrm{b},\pm}$, centered at p,

$$\chi_{p,\pm} \equiv (t_{x_0}, y_q) : U_{p,\pm} \equiv \mathbb{R}^\pm \times B_L(q, r) \to \mathbb{R} \times B\,.$$

This shows that the injectivity bi-radius of $\mathring{\mathcal{F}}_\pm$ with $g_{\mathrm{b},\pm}$ is positive. □

By Proposition 4.2.1 and according to Sect. 3.4.1, $\mathring{M}_\pm$ is of bounded geometry (the property (A) of Sect. 2.5.20).

Proposition 4.2.2 *$M_\pm$ satisfies the property (B) of Sect. 2.5.20.*

Proof According to the proof of Proposition 4.2.1, it is easy to check that $(\widetilde{M}_\pm, \tilde{g}_{\mathrm{b},\pm})$ satisfies (B) on the whole of $\widetilde{M}_\pm$ with $\tilde{\rho}\partial_{\tilde{\rho}}$ and the extensions $B' = (B, 0) \in \mathfrak{X}(\widetilde{M}_\pm)$ of vector fields $B \in \mathfrak{X}(\widetilde{L})$. It follows that $(M_\pm, g_{\mathrm{b},\pm})$ also satisfies (B) on the whole of $M_\pm$ with $\rho\partial_\rho$ and the extensions $A' \in \mathfrak{X}(M_\pm)$ of vector fields $A \in \mathfrak{X}(L)$ defined as follows. For every $A \in \mathfrak{X}(L)$, let $\widetilde{A}$ denote its lift to $\widetilde{L}$. Then $\widetilde{A}' = (\widetilde{A}, 0) \in \mathfrak{X}(\widetilde{M}_\pm)$ is Γ-invariant, and therefore it is $\pi_{M_\pm}$-projectable to a vector field $A' \in \mathfrak{X}(M_\pm)$. □

Proposition 4.2.3 *We have* $d(\ln\rho) \in C^\infty_{ub}(\mathring{M}_\pm; \Lambda^1)$.

Proof On the one hand, by the compactness of L and the definition of $g_{\mathrm{b},\pm}$, we have $\eta \in C^\infty_{\mathrm{ub}}(\mathring{M}_\pm; \Lambda^{0,1})$. On the other hand, $\rho^{-1}\omega \in C^\infty_{\mathrm{ub}}(\mathring{M}_\pm; \Lambda^{1,0})$ by Proposition 4.2.1, since $\omega_{\mathrm{b},\pm} = \operatorname{sign}(\varkappa)\rho^{-1}\omega$ is the $g_{\mathrm{b},\pm}$-transverse volume form. Hence $d(\ln\rho) = \eta \pm |\varkappa|\rho^{-1}\omega \in C^\infty_{\mathrm{ub}}(\mathring{M}_\pm; \Lambda^1)$ by (4.2.14). □

Let $|\cdot| : \Gamma \to \mathbb{N}_0$ be the word length function given by a finite generating set. There is some $c_0 > 0$ so that, for all $\gamma \in \Gamma$,

$$|\ln a_\gamma| \le c_0\,|\gamma|\,. \tag{4.2.19}$$

By (2.9.16), (4.2.3) and (4.2.18), for all $\tilde{p} \in \mathring{\widetilde{M}}_\pm$ and $\gamma \in \Gamma$,

$$c_1^{-1}|\gamma| \le d_{\mathring{\widetilde{\mathcal{F}}}_\pm}\big(\gamma^{-1}\cdot\xi^{h_\pm(\gamma)}(\tilde{p}), \tilde{p}\big) \le c_1|\gamma|\,,$$

using the holonomy homomorphism $h_\pm : \Gamma \to \mathbb{R}$ of $\mathcal{F}^1_\pm \equiv \mathring{\mathcal{F}}_\pm$ (Sect. 4.2.2).

Lemma 4.2.4 *There are* $C > 0$ *and* $c \ge 1$ *so that, for all* $\tilde{y}, \tilde{y}' \in \widetilde{L}$ *and* $x \in \mathbb{R}^\times$,

$$d_{\widetilde{L}}(\tilde{y}, \tilde{y}') \ge C \ln\frac{\rho([x, \tilde{y}])}{c\rho([x, \tilde{y}'])}\,.$$

Proof Let $\boldsymbol{F} \subset \widetilde{L}$ be a fundamental domain. Without loss of generality, we can assume $\tilde{y} \in \boldsymbol{F}$. Take some $\gamma \in \Gamma$ such that $\gamma\cdot\tilde{y}' \in \boldsymbol{F}$. Then

$$\rho([x, \tilde{y}]) = e^{F(\tilde{y})}x\,, \quad \rho([x, \tilde{y}']) = \rho([a_\gamma x, \gamma\cdot\tilde{y}']) = e^{F(\gamma\cdot\tilde{y}')}a_\gamma x\,.$$

There is some $C_0 \ge 1$ such that, for all $\tilde{y}_1, \tilde{y}_2 \in \boldsymbol{F}$,

$$C_0^{-1}e^{F(\tilde{y}_1)} \le e^{F(\tilde{y}_2)} \le C_0 e^{F(\tilde{y}_1)}\,.$$

So, using (2.9.17) with $\boldsymbol{K} = \boldsymbol{F}^2$ and (4.2.19),

$$\frac{\rho([x, \tilde{y}])}{\rho([x, \tilde{y}'])} = \frac{e^{F(\tilde{y})}}{e^{F(\gamma\cdot\tilde{y}')}a_\gamma} \le C_0 a_{\gamma^{-1}} \le C_0 e^{c_0|\gamma^{-1}|} \le C_0 e^{c_0 c_1 (d_{\widetilde{L}}(\tilde{y}', \tilde{y}) + c_2)}\,.$$

□

Corollary 4.2.5 *For $R, \epsilon > 0$, we have* $\operatorname{Pen}_{\mathcal{F}}(T_\epsilon, R) \subset T_{ce^{R/C}\epsilon}$.

Lemma 4.2.6 *For $p \in \mathring{M}_\pm$ and $t \in \mathbb{R}^\times$, if $\phi^t(L_p) = L_p$, then $d_{\mathcal{F}}(\phi^t(p), p) \geq \operatorname{inj}_L$.*

Proof We have $p = [x_0, \tilde{q}]$ for some $x_0 \in \mathbb{R}^\pm$ and $\tilde{q} \in \widetilde{L}$, and let $q = \pi(\tilde{q}) \in L$. For $r = \operatorname{inj}_L$, take normal charts $\tilde{y}_{\tilde{q}} : B_{\widetilde{L}}(\tilde{q}, r) \to B$ and $y_q : B_L(q, r) \to B$ like in the proof of Proposition 4.2.1. We have the foliated chart of $\widetilde{\mathcal{F}}$,

$$\tilde{\chi}_{\tilde{q}} = (x, \tilde{y}_{\tilde{q}}) : \widetilde{U}_{\tilde{q}} = \mathbb{R} \times B_{\widetilde{L}}(\tilde{q}, r) \to \mathbb{R} \times B ,$$

which induces via π_M a foliated chart of $\mathcal{F}$,

$$\chi_q = (x, y_q) : U_q = \pi_M(\widetilde{U}_{\tilde{q}}) \equiv \mathbb{R} \times B_L(q, r) \to \mathbb{R} \times B .$$

On the one hand, if $\phi^t(p) \in U_q$, then $\chi_q \phi^t(p) = (e^{\varkappa t} x_0, \tilde{\phi}^t_{x_0}(\tilde{q}))$ by (4.2.1), with $e^{\varkappa t} x_0 \neq x_0$ because $t \neq 0$. So p and $\phi^t(p)$ lie in different plaques of (U_q, χ_q). On the other hand, if $\phi^t(p) \notin U_q$, then a fortiori $\phi^t(p)$ is not in the plaque of (U_q, χ_q) through p. In any case, $d_{\mathcal{F}}(\phi^t(p), p) \geq r$ because the plaque through p is $B_{L_p}(p, r) \equiv B_{\widetilde{L}}(\tilde{q}, r)$. □

Proposition 4.2.7 *If $Z_\pm \in \mathfrak{X}_{ub}(M^1_\pm, \mathcal{F}^1_\pm)$, then, for any compact $I \subset \mathbb{R}$, there are $c_1, c_2 > 0$ such that, for all $p \in \mathring{M}_\pm$ and $\gamma \in \Gamma$ with $h_\pm(\gamma) \in I$,*

$$d_{\mathcal{F}}\big(\phi^{h_\pm(\gamma)}(p), p\big) \geq c_1^{-1}|\gamma| - c_2 .$$

Proof Since I is compact and $\phi_\pm$ is of $\mathbb{R}$-local bounded geometry on $\mathring{M}_\pm$ (Sect. 2.4.7), there is some $R > 0$ such that $d_{\widetilde{L}}(\tilde{\phi}^t_x(\tilde{y}), \tilde{y}) \leq R$ for all $x \in \mathbb{R}^\pm$, $t \in I$ and $\tilde{y} \in \widetilde{L}$. Given any fundamental domain $\boldsymbol{F} \subset \widetilde{L}$, let $\boldsymbol{K} = \overline{\operatorname{Pen}_{\widetilde{L}}(\boldsymbol{F}, R)} \times \boldsymbol{F}$. By (2.9.17), there are $c_1 \geq 1$ and $c_2 \geq 0$ such that

$$d_{\widetilde{L}}(\gamma \cdot \tilde{\phi}^t_x(\tilde{y}), \tilde{y}') \geq c_1^{-1}|\gamma| - c_2 \tag{4.2.20}$$

for all $x \in \mathbb{R}^\pm$, $t \in I$, $\tilde{y}, \tilde{y}' \in \boldsymbol{F}$ and $\gamma \in \Gamma$, because $(\tilde{\phi}^t_x(\tilde{y}), \tilde{y}') \in \boldsymbol{K}$.

Any $p \in \mathring{M}_\pm$ is of the form $p = [x, \tilde{y}]$ for some $x \in \mathbb{R}^\pm$ and $\tilde{y} \in \boldsymbol{F}$. Let $\gamma \in \Gamma$ with $t := h_\pm(\gamma) = \varkappa^{-1} \ln a_\gamma \in I$. Then $\phi^t(L_p) = L_p$ (Sect. 4.2.1), and, by (4.2.20),

$$\begin{aligned} d_{\mathcal{F}}(\phi^t(p), p) &= d_{\mathcal{F}}([e^{\varkappa t}x, \tilde{\phi}^t_x(\tilde{y})], [x, \tilde{y}]) = d_{\mathcal{F}}([a_\gamma x, \tilde{\phi}^t_x(\tilde{y})], [x, \tilde{y}]) \\ &= d_{\mathcal{F}}([x, \gamma^{-1} \cdot \tilde{\phi}^t_x(\tilde{y})], [x, \tilde{y}]) = d_{\widetilde{L}}(\gamma^{-1} \cdot \tilde{\phi}^t_x(\tilde{y}), \tilde{y}) \geq c_1^{-1}|\gamma| - c_2 . \end{aligned}$$

□

Corollary 4.2.8 *If $Z_\pm \in \mathfrak{X}_{ub}(M^1_\pm, \mathcal{F}^1_\pm)$, then, for any compact $I \subset \mathbb{R}^\times$, there is some $c_3 > 0$ such that, for all $p \in \mathring{M}_\pm$ and $\gamma \in \Gamma$ with $h_\pm(\gamma) \in I$,*

$$d_{\mathcal{F}}\big(\phi^{h_\pm(\gamma)}(p), p\big) \geq c_3|\gamma|\,.$$

Proof By Lemmas 4.2.6 and 4.2.7, the result follows taking $c_3 > 0$ such that, for all $\gamma \in \Gamma$,

$$c_3|\gamma| \leq \begin{cases} \mathrm{inj}_L & \text{if } |\gamma| \leq c_1c_2 \\ c_1^{-1}|\gamma| - c_2 & \text{if } |\gamma| > c_1c_2\,. \end{cases}$$

□

Proposition 4.2.9 *If $Z_\pm \in \mathfrak{X}_{ub}(M^1_\pm, \mathcal{F}^1_\pm)$, then, for any compact $I \subset \mathbb{R}$, there exists some $c' > 0$ such that $\phi^t(T_\epsilon) \subset T_{c'\epsilon}$ for all $t \in I$ and $\epsilon > 0$.*

Proof Take some $R > 0$ like in the proof of Proposition 4.2.7. Let $\tilde{y} \in \widetilde{L}$ and $x \in \mathbb{R}^\pm$ such that $\rho([x, \tilde{y}]) < \epsilon$. If $\boldsymbol{c} : [0, 1] \to \widetilde{L}$ is a minimizing geodesic segment from $\tilde{y}$ to $\tilde{\phi}^t_x(\tilde{y})$, then

$$F(\tilde{\phi}^t_x(\tilde{y})) - F(\tilde{y}) = \int_0^1 \boldsymbol{c}^*dF = \int_0^1 \boldsymbol{c}^*\tilde{\eta} \leq \|\tilde{\eta}\|_{L^\infty} d_{\widetilde{L}}(\tilde{y}, \tilde{\phi}^t_x(\tilde{y})) \leq \|\eta\|_{L^\infty} R\,.$$

Take also some $c'_1 > 0$ such that $e^{\varkappa t} \leq c'_1$ for all $t \in I$. Then

$$\rho(\phi^t([x, \tilde{y}])) = e^{F(\tilde{\phi}^t_x(\tilde{y})) - F(\tilde{y}) + \varkappa t}\rho([x, \tilde{y}]) \leq e^{\|\eta\|_{L^\infty} R}c'_1\epsilon\,.$$

□

Proposition 4.2.10 *Suppose Γ is nontrivial. For any $\epsilon > 0$, there is some $0 < \epsilon' < \epsilon$ such that, for all leaf L' of $\mathcal{F}$, if a connected component W of $L' \cap T_\epsilon$ meets $T_{\epsilon'}$, then $L \subset \overline{W}$.*

Proof Let $\boldsymbol{F} \subset \widetilde{L}$ be a fundamental domain. We can choose $0 < \epsilon' < \epsilon$ such that $e^{F(\tilde{y}) - F(\tilde{y}')}\epsilon' < \epsilon$ for all $\tilde{y}, \tilde{y}' \in \boldsymbol{F}$. Let W be a connected component of $L' \cap T_\epsilon$ that meets $T_{\epsilon'}$ at some point $[x, \tilde{y}]$. We can assume $\tilde{y} \in \boldsymbol{F}$. For every $\tilde{y}' \in \boldsymbol{F}$, we have $[x, \tilde{y}'] \in L'$ and

$$|\rho([x, \tilde{y}'])| = e^{F(\tilde{y}')}|x| = e^{F(\tilde{y}') - F(\tilde{y})}|\rho([x, \tilde{y}])| < e^{F(\tilde{y}') - F(\tilde{y})}\epsilon' < \epsilon\,.$$

So $\pi_M(\{x\} \times \boldsymbol{F}) \subset W$ because $\boldsymbol{F}$ is connected. Since Γ is nontrivial and h is injective, there is some $\gamma \in \Gamma$ such that $\gamma \cdot \boldsymbol{F} \cap \boldsymbol{F} \neq \emptyset$ and $a_\gamma > 1$. Then $W_0 := \bigcup_{m=0}^\infty \gamma^m \cdot \boldsymbol{F}$ is connected in $\widetilde{L}$. Moreover, by (4.2.4), for all $m \in \mathbb{N}_0$ and $\tilde{y}' \in \boldsymbol{F}$,

$$|\rho([x, \gamma^m\tilde{y}'])| = e^{F(\gamma^m\tilde{y}')}|x| = a_\gamma^{-m}e^{F(\tilde{y}')}|x| = a_\gamma^{-m}|\rho([x, \tilde{y}'])| < a_\gamma^{-m}\epsilon\,,$$

which is $< \epsilon$ and converges to 0 as $m \to \infty$. Since $[x, \gamma^m \tilde{y}'] \in \varpi^{-1}([\tilde{y}'])$, it follows that $[x, \gamma^m \tilde{y}'] \to [0, \tilde{y}'] \equiv [\tilde{y}']$ as $m \to \infty$. Hence $\pi_M(\{x\} \times W_0) \subset W$ and $L \subset \overline{\pi_M(\{x\} \times W_0)}$. □

4.2.8 The b-Metrics $g_{c,\pm}$

Using (4.2.13), we can also define the metric $g_{c,\pm} \equiv (\varkappa\rho)^{-2}(d\rho)^2 + g_L$ on $\mathring{M}_\pm$ and its lift $\tilde{g}_{c,\pm} \equiv (\varkappa\tilde{\rho})^{-2}(d\rho)^2 + g_{\tilde{L}}$ on $\mathring{\widetilde{M}}_\pm$. These are restrictions to the interiors of b-metrics, also denoted by $g_{c,\pm}$ and $\tilde{g}_{c,\pm}$. The b-metrics $\varkappa^2 g_{c,\pm}$ and $\varkappa^2 \tilde{g}_{c,\pm}$ are exact and cylindrical around the boundary (Sect. 2.5.1); in particular, the level hypersurfaces of ρ and $\tilde{\rho}$ are totally geodesic for $g_{c,\pm}$ and $\tilde{g}_{c,\pm}$.

Proposition 4.2.11 *The metrics $g_{c,\pm}$ and $g_{b,\pm}$ are quasi-isometric on $\mathring{M}_\pm$; more precisely,*

$$|\cdot|_{g_{c,\pm}} \le \sqrt{2(1 + \varkappa^{-2}\|\eta\|_{L^\infty})}\,|\cdot|_{g_{b,\pm}}\,, \quad |\cdot|_{g_{b,\pm}} \le \sqrt{2}\,|\cdot|_{g_{c,\pm}}\,.$$

Proof Take any $p \in \mathring{M}_\pm$ and $u \in T_p\mathring{M}_\pm \equiv T_pM = \boldsymbol{H}_p \oplus \boldsymbol{V}_p$, and let $v = \boldsymbol{V}u$ and $w = \boldsymbol{H}u$. Then

$$|u|^2_{g_{b,\pm}} = \rho^{-2}\omega(w)^2 + |v|^2_{g_L}\,.$$

By (4.2.14) and since $\omega(X) = 1$, it follows that u is the sum of the vectors

$$v \mp |\varkappa|^{-1}\rho\eta(v)X_p \in \ker(d\rho)_p\,, \quad w \pm |\varkappa|^{-1}\rho\eta(v)X_p \in \boldsymbol{H}_p = \ker\varpi_{*p}\,,$$

obtaining

$$\begin{aligned} |u|^2_{g_{c,\pm}} &= (\varkappa\rho)^{-2}\big(\pm|\varkappa|\omega(w) + \rho\eta(v)\big)^2 + |v|^2_{g_L} \\ &= |u|^2_{g_{b,\pm}} + \varkappa^{-2}\eta(v)^2 \pm 2(|\varkappa|\rho)^{-1}\omega(w)\eta(v)\,. \end{aligned}$$

Then

$$|u|^2_{g_{c,\pm}} \le |u|^2_{g_{b,\pm}} + 2\varkappa^{-2}\eta(v)^2 + \rho^{-2}\omega(w)^2 \le 2(1 + \varkappa^{-2}\|\eta\|_{L^\infty})|u|^2_{g_{b,\pm}}\,,$$

$$|u|^2_{g_{b,\pm}} \le |u|^2_{g_{c,\pm}} + \rho^{-2}\omega(w)^2 \le 2|u|^2_{g_{c,\pm}}\,.$$

□

4.2.9 Vector Fields

Assume again that $\mathring{\widetilde{M}}_{\pm}$ and $\mathring{M}_{\pm}$ are endowed with $\tilde{g}_{\mathrm{b},\pm}$ and $g_{\mathrm{b},\pm}$. Recall that any $A \in \mathfrak{X}(M, \mathcal{F})$ induces a vector field $A_{\pm} \in \mathfrak{X}(M_{\pm}, \mathcal{F}_{\pm})$, whose restriction to $\mathring{M}_{\pm}$ is also denoted by $A_{\pm}$ (Sects. 4.1.2 and 4.2.1).

Proposition 4.2.12 $Y_{\pm} \in \mathfrak{X}_{ub}(\mathring{M}_{\pm}, \mathcal{F}_{\pm})$.

Proof This follows from the proof of Proposition 4.2.1 because ∂_t corresponds to $\varkappa x \partial_x$ by the change of coordinate $t = \varkappa^{-1} \ln |x|$. □

Lemma 4.2.13 *If $V \in \mathfrak{X}_c(\mathcal{F})$, then $V_{\pm} \in \mathfrak{X}_{ub}(\mathring{\mathcal{F}}_{\pm})$.*

Proof Consider the normal foliated charts of $\mathring{\mathcal{F}}_{\pm}$, $\chi_{p,\pm} = (t_{x_0}, y_q) = (t, \underset{\sim}{y})$ on $U_{p,\pm}$, like in the proof of Proposition 4.2.1, and the foliated charts of $\mathcal{F}$, $\chi_q = (x, y_q) = (x, y)$ on U_q, like in the proof of Lemma 4.2.6. Then $U_q = \varpi^{-1}(B_L(q, r))$, $U_{p,\pm} = U_q \cap \mathring{M}_{\pm}$ and $x = e^{\varkappa t} x_0$. Let $\partial_i = \partial_{y^i}$ and $\partial_I = \partial_{i_1} \cdots \partial_{i_m}$ for any multi-index $I = (i_1, \dots, i_m)$. Take a partition of unity subordinated to a finite open cover of the compact manifold L by balls $B_L(q, r)$ ($q \in L$). By using the ϖ-lift of this partition of unity to M, it easily follows that we can assume V is supported in some U_q. Thus we can write $V = f^i(x, y)\partial_i$ on U_q for functions $f^i \in C^\infty_{\mathrm{c}}(\mathbb{R} \times B) \equiv C^\infty_{\mathrm{c}}(U_q)$, and write $V_{\pm} = h^i(t, y)\partial_i$ on $U_{p,\pm}$ for functions $h^i \in C^\infty(\mathbb{R}_{\pm} \times B)$. We have

$$\partial_I h^i(t, y) = \partial_I f^i(e^{\varkappa t} x_0, y) . \tag{4.2.21}$$

Claim 4.2.14 For $l \leq k$ in $\mathbb{N}$, there are $c_{k,l} \in \mathbb{N}$ such that, on $U_{p,\pm}$,

$$\partial_t^k = \sum_l c_{k,l} x^l \partial_x^l = \sum_l (\pm 1)^l c_{k,l} \rho^l X^l .$$

To simplify the notation, we define $c_{k,l}$ for all $k, l \in \mathbb{Z}$ by setting $c_{k,l} = 0$ if $\min\{k, l\} < 0$, $c_{0,0} = 1$, and $c_{k,l} = \varkappa(l c_{k-1,l} + c_{k-1,l-1})$ if $\max\{k, l\} > 0$. Note that $c_{k,l} = 0$ if $l \leq 0 < k$ or $l > k$.

The first equality of Claim 4.2.14 follows by induction on k. The case $k = 1$ is true because $\partial_t = \varkappa x \partial_x$. If $k > 1$ and the first equality holds for $k - 1$, then

$$\begin{aligned}\partial_t^k &= \varkappa x \partial_x \sum_l c_{k-1,l} x^l \partial_x^l = \sum_l \varkappa c_{k-1,l} \big(x^{l+1} \partial_x^{l+1} + x[\partial_x, x^l]\partial_x^l\big) \\ &= \sum_l \varkappa c_{k-1,l} \big(x^{l+1} \partial_x^{l+1} + l x^l \partial_x^l\big) = \sum_l \varkappa(l c_{k-1,l} + c_{k-1,l-1}) x^l \partial_x^l .\end{aligned}$$

The second equality of Claim 4.2.14 holds because

$$\tilde{\rho}^l \widetilde{X}^l = e^{lF(y)} |x|^l (e^{-F(y)} \partial_x)^l = |x|^l \partial_x^l = (\pm 1)^l x^l \partial_x^l .$$

By (4.2.21) and Claim 4.2.14,

$$\partial_t^{k+1}\partial_I h^i(t,y) = \sum_{l=1}^{k}(\pm 1)^l c_{k,l}\rho^l X^l \partial_I f^i(e^{\varkappa t}x_0, y)\,.$$

Thus every function $|\partial_t^{k+1}\partial_I h^i|$ is uniformly bounded on $\mathbb{R}_\pm \times B$ because $f^i \in C_c^\infty(\mathbb{R}\times B) \equiv C_c^\infty(U_q)$ and $X \in \mathfrak{X}(M)$. □

Proposition 4.2.15 *For any $\epsilon > 0$, there is some $A \in \mathfrak{X}_{com}(M,\mathcal{F})$ such that $A_\pm \in \mathfrak{X}_{ub}(\mathring{M}_\pm, \mathring{\mathcal{F}}_\pm)$ and $A = Z$ on T_ϵ.*

Proof Let $\widetilde{V} = (0, \widetilde{Z}_x) \in \mathfrak{X}(\widetilde{\mathcal{F}})$, which projects to a vector field $V \in \mathfrak{X}(\mathcal{F})$ by (4.2.2). For any $\lambda \in C_c^\infty(M)$ such that $0 \le \lambda \le 1$ and $\lambda = 1$ on T_ϵ, we have $V' := \lambda V \in \mathfrak{X}_c(\mathcal{F})$. Then $V'_\pm \in \mathfrak{X}_{ub}(\mathring{\mathcal{F}}_\pm)$ by Lemma 4.2.13, $A := Y + V' \in \mathfrak{X}_{com}(M,\mathcal{F})$, $A = Z$ on T_ϵ, and $A_\pm = Y_\pm + V'_\pm \in \mathfrak{X}_{ub}(\mathring{M}_\pm, \mathring{\mathcal{F}}_\pm)$ by Proposition 4.2.12. □

4.3 Global Objects on Foliations with Simple Foliated Flows

Consider the notation of Sects. 3.1.12 and 4.1.2, where M is compact, $\mathcal{F}$ is transversely oriented, and ϕ is transversely simple.

4.3.1 Tubular Neighborhoods of M^0

In the following, for $L \in \pi_0 M^0$ (the set of leaves in M^0), we have corresponding objects $\hat{h}_L$, h_L, Γ_L, $\pi_L : \widetilde{L} \to L$, $\varkappa_L$ and $a_{L,\gamma}$ (Sect. 4.1.2). Consider also the corresponding suspension foliated manifold, $(M'_L, \mathcal{F}'_L)$, and all other associated objects (Sects. 4.2.1 to 4.2.8). A prime and the subscript "L" is added to their notation; for instance, we have $\xi'_L = \{\xi'^t_L\}$, Y'_L, M'^0_L, M'^1_L, $\mathcal{F}'^1_L$, ϖ'_L, ρ'_L, $T'_{L,\epsilon}$, $T'^1_{L,\epsilon}$, X'_L, ω'_L, η'_L, $g_{M'_L}$ and $g_{\mathcal{F}'_L}$. The corresponding disjoint unions or combinations, with L running in $\pi_0 M^0$, are denoted by M', $\mathcal{F}'$, $\xi' = \{\xi'^t\}$, Y', M'^0, M'^1, $\mathcal{F}'^1$, ϖ', ρ', T'_ϵ, T'^1_ϵ, X', ω', η', $g_{M'}$ and $g_{\mathcal{F}'}$, removing the subscript "L".

By the Reeb's local stability, if $\epsilon > 0$ is small enough, there is a tubular neighborhood of every L in M, $\varpi_L : T_{L,\epsilon} \to L$, such that $T_{L,\epsilon}$ is diffeomorphic to $T'_{L,\epsilon}$, with ϖ_L and $\mathcal{F}|_{T_{L,\epsilon}}$ corresponding to ϖ'_L and $\mathcal{F}'_L|_{T'_{L,\epsilon}}$; we simply write $\varpi_L \equiv \varpi'_L$ and $\mathcal{F} \equiv \mathcal{F}'_L$ on $T_{L,\epsilon} \equiv T'_{L,\epsilon}$. We can assume the closures $\overline{T_{L,\epsilon}}$ are disjoint one another. Then the combination of the maps $\varpi_{L,\epsilon}$ is a tubular neighborhood of M^0 in M,

$$\varpi \equiv \varpi' : T_\epsilon := \bigcup_L T_{L,\epsilon} \equiv T'_\epsilon \to M^0 \equiv M'^0\,.$$

4.3.2 Collar Neighborhoods of Every ∂M_l

Given any connected component M_l^1 of M^1, consider only leaves $L \in \pi_0(M^0 \cap \overline{M_l^1}) \equiv \pi_0(\partial M_l)$. The notation $(M'_{L,l}, \mathcal{F}'_{L,l})$ is used for $(M'_{L,+}, \mathcal{F}'_{L,+})$ (resp., $(M'_{L,-}, \mathcal{F}'_{L,-})$) if the transverse orientation of $\mathcal{F}_l$ along L points inwards (resp., outwards), like the transverse orientation along L of $\mathcal{F}'_{L,+}$ (resp., $\mathcal{F}'_{L,-}$). This kind of change is applied to the rest of notation concerning these foliated manifolds with boundary (Sects. 4.2.2, 4.2.5, 4.2.6, 4.2.7, 4.2.8 and 4.2.9). For instance, we obtain $\xi'_{L,l} = \{\xi'^t_{L,l}\}$, $Y'_{L,l}$, $\varpi'_{L,l}$, $\rho'_{L,l}$, $\nu'_{L,l}$, $T'_{L,l,\epsilon}$, $\omega'_{\mathrm{b},L,l}$, $\eta'_{L,l}$, $g'_{\mathrm{b},L,l}$ and $g'_{\mathrm{c},L,l}$. Similarly, we have $(M'^1_{L,l}, \mathcal{F}'^1_{L,l}) \equiv (\mathring{M}'_{L,l}, \mathring{\mathcal{F}}'_{L,l})$, whose Molino's description involves $\widetilde{M}'^1_{L,l}$, $\widetilde{\mathcal{F}}'^1_{L,l}$, $h'_{L,l} : \Gamma_L \to \mathbb{R}$ and $D'_{L,l} : \widetilde{M}'^1_{L,l} \to \mathbb{R}$. We have $\mathring{T}'_{L,l,\epsilon} \equiv T'_{L,\epsilon} \cap M'^1_l =: T'^1_{L,l,\epsilon}$. The corresponding disjoint unions or combinations, with L running in $\pi_0(\partial M_l)$, are denoted by M'_l, $\mathcal{F}'_l$, $\xi'_l = \{\xi'^t_l\}$, Y'_l, ϖ'_l, ρ'_l, $\nu'_{L,l}$, $T'_{l,\epsilon}$, $\omega'_{\mathrm{b},l}$, η'_l, $g'_{\mathrm{b},l}$ and $g'_{\mathrm{c},l}$, deleting the subscript "L". In the same way, we have M'^1_l, $\mathcal{F}'^1_l$ and $T'^1_{l,\epsilon} = T'_\epsilon \cap M'^1_l \equiv \mathring{T}'_{l,\epsilon}$.

Next, we delete "l" from this notation and use boldface for the corresponding disjoint unions or combinations for all l, obtaining $\boldsymbol{M}'$, $\boldsymbol{\mathcal{F}}'$, $\boldsymbol{\varpi}'$, $\boldsymbol{\rho}'$, $\boldsymbol{\nu}'$, $\boldsymbol{T}'_\epsilon$, $\boldsymbol{\omega}'_{\mathrm{b}}$, $\boldsymbol{\eta}'$, $\boldsymbol{g}'_{\mathrm{b}}$ and $\boldsymbol{g}'_{\mathrm{c}}$.

On the other hand, $\varpi_L : T_{L,\epsilon} \to L$ induces a collar neighborhood $\varpi_{L,l} : T_{L,l,\epsilon} \to L$ of the boundary component L of M_l, and the identity $T_{L,\epsilon} \equiv T'_{L,\epsilon}$ induces an identity $T_{L,l,\epsilon} \equiv T'_{L,l,\epsilon}$, and we have $\varpi_{L,l} \equiv \varpi'_{L,l}$ and $\mathcal{F}_l \equiv \mathcal{F}'_{L,l}$ on $T_{L,l,\epsilon} \equiv T'_{L,l,\epsilon}$. Moreover $T'^1_{L,l,\epsilon} \equiv T^1_{L,l,\epsilon} := T_{L,\epsilon} \cap M^1_l \equiv \mathring{T}_{L,l,\epsilon}$ and $T'^1_{L,l,\epsilon} \equiv T^1_{l,\epsilon} := T_\epsilon \cap M^1_l \equiv \mathring{T}_{L,l,\epsilon}$.

The combination of the maps $\varpi_{L,l}$, with L running in $\pi_0 M_l$, is a collar neighborhood $\varpi_l \equiv \varpi'_l : T_{l,\epsilon} \equiv T'_{l,\epsilon} \to \partial M_l \equiv \partial M'_l$ of the boundary in M_l, where $\mathcal{F}_l \equiv \mathcal{F}'_l$. In turn, the combination of the maps ϖ_l is a collar neighborhood

$$\boldsymbol{\varpi} \equiv \boldsymbol{\varpi}' : \boldsymbol{T}_\epsilon := \bigsqcup_l T_{l,\epsilon} \equiv \boldsymbol{T}'_\epsilon \to \partial \boldsymbol{M} \equiv M^0 \sqcup M^0$$

of the boundary in $\boldsymbol{M}$, and we have $\boldsymbol{\mathcal{F}} \equiv \boldsymbol{\mathcal{F}}'$ on $\boldsymbol{T}_\epsilon \equiv \boldsymbol{T}'_\epsilon$.

4.3.3 Globalization

For fixed $0 < \epsilon < \epsilon_0$ small enough, we can construct the following objects with standard arguments, using a partition of unity subordinated to the open cover $\{T_{\epsilon_0}, M \setminus \overline{T_\epsilon}\}$ of M:

(E) For any $A' \in \mathfrak{X}_{\mathrm{com}}(M', \mathcal{F}')$ with $\overline{A'} = \overline{Y'}$, there is some $A \in \mathfrak{X}_{\mathrm{com}}(M, \mathcal{F})$ with $\overline{A} = \overline{Z}$, $A \equiv A'$ on $T_\epsilon \equiv T'_\epsilon$ and $A = Z$ on $M \setminus T_{\epsilon_0}$. Moreover A induces a vector field $A_l \in \mathfrak{X}(M_l, \mathcal{F}_l)$ (Sect. 4.1.2), whose restriction to $\mathring{M}_l \equiv M^1_l$ is denoted in the same way. In particular, this applies to $Y' \in \mathfrak{X}_{\mathrm{com}}(M', \mathcal{F}')$, obtaining $Y \in \mathfrak{X}(M, \mathcal{F})$ with flow $\xi = \{\xi^t\}$ and $Y_l \in \mathfrak{X}(M_l, \mathcal{F}_l)$ with flow $\xi_l = \{\xi_l^t\}$. We have $\mathrm{Fix}(\xi) = M^0$, and the orbits of ξ agree with the fibers of ϖ on $T_\epsilon \cap M^1$. Thus ξ has no closed orbit in $T_\epsilon \cap M^1$.

(F) Some $Z' \in \mathfrak{X}_{\mathrm{com}}(M', \mathcal{F}')$, with flow $\phi' = \{\phi'^t\}$, such that $\overline{Z'} = \overline{Y'}$, $Z' \equiv Z$ on $T_\epsilon \equiv T'_\epsilon$, and $Z' = Y'$ on $M' \setminus T'_{\epsilon_0}$. This Z' induces vector fields $Z'_{L,l} \in \mathfrak{X}(M'_{L,l}, \mathcal{F}'_{L,l})$ with flow $\phi'_{L,l} = \{\phi'^t_{L,l}\}$, and $Z'_l \in \mathfrak{X}(M'_l, \mathcal{F}'_l)$ with flow $\phi'_l = \{\phi'^t_l\}$.

(G) A bundle-like metric $g_{\mathrm{b},l}$ of every $\mathcal{F}^1_l \equiv \mathring{\mathcal{F}}_l$ on $M^1_l \equiv \mathring{M}_l$ such that $g_{\mathrm{b},l} \equiv g'_{\mathrm{b},l}$ on $T^1_{l,\epsilon} \equiv T'^1_{l,\epsilon}$. Thus $g_{\mathrm{b},l}$ is the restriction to $\mathring{M}_l$ of a b-metric on M_l, also denoted by $g_{\mathrm{b},l}$. Let $\omega_{\mathrm{b},l}$ be the $g_{\mathrm{b},l}$-transverse volume form, defining the transverse orientation given by $\overline{Z_l}$; thus $\omega_{\mathrm{b},l} \equiv \omega'_{\mathrm{b},l}$ on $T^1_{l,\epsilon} \equiv T'^1_{l,\epsilon}$. Since $\omega'_{\mathrm{b},l}(Y'_l) = \omega'_{\mathrm{b},l}(Z'_l) = 1$ (Sect. 4.2.5), we can assume $\omega_{\mathrm{b},l}(Y_l) = \omega_{\mathrm{b},l}(Z_l) = 1$.

(H) A Riemannian metric $g_{\mathrm{c},l}$ on every $M^1_l \equiv \mathring{M}_l$ such that $g_{\mathrm{c},l} \equiv g'_{\mathrm{c},l}$ on $T^1_{l,\epsilon} \equiv T'^1_{l,\epsilon}$. Thus $g_{\mathrm{c},l}$ is the restriction to $\mathring{M}_l$ of a b-metric on M_l, also denoted by $g_{\mathrm{c},l}$, and the b-metric $\varkappa_L^2 g_{\mathrm{c},l}$ is exact and cylindrical around every boundary component L of M_l.

(I) A Riemannian metric g_M on M such that $g_M \equiv g_{M'}$ on $T_\epsilon \equiv T'_\epsilon$, $g_M = g_{\mathrm{b},l}$ on every $M^1_l \setminus T_{\epsilon_0}$, and g_M defines the same orthogonal complement of $T\mathcal{F}$ as $g_{\mathrm{b},l}$ on every M^1_l. We consider the bigrading of ΛM defined by the g_M-orthogonal complement of the leaves (Sect. 3.2).

(J) A leafwise Riemannian metric $g_{\mathcal{F}}$ of $\mathcal{F}$ such that $g_{\mathcal{F}} \equiv g_{\mathcal{F}'}$ on $T_\epsilon \equiv T'_\epsilon$. We can assume it is induced by g_M on M, and by $g_{\mathrm{b},l}$ and $g_{\mathrm{c},l}$ on every M^1_l. It induces a leafwise metric $g_{\mathcal{F}_l}$ for every $\mathcal{F}_l$.

(K) Differential forms, $\omega \in C^\infty(M; \Lambda^{1,0})$ and $\eta \in C^\infty(M; \Lambda^{0,1})$, such that ω is the transverse volume form of $\mathcal{F}$ with respect to g_M, and $d\omega = \eta \wedge \omega$. Thus $\ker \omega = T\mathcal{F}$, $\eta = 0$ on $M \setminus T_{\epsilon_0}$, and they extend the forms ω and η we had on T_ϵ. For every $L \in \pi_0 M^0$, we may use the notation $\eta_L = \eta|_L$ and $\tilde\eta_L = \pi_L^* \eta_L = d_{\tilde{\mathcal{F}}} F_L$ for some $F_L \in C^\infty(\widetilde{L})$. Moreover, $\eta = \eta_0$ on T_ϵ with the notation of Sect. 3.3.1 because this is true for every $\mathcal{F}'_L$.

(L) A defining function $\rho \equiv \rho'$ of M^0 in $T_{\epsilon_0} \equiv T'_{\epsilon_0}$.

(M) A boundary-defining function $\rho = \rho_l$ on every M_l such that $\rho_l \equiv \rho'_l$ on $T_{l,\epsilon} \equiv T'_{l,\epsilon}$, and $\rho_l = 1$ on $M^1_l \setminus T^1_{l,\epsilon_0}$. The level hypersurfaces of ρ_l in $T^1_{l,\epsilon}$ are totally geodesic with respect to $g_{\mathrm{c},l}$. Let $\nu = \nu_l$ be the unique smooth trivialization of ${}_+N\partial M_l$ with $d\rho_l(\nu_l) = 1$ (Sect. 2.5.1). Thus $\nu_l \equiv \nu'_l$ via $T_{l,\epsilon} \equiv T'_{l,\epsilon}$.

From Proposition 4.2.1 to 4.2.3, 4.2.12 and 4.2.15, it easily follows that $\mathcal{F}^1_l$ is of bounded geometry, $(M_l, g_{\mathrm{b},l})$ satisfies the properties (A) and (B) of Sect. 2.5.18, $d(\ln \rho_l) \in C^\infty_{\mathrm{ub}}(\mathring{M}_l; T^*\mathring{M}_l)$, and $Y_l \in \mathfrak{X}_{\mathrm{ub}}(M^1_l, \mathcal{F}^1_l)$ with respect to $g_{\mathrm{b},l}$, and we can assume $Z'_{L,l} \in \mathfrak{X}_{\mathrm{ub}}(M'^1_{L,l}, \mathcal{F}'^1_{L,l})$ with respect to $g'_{\mathrm{b},L,l}$. So $Z_l \in \mathfrak{X}_{\mathrm{ub}}(M^1_l, \mathcal{F}^1_l)$ with respect to $g_{\mathrm{b},l}$. By Proposition 4.2.11 and since $M^1_l \setminus T^1_{l,\epsilon}$ is compact, we also get that the metrics $g_{\mathrm{b},l}$ and $g_{\mathrm{c},l}$ are quasi-isometric on M^1_l; this also follows because both of these metrics are restrictions to $\mathring{M}_l$ of b-metrics on the compact manifold with boundary M_l.

By (4.2.16), we have $\omega = \operatorname{sign}(\omega(Z_l))\,\rho_l\omega_{\mathrm{b},l}$ on $\mathring{M}_l \cap \boldsymbol{T}_\epsilon \equiv M_l^1 \cap T_\epsilon$. This equality is also true on $M_l^1 \setminus T_{\epsilon_0}$, where $g_M = g_{\mathrm{b},l}$ and $\rho_l = 1$. Indeed, we can choose ρ_l so that this equality holds on the whole of $\mathring{M}_l \equiv M_l^1$. So

$$d\omega = \operatorname{sign}(\omega(Z_l))\,d\rho_l \wedge \omega_{\mathrm{b},l} = d\rho_l \wedge \rho_l^{-1}\omega = d(\ln \rho_l) \wedge \omega$$

on $\mathring{M}_l \equiv M_l^1$, yielding

$$\eta_0 = d_{0,1}(\ln \rho_l) \equiv d_{\mathcal{F}_l}(\ln \rho_l)\,. \tag{4.3.1}$$

Taking combinations of the above objects on the manifolds M_l, we get a boundary-defining function $\boldsymbol{\rho}$ on $\boldsymbol{M}$, a trivialization $\boldsymbol{\nu}$ of $_+N\partial\boldsymbol{M}$, real 1-forms $\boldsymbol{\omega}_{\mathrm{b}}$ and $\boldsymbol{\eta}$, and b-metrics $\boldsymbol{g}_{\mathrm{b}}$ and $\boldsymbol{g}_{\mathrm{c}}$. They agree with $\boldsymbol{\rho}'$, $\boldsymbol{\nu}'$, $\boldsymbol{\omega}_{\mathrm{b}}'$, $\boldsymbol{\eta}'$, $\boldsymbol{g}_{\mathrm{b}}'$ and $\boldsymbol{g}_{\mathrm{c}}'$ on $\boldsymbol{T}_\epsilon \equiv \boldsymbol{T}_\epsilon'$.

4.3.4 The Components of M^1

Recall that every $\mathcal{F}_l^1 \equiv \mathring{\mathcal{F}}_l$ on $M_l^1 \equiv \mathring{M}_l$ is a transversely complete $\mathbb{R}$-Lie foliation, where this transverse structure is defined by $Z_l \in \mathfrak{X}_{\mathrm{com}}(M_l^1, \mathcal{F}_l^1)$. Of course, the transverse orientation of $\mathcal{F}_l^1$ defined by Z_l may not agree with the original transverse orientation of $\mathcal{F}$.

The Fedida's description of $\mathcal{F}_l^1$ is given by a regular covering $\pi_l : \widetilde{M}_l^1 \to M_l^1$ with group of deck transformations Γ_l, a holonomy monomorphism $h_l : \Gamma_l \to \mathbb{R}$ and a developing map $D_l : \widetilde{M}_l^1 \to \mathbb{R}$ (Sects. 3.1.9 and 3.1.11). Note that Γ_l has finite rank because $M_l^1 \equiv \mathring{M}_l$ and M_l is compact. Recall that the action of any $\gamma \in \Gamma_l$ on $\widetilde{M}_l^1$ is denoted by $\tilde{p} \mapsto \gamma \cdot \tilde{p}$ or by T_γ.

Let Y_l and $\xi_l = \{\xi_l^t\}$ denote the restrictions of Y and ξ to every M_l^1. Let $\widetilde{\mathcal{F}}_l^1$, $\widetilde{Y}_l$, $\widetilde{Z}_l$, $\tilde{\xi}_l = \{\tilde{\xi}_l^t\}$ and $\tilde{\phi}_l = \{\tilde{\phi}_l^t\}$ be the lifts to $\widetilde{M}_l^1$ of $\mathcal{F}_l^1$, Z_l and ϕ_l, respectively. Recall from Sect. 3.4.7 that $\widetilde{Z}_l$ is Γ_l-invariant and D_l-projectable, and $\tilde{\phi}_l$ is Γ_l-equivariant. Moreover we can assume $D_{l*}\widetilde{Z}_l = \partial_x$, where x denotes the canonical global coordinate of $\mathbb{R}$, and therefore $\tilde{\phi}_l$ corresponds via D_l to the flow $\bar{\phi}_l$ on $\mathbb{R}$ defined by $\bar{\phi}_l^t(x) = t + x$. So D_l restricts to diffeomorphisms between the orbits of $\tilde{\phi}_l$ and $\mathbb{R}$.

Proposition 4.3.1 *Given any leaf L_l of $\mathcal{F}_l^1$, there is a left action of Γ_l on L_l and there is an identity $\widetilde{M}_l^1 \equiv \mathbb{R} \times L_l$ such that:*

(i) *D_l is the left-factor projection;*
(ii) *$\widetilde{Y}_l \equiv (\partial_x, 0)$ and $\tilde{\xi}_l^t(x, y) = (t + x, y)$;*
(iii) *the action of Γ_l on $\widetilde{M}_l^1$ is given by $\gamma \cdot (x, y) = (h_l(\gamma) + x, \gamma \cdot y)$; and*

(iv) *there is some compact* $K_l \subset M_l^1$ *so that, if* $\gamma \cdot y = y$ *for some* $\gamma \in \Gamma_l$ *and* $y \in L_l \setminus K_l$, *then* $\gamma = e$.

Proof Since $\widetilde{Y}_l$ is projectable by D_l to ∂_x because $\widetilde{\widetilde{Y}}_l = \widetilde{\widetilde{Z}}_l$, it follows that D_l also restricts to diffeomorphisms of the $\tilde{\xi}_l$-orbits to $\mathbb{R}$. So, given any leaf $\widetilde{L}_l$ of $\widetilde{\mathcal{F}}_l^1$ over L_l, we get $\widetilde{M}_l^1 \equiv \mathbb{R} \times \widetilde{L}_l \equiv \mathbb{R} \times L_l$ such that (i) and (ii) hold.

The action of every $\gamma \in \Gamma_l$ on $(x, y) \in \mathbb{R} \times L_l \equiv \widetilde{M}_l^1$ can be written as $\gamma \cdot (x, y) = (h_l(\gamma) + x, T_\gamma(x, y))$ for some smooth map $T_\gamma : \mathbb{R} \times L_l \to L_l$. Then, since the flow $\tilde{\xi}_l^t$ is Γ_l-equivariant, it easily follows that $T_\gamma(x, y) = T_\gamma(t + x, y)$. So $T_\gamma(x, y)$ is independent of x, and therefore it can be written as $\gamma \cdot y$. It is easy to check that this defines a left Γ_l-action on L_l, and (iii) follows.

Let us prove (iv). If $\gamma \cdot y = y$ for some $\gamma \in \Gamma_l \setminus \{e\}$ and $y \in L_l$, then we easily compute $\gamma \cdot \tilde{\xi}^t(x, y) = \tilde{\xi}^{h_l(\gamma)+t}(x, y)$ for all $x, t \in \mathbb{R}$. Thus the $\tilde{\xi}^t$-orbit of (x, y) is invariant by the action of γ, and therefore the ξ^t-orbit of $[x, y]$ is closed because $\gamma \neq e$. Since $Y \equiv Y'$ on $T_\epsilon \equiv T_\epsilon'$, it follows that $y \in L_l \setminus T_\epsilon$, and $M_l^1 \setminus T_\epsilon$ is compact in M_l^1. □

Remark 4.3.2 In Proposition 4.3.1, the projection $L_l \to \Gamma_l \backslash L_l$ may not be a covering map, and therefore $(M_l^1, \mathcal{F}_l^1)$ may not be given by a suspension. According to its proof, a point $y \in L_l$ is fixed by some $\gamma \in \Gamma_l \setminus \{e\}$ just when $\mathbb{R} \times \{y\}$ projects to a closed orbit of ξ^t in M_l^1 whose group of periods contains $h_l(\gamma)$.

According to Proposition 4.3.1, we may use the notation $[x, y] = \pi_l(x, y) \in M_l^1$ for $(x, y) \in \mathbb{R} \times L_l \equiv \widetilde{M}_l^1$, and the action of every $\gamma \in \Gamma_l$ on L_l may be also denoted by T_γ. Like in (4.2.1) and (4.2.2), we get

$$\tilde{\phi}_l^t(x, y) = (t + x, \tilde{\phi}_{l,x}^t(y)) , \quad \widetilde{Z}_l = (\partial_x, \widetilde{Z}_{l,x}) , \tag{4.3.2}$$

for some smooth families, $\{\, \tilde{\phi}_{l,x}^t \mid x, t \in \mathbb{R} \,\} \subset \mathrm{Diffeo}(L_l)$ and $\{\, \widetilde{Z}_{l,x} \mid x \in \mathbb{R} \,\} \subset \mathfrak{X}(L_l)$, such that

$$T_\gamma \tilde{\phi}_{l,x}^t = \tilde{\phi}_{l,h_l(\gamma)+x}^t T_\gamma , \quad T_{\gamma *} \widetilde{Z}_x = \widetilde{Z}_{h_l(\gamma)+x} .$$

Let c be a closed orbit of ϕ_l with period t_0, and let $p = [x, y] \in c$ and $\tilde{p} = (x, y) \in \widetilde{M}_l^1 \equiv \mathbb{R} \times L_l$. Then $k = t_0/\ell(c) \in \mathbb{Z}$ and there is a unique $\gamma_0 \in \Gamma_l$ such that $\tilde{\phi}_l^{t_0}(\tilde{p}) = \gamma_0 \cdot \tilde{p}$. Using (4.3.2) and Proposition 4.3.1 (iii), it easily follows that $t_0 = h_l(\gamma_0)$ and $\tilde{\phi}_{l,x}^{t_0}(y) = \gamma_0 \cdot y$; i.e., y is a fixed point of the diffeomorphism $T_{\gamma_0}^{-1} \tilde{\phi}_{l,x}^{t_0}$ of L_l. Moreover y is simple if and only if c is simple, and, in this case, $\epsilon_y(T_{\gamma_0} \tilde{\phi}_{l,x}^{t_0}) = \epsilon_c(k, \phi) = \epsilon_c(k)$.

We have $\tilde{\omega}_{b,l} := \pi_l^* \omega_{b,l} = D_l^* dx \equiv dx$ because $D_{l*} \widetilde{Z}_l = \partial_x$ and $\omega_{b,l}(Z_l) = 1$ (Sect. 4.3.3).

4.3.5 Metric Properties of the Components of M^1

With the notation of Sects. 4.3.2 to 4.3.4, for leaves $L \subset M^0 \cap \overline{M_l^1}$ and $0 < \epsilon' \le \epsilon$, the open subsets

$$\widetilde{T}'^1_{L,l,\epsilon'} = \pi^{-1}_{M'_{L,l}}(T'^1_{L,l,\epsilon'}) \subset \widetilde{M}'^1_{L,l} , \quad \widetilde{T}^1_{L,l,\epsilon'} = \pi_l^{-1}(T^1_{L,l,\epsilon'}) \subset \widetilde{M}^1_l ,$$

are invariant by Γ_L and Γ_l, respectively. Let $\tilde{\rho}_l = \pi_l^* \rho_l$ and $M^1_{l,\epsilon'} = M^1_l \setminus T_{l,\epsilon'}$, which is a connected compact smooth submanifold with boundary of M^1_l. Then $\widetilde{T}^1_{l,\epsilon'} := \pi_l^{-1}(T^1_{l,\epsilon'}) = \{\tilde{\rho}_l < \epsilon'\}$ is a Γ_l-invariant open subspace of $\widetilde{M}^1_l$, and $\pi_l : \widetilde{M}^1_{l,\epsilon'} := \widetilde{M}^1_l \setminus \widetilde{T}^1_{l,\epsilon'} \to M^1_{l,\epsilon'}$ is a regular Γ_l-covering.

Let d_l denote the length-metric on M^1_l defined by $g_{\mathrm{b},l}$. Let $\tilde{g}_{\mathrm{b},l}$ and $\tilde{g}_{\mathrm{c},l}$ be the lifts to $\widetilde{M}^1_l$ of $g_{\mathrm{b},l}$ and $g_{\mathrm{c},l}$. Both of them induce the same leafwise metric $g_{\widetilde{\mathcal{F}}_l}$ of $\widetilde{\mathcal{F}}_l$, which is the lift of $g_{\mathcal{F}_l}$. Let $\tilde{d}_l$ and $\tilde{d}_{l,\epsilon'}$ denote the length-metrics on $\widetilde{M}^1_l$ and $\widetilde{M}^1_{l,\epsilon'}$ defined by $\tilde{g}_{\mathrm{b},l}$. Similarly, let $\tilde{d}_{\mathrm{c},l}$ and $\tilde{d}_{\mathrm{c},l,\epsilon'}$ be the length-metrics on $\widetilde{M}^1_l$ and $\widetilde{M}^1_{l,\epsilon'}$ defined by $\tilde{g}_{\mathrm{c},l}$. Since $g_{\mathrm{b},l}$ and $g_{\mathrm{c},l}$ are quasi-isometric (Sect. 4.3.3), the metrics $\tilde{g}_{\mathrm{b},l}$ and $\tilde{g}_{\mathrm{c},l}$ are also quasi-isometric. Therefore there is some $C_1 \ge 1$ such that, for all $\tilde{p}, \tilde{q} \in \widetilde{M}^1_l$,

$$C_1^{-1} \tilde{d}_l(\tilde{p}, \tilde{q}) \le \tilde{d}_{c,l}(\tilde{p}, \tilde{q}) \le C_1 \tilde{d}_l(\tilde{p}, \tilde{q}) . \tag{4.3.3}$$

On the other hand, $\tilde{d}_l \le \tilde{d}_{l,\epsilon'}$ on $\widetilde{M}^1_{l,\epsilon'}$.

Lemma 4.3.3 *We have* $\tilde{d}_{c,l} = \tilde{d}_{c,l,\epsilon'}$ *on* $\widetilde{M}^1_{l,\epsilon'}$.

Proof It is enough to show that any $\tilde{g}_{\mathrm{c},l}$-geodesic segment with end-points in $\widetilde{M}^1_{l,\epsilon'}$ is contained in $\widetilde{M}^1_{l,\epsilon'}$ ($\widetilde{M}^1_{l,\epsilon'}$ is $\tilde{g}_{\mathrm{c},l}$-convex). This follows easily using that the level hypersurfaces of $\tilde{\rho}_l$ are $\tilde{g}_{\mathrm{c},l}$-totally geodesic because the level hypersurfaces of ρ_l in $T^1_{l,\epsilon}$ are $g_{\mathrm{c},l}$-totally geodesic ((M) of Sect. 4.3.3). □

Let $|\cdot| = |\cdot|_l : \Gamma_l \to \mathbb{N}_0$ and $|\cdot| = |\cdot|_L : \Gamma_L \to \mathbb{N}_0$ be the word length functions induced by any choice of finite sets of generators of Γ_l and Γ_L. By the compactness of $M^1_{l,\epsilon'}$, there is some $C_2 = C_2(\epsilon') \ge 1$ such that, for all $\gamma \in \Gamma_l$ and $\tilde{p} \in \widetilde{M}^1_{l,\epsilon'}$,

$$C_2^{-1} |\gamma| \le \tilde{d}_{l,\epsilon'}(\tilde{p}, \gamma \cdot \tilde{p}) \le C_2 |\gamma| . \tag{4.3.4}$$

Since $g_{\mathrm{b},l}$ and $g_{\mathrm{c},l}$ are quasi-isometric on M^1_l, it follows from (4.3.3), (4.3.4) and Lemma 4.3.3 that there is some $C_3 = C_3(\epsilon') \ge 1$ such that, for all $\gamma \in \Gamma_l$ and $\tilde{p} \in \widetilde{M}^1_{l,\epsilon'}$,

$$C_3^{-1} |\gamma| \le \tilde{d}_l(\tilde{p}, \gamma \cdot \tilde{p}) \le C_3 |\gamma| . \tag{4.3.5}$$

Remark 4.3.4 For any leaf $L \subset M^0 \cap \overline{M_l^1}$, the given descriptions of $\mathcal{F}$ on $T_{L,\epsilon}^1$ and M_l^1 have the following relation, whose proof is omitted because it will not be used. There is a monomorphism $H_{L,l} : \Gamma_L \to \Gamma_l$ such that, for every connected component $\widetilde{T}_{L,l,\epsilon',0}^1$ of $\widetilde{T}_{L,l,\epsilon'}^1$, the identity $T_{L,l,\epsilon'}^{\prime 1} \equiv T_{L,l,\epsilon'}^1$ can be lifted to an $H_{L,l}$-equivariant identity $\widetilde{T}_{L,l,\epsilon'}^{\prime 1} \equiv \widetilde{T}_{L,l,\epsilon',0}^1$, which is locally equivariant with respect to the local flows defined by $\tilde{\xi}_L'$ on $\widetilde{T}_{L,\epsilon',\pm}^{\prime 1}$ and $\tilde{\xi}_l$ on $\widetilde{T}_{L,l,\epsilon',0}^1$, and so that D_l corresponds to $D_{L,l}'$.

Chapter 5
Conormal Leafwise Reduced Cohomology

5.1 Conormal Sequence of Leafwise Currents

Let $\mathcal{F}$ be a transversely orientable smooth foliation of codimension one on a closed manifold M satisfying the conditions (C) and (D) of Sect. 4.1.2. Then M^0 is determined by $\mathcal{F}$ in the cases (d)–(f) of Sect. 4.1.2, whereas M^0 must be also given in the case (c). The compactness condition on M is assumed for the sake of simplicity, but all concepts, results and arguments of this section have straightforward extensions to the case where M is not compact and M^0 is compact, using compactly supported versions or versions without support restrictions of the spaces of leafwise currents that will be considered. The compactly supported versions, in the non-compact case, will be used in the arguments.

Since $\mathrm{Diff}^1(\mathcal{F}; \Lambda\mathcal{F}) \subset \mathrm{Diff}^1(M, M^0; \Lambda\mathcal{F})$, the graded LCHS

$$I(\mathcal{F}) = I\Lambda^\bullet(\mathcal{F}) := I(M, M^0; \Lambda\mathcal{F})$$

becomes a topological complex with $d_\mathcal{F}$ (Sects. 2.2.7 and 3.2.1). If we take coefficients in some leafwise flat vector bundle E, then the notation $I(\mathcal{F}; E)$ will be used, and all other notations will be modified in the same way. We may even consider $I(\mathcal{F}; E)$ for an arbitrary vector bundle E, missing the leafwise differential map $d_\mathcal{F}$.

The topological complex $(I(\mathcal{F}), d_\mathcal{F})$ produces the *conormal leafwise cohomology* and *conormal leafwise reduced cohomology* of $\mathcal{F}$ (or of $(\mathcal{F}, M^0)$ when M^0 is not determined by $\mathcal{F}$), denoted by $H^\bullet I(\mathcal{F})$ and $\bar{H}^\bullet I(\mathcal{F})$, which are LCSs (Sect. 2.1.13). The image and kernel of $d_\mathcal{F}$ in $I(\mathcal{F})$ are denoted by $BI(\mathcal{F})$ and $ZI(\mathcal{F})$, and we write $\bar{B}I(\mathcal{F}) = \overline{BI(\mathcal{F})}$.

The LCHSs

$$I^{(s)}(\mathcal{F}) = I^{(s)}\Lambda^\bullet(\mathcal{F}) := I^{(s)}(M, M^0; \Lambda\mathcal{F}) \quad (s \in \mathbb{R})$$

J. A. Álvarez López et al., *A Trace Formula for Foliated Flows*, Lecture Notes in Mathematics 2387, https://doi.org/10.1007/978-3-032-15413-2_5

also become topological complexes with $d_{\mathcal{F}}$ (Sect. 2.2.7). The notation $H^\bullet I^{(s)}(\mathcal{F})$, $\bar{H}^\bullet I^{(s)}(\mathcal{F})$, $BI^{(s)}(\mathcal{F})$, $ZI^{(s)}(\mathcal{F})$ and $\bar{B}I^{(s)}(\mathcal{F})$ is used as before. We have continuous inclusion maps (Sect. 2.2.2)

$$j_s : I^{(s)}(\mathcal{F}) \hookrightarrow I(\mathcal{F}) , \quad j_{s,s'} : I^{(s)}(\mathcal{F}) \hookrightarrow I^{(s')}(\mathcal{F}) \quad (s' \le s) . \tag{5.1.1}$$

The induced homomorphism in cohomology and reduced cohomology are denoted by j_{s*}, $j_{s,s'*}$, $\bar{j}_{s*}$ and $\bar{j}_{s,s'*}$. The homomorphism $j_{s,s'*}$ and $\bar{j}_{s*}$ form inductive spectra, giving rise to inductive limits as $s \downarrow -\infty$. The maps j_{s*} and $\bar{j}_{s*}$ induce canonical continuous linear isomorphisms (Sect. 5.2),

$$\begin{cases} \tilde{j}_* := \varinjlim j_{s*} : \widetilde{H}^\bullet I(\mathcal{F}) := \varinjlim H^\bullet I^{(s)}(\mathcal{F}) \xrightarrow{\cong} H^\bullet I(\mathcal{F}) , \\ \hat{j}_* := \varinjlim \bar{j}_{s*} : \widehat{H}^\bullet I(\mathcal{F}) := \varinjlim \bar{H}^\bullet I^{(s)}(\mathcal{F}) \xrightarrow{\cong} \bar{H}^\bullet I(\mathcal{F}) . \end{cases} \tag{5.1.2}$$

The canonical maps of the steps to the inductive limits are denoted by

$$\tilde{j}_{s*} : H^\bullet I^{(s)}(\mathcal{F}) \to \widetilde{H}^\bullet I(\mathcal{F}) , \quad \hat{j}_{s*} : \bar{H}^\bullet I^{(s)}(\mathcal{F}) \to \widehat{H}^\bullet I(\mathcal{F}) .$$

The graded LCHSs,

$$J(\mathcal{F}) = J\Lambda^\bullet(\mathcal{F}) := J(M, M^0; \Lambda\mathcal{F}) , \quad K(\mathcal{F}) = K\Lambda^\bullet(\mathcal{F}) := K(M, M^0; \Lambda\mathcal{F}) ,$$

also become topological complexes with $d_{\mathcal{F}}$ (Sect. 2.6.13). The above kind of notation is also used for the induced spaces: $BJ(\mathcal{F})$, $ZJ(\mathcal{F})$, $\bar{B}J(\mathcal{F})$ and $H^\bullet J(\mathcal{F})$, $\bar{H}^\bullet J(\mathcal{F})$, and the same for $K(\mathcal{F})$.

Similarly, we have topological complexes $J^{(s)}(\mathcal{F})$, $J^m(\mathcal{F})$ and $K^{(s)}(\mathcal{F})$ ($s, m \in \mathbb{R}$) with $d_{\mathcal{F}}$ (Sect. 2.6.13). The analogs of the inclusion maps (5.1.1) for the spaces $J^{(s)}(\mathcal{F})$ and $K^{(s)}(\mathcal{F})$ are denoted in the same way. The induced homomorphisms in cohomology and reduced cohomology form inductive spectra. Their inductive limits, denoted by $\widetilde{H}^\bullet K(\mathcal{F})$, $\widehat{H}^\bullet K(\mathcal{F})$, $\widetilde{H}^\bullet J(\mathcal{F})$ and $\widehat{H}^\bullet J(\mathcal{F})$, satisfy analogs of (5.1.2) (proved with the same arguments). In fact, in the case of $K(\mathcal{F})$, we have canonical TVS-identities (Corollary 5.3.2),

$$\begin{cases} H^\bullet K(\mathcal{F}) \equiv \bar{H}^\bullet K(\mathcal{F}) , \quad H^\bullet K^{(s)}(\mathcal{F}) \equiv \bar{H}^\bullet K^{(s)}(\mathcal{F}) , \\ \widetilde{H}^\bullet K^{(s)}(\mathcal{F}) \equiv H^\bullet K(\mathcal{F}) , \quad \widehat{H}^\bullet K^{(s)}(\mathcal{F}) \equiv \bar{H}^\bullet K(\mathcal{F}) . \end{cases} \tag{5.1.3}$$

There are also continuous inclusion maps (Sect. 2.6.7)

$$\begin{cases} j_m : J^m(\mathcal{F}) \hookrightarrow J(\mathcal{F}) , \quad j_{m,m'} : J^m(\mathcal{F}) \hookrightarrow J^{m'}(\mathcal{F}) \quad (m' \le m) , \\ j_{s,m} : J^{(s)}(\mathcal{F}) \hookrightarrow J^m(\mathcal{F}) \quad (m < s - n/2 - 1) , \\ j_{m,s} : J^m(\mathcal{F}) \hookrightarrow J^{(s)}(\mathcal{F}) \quad (s \le m, 0) , \end{cases} \tag{5.1.4}$$

denoted like in (5.1.1) with some abuse of notation. The homomorphisms induced by the maps $j_{m,m'}$ in cohomology and reduced cohomology form inductive spectra whose inductive limits as $m \downarrow -\infty$ agree with the previous ones for $J(\mathcal{F})$, and the maps j_m induce a continuous linear isomorphism analogous to (5.1.2).

There are similar constructions for the spaces of the symbol-order filtration of $I(\mathcal{F})$ and $K(\mathcal{F})$, with similar properties, but they will not be used here.

The *leafwise conormal exact sequence* of $\mathcal{F}$ is the bottom row of (2.6.41) with $\Lambda\mathcal{F}$,

$$0 \to K(\mathcal{F}) \xrightarrow{\iota} I(\mathcal{F}) \xrightarrow{R} J(\mathcal{F}) \to 0 . \tag{5.1.5}$$

Besides being exact in the category of continuous linear maps between LCSs, it is compatible with $d_{\mathcal{F}}$. The exactness of the induced sequences,

$$0 \to H^\bullet K(\mathcal{F}) \xrightarrow{\iota_*} H^\bullet I(\mathcal{F}) \xrightarrow{R_*} H^\bullet J(\mathcal{F}) \to 0 , \tag{5.1.6}$$

$$0 \to H^\bullet K(\mathcal{F}) \xrightarrow{\bar\iota_*} \bar H^\bullet I(\mathcal{F}) \xrightarrow{\bar R_*} \bar H^\bullet J(\mathcal{F}) \to 0 , \tag{5.1.7}$$

will be proved in Sect. 5.5; in particular, this shows Theorem 1.3.3.

Concerning notation, the subscript "s" may be added to the notation of cochain maps between the topological complexes $K^{(s)}(\mathcal{F})$, $I^{(s)}(\mathcal{F})$ or $J^{(s)}(\mathcal{F})$, like

$$\iota_s = \iota : K^{(s)}(\mathcal{F}) \to I^{(s)}(\mathcal{F}) , \quad R_s = R : I^{(s)}(\mathcal{F}) \to J^{(s)}(\mathcal{F}) . \tag{5.1.8}$$

The subscript "s" may be also added to the elements of their cohomologies or reduced cohomologies: $[\alpha]_s \in H^\bullet I^{(s)}(\mathcal{F})$ and $\overline{[\alpha]}_s \in \bar H^\bullet I^{(s)}(\mathcal{F})$ for $\alpha \in ZI^{(s)}(\mathcal{F})$.

5.2 Injective Limits in Cohomology and Reduced Cohomology

The purpose of this section is to prove that the maps (5.1.2) are isomorphisms. The details are given for the case of $\bar H^\bullet I(\mathcal{F})$. Some remarks indicate how to modify the arguments to show the simpler case of $H^\bullet I(\mathcal{F})$.

5.2.1 *Injectivity of $\hat\jmath_*$*

Take any element in $\ker \hat\jmath_*$, which is of the form $\hat\jmath_{s*}(\overline{[\alpha]}_s)$ for some $\overline{[\alpha]}_s \in \bar H^\bullet I^{(s)}(\mathcal{F})$. Then there is some net $\varphi_l \in I(\mathcal{F})$ such that $\alpha = \lim_l d_{\mathcal{F}}\varphi_l$ in $I(\mathcal{F})$. We can assume $\varphi_l \in C^\infty(M; \Lambda\mathcal{F})$ by the density of $C^\infty(M; \Lambda\mathcal{F})$ in $I(\mathcal{F})$ (Sect. 2.2.2). The set $\{\alpha, d_{\mathcal{F}}\varphi_l\}_l$ is compact in $I(\mathcal{F})$. Then $\{\alpha, d_{\mathcal{F}}\beta_l\}_l$ is contained and compact in some step $I^{(s')}(\mathcal{F})$ ($s' \le s$) because $I(\mathcal{F})$ is compactly retractive (Sect. 2.2.2). Thus $\alpha = \lim_l d_{\mathcal{F}}\varphi_l$ in $I^{(s')}(\mathcal{F})$; otherwise, using that $\{\alpha, d_{\mathcal{F}}\varphi_l\}_l$ is compact in $I^{(s')}(\mathcal{F})$, it is easy to find a subnet $d_{\mathcal{F}}\varphi_{l_k}$ convergent to some $\beta \neq \alpha$ in $I^{(s')}(\mathcal{F})$, which contradicts the continuity of $j_{s'} : I^{(s')}(\mathcal{F}) \to I(\mathcal{F})$ and the convergence $d_{\mathcal{F}}\varphi_l \to \alpha$ in $I(\mathcal{F})$. (Indeed, we can assume $d_{\mathcal{F}}\phi_l$ is a sequence because $I^{(s')}(\mathcal{F})$ is a Fréchet space.) So $\overline{[\alpha]}_{s'} = 0$ in $\bar{H}^\bullet I^{(s')}(\mathcal{F})$, and therefore $\hat{j}_{s*}(\overline{[\alpha]}_s) = \hat{j}_{s'*}(\overline{[\alpha]}_{s'}) = 0$.

Remark 5.2.1 To prove injectivity of $\tilde{j}_*$, take some $\tilde{j}_{s*}(\overline{[\alpha]}_s)$ in $\ker \tilde{j}_*$. Now modify the above argument by using cohomology classes, and taking an element $\varphi \in I(\mathcal{F})$ with $d_{\mathcal{F}}\varphi = \alpha$ instead of a net φ_l. Then φ and α are in some step $I^{(s')}(\mathcal{F})$ ($s' \le s$), yielding $[\alpha]_{s'} = 0$ in $H^\bullet I^{(s')}(\mathcal{F})$, and therefore $\hat{j}_{s*}([\alpha]_s) = \hat{j}_{s'*}([\alpha]_{s'}) = 0$.

5.2.2 *Surjectivity of $\hat{j}_*$*

For any $\overline{[\alpha]} \in \bar{H}^\bullet I(\mathcal{F})$, there is some s such that $\alpha \in I^{(s)}(\mathcal{F})$, and therefore $\alpha \in ZI^{(s)}(\mathcal{F})$. Hence the element $\overline{[\alpha]}_s \in \bar{H}I^{(s)}(\mathcal{F})$ is defined, and the element $\hat{j}_{s*}(\overline{[\alpha]}_s) \in \widehat{H}^\bullet I^{(s)}(\mathcal{F})$ is mapped to $\overline{[\alpha]}$ by $\hat{j}_*$.

Remark 5.2.2 To prove the surjectivity of $\tilde{j}_*$, simply modify the argument by using cohomology classes instead of reduced cohomology classes.

5.3 Description of $H^\bullet K(\mathcal{F})$

Consider also the notation of Sect. 4.3.3. For every $z \in \mathbb{C}$, we have the Witten's complex $d_z = d + z\,\eta\wedge$ on $C^\infty(M^0; \Lambda)$, whose cohomology is denoted by $H_z^\bullet(M^0)$ (Sect. 2.9.1). Consider also the trivialization of the flat line bundle $\Omega^z NM^0 = \Omega^z N\mathcal{F}|_{M^0}$ defined by $|\omega|^z$. Then, by (2.9.5) and since $d\omega = \eta \wedge \omega$ ((K) of Sect. 4.3.3),

$$C^{\pm\infty}(M^0; \Lambda \otimes \Omega^z NM^0) \equiv C^{\pm\infty}(M^0; \Lambda) \otimes \mathbb{R}|\omega|^z \equiv C^{\pm\infty}(M^0; \Lambda)\,,$$

$$d \equiv d_z \otimes 1 \equiv d_z\,, \quad H^\bullet(M^0; \Omega^z NM^0) \equiv H_z(M^0)\,.$$

These identities will be applied without further comment. By Reeb's local stability, the following result follows from the case of a suspension foliation, which will be proved in Sect. 5.6.1 (Corollary 5.6.2).

Proposition 5.3.1 *We have identities of topological complexes,*

$$K(\mathcal{F}) \equiv \bigoplus_k C^\infty(M^0; \Lambda) \equiv \bigoplus_k C^\infty(M^0; \Lambda \otimes \Omega^{-k-1} NM^0)\,,$$

$$d_{\mathcal{F}} \equiv \bigoplus_k d_{-k-1} \equiv \bigoplus_k d\,,$$

where k runs in $\mathbb{N}_0$. Moreover the subcomplex $K^{(s)}(\mathcal{F}) \subset K(\mathcal{F})$ corresponds to the finite direct sum with $k < -s - 1/2$.

Corollary 5.3.2 *We have TVS-identities,*

$$H^\bullet K(\mathcal{F}) \equiv \bigoplus_k H^\bullet_{-k-1}(M^0) \equiv \bigoplus_k H^\bullet(M^0, \Omega^{-k-1} N M^0)\,.$$

Moreover $H^\bullet K^{(s)}(\mathcal{F})$ is the topological vector subspace of $H^\bullet K(\mathcal{F})$ given by the finite direct sum with $k < -s - 1/2$. In particular, (5.1.3) *is satisfied.*

Remark 5.3.3 The differential complexes on M^0 used in Proposition 5.3.1 obviously split into direct sums of the same complexes given by leaves $L \subset M^0$. The same applies to their cohomologies in Corollary 5.3.2.

Remark 5.3.4 Like in Proposition 5.3.1, the isomorphism (2.6.24) gives

$$C^{-\infty}_{M^0}(M; \Lambda\mathcal{F}) \equiv \bigoplus_k C^{-\infty}(M^0; \Lambda) \equiv \bigoplus_k C^{-\infty}(M^0; \Lambda \otimes \Omega^{-k-1} N M^0)\,.$$

5.4 Description of $\bar{H}^\bullet J(\mathcal{F})$

With the notation of Sects. 3.1.12 and 4.1.2, by (2.6.30) and (2.6.31), for $m \in \mathbb{R}$,

$$J^m(\mathcal{F}) \cong \rho^m H^\infty_{\mathrm{b}}(\boldsymbol{M}; \Lambda\boldsymbol{\mathcal{F}}) \equiv \rho^{m+\frac{1}{2}} H^\infty(\mathring{\boldsymbol{M}}; \Lambda\mathring{\boldsymbol{\mathcal{F}}})\,, \tag{5.4.1}$$

$$J(\mathcal{F}) \cong \bigcup_m \rho^m H^\infty_{\mathrm{b}}(\boldsymbol{M}; \Lambda\boldsymbol{\mathcal{F}}) = \bigcup_m \rho^m H^\infty(\mathring{\boldsymbol{M}}; \Lambda\mathring{\boldsymbol{\mathcal{F}}})\,, \tag{5.4.2}$$

as topological complexes with $d_{\mathcal{F}}$, $d_{\boldsymbol{\mathcal{F}}}$ or $d_{\mathring{\boldsymbol{\mathcal{F}}}}$, using the b-metric $\boldsymbol{g}$ to define $H^\infty_{\mathrm{b}}(\boldsymbol{M}; \Lambda\boldsymbol{\mathcal{F}})$, and using $\boldsymbol{g}|_{\mathring{\boldsymbol{M}}}$ to define $H^\infty(\mathring{\boldsymbol{M}}; \Lambda\mathring{\boldsymbol{\mathcal{F}}})$.

On the other hand, since $\eta = d_{\boldsymbol{\mathcal{F}}}(\ln\rho)$ on $\mathring{\boldsymbol{M}}$ by (4.3.1), we get isomorphisms of topological complexes,

$$\rho^{-m-\frac{1}{2}} : \left(\rho^{m+\frac{1}{2}} H^\infty(\mathring{\boldsymbol{M}}; \Lambda\mathring{\boldsymbol{\mathcal{F}}}), d_{\mathring{\boldsymbol{\mathcal{F}}}}\right) \xrightarrow{\cong} \left(H^\infty(\mathring{\boldsymbol{M}}; \Lambda\mathring{\boldsymbol{\mathcal{F}}}), d_{\mathring{\boldsymbol{\mathcal{F}}}, m+\frac{1}{2}}\right)\,, \tag{5.4.3}$$

by the leafwise version of (2.9.4) (Sect. 3.5).

By (5.4.1) and (5.4.3), and the analog of (3.4.16) for $\Delta_{\mathring{\mathcal{F}},m+\frac{1}{2}}$ in $H^\infty(\mathring{M};\Lambda\mathring{\mathcal{F}})$ (Sect. 3.5), we get induced TVS-isomorphisms

$$\bar{H}^\bullet J^m(\mathcal{F}) \cong \bar{H}^\bullet\big(\rho^{m+\frac{1}{2}}H^\infty(\mathring{M};\Lambda\mathring{\mathcal{F}}),d_{\mathring{\mathcal{F}}}\big) \tag{5.4.4}$$

$$\cong \bar{H}^\bullet\big(H^\infty(\mathring{M};\Lambda\mathring{\mathcal{F}}),d_{\mathring{\mathcal{F}},m+\frac{1}{2}}\big) \tag{5.4.5}$$

$$\cong \ker\Delta_{\mathring{\mathcal{F}},m+\frac{1}{2}}\ . \tag{5.4.6}$$

By the analog of (5.1.2) for $J(\mathcal{F})$ and (5.4.6), the LCHS $\bar{H}^\bullet J(\mathcal{F})$ is an inductive limit of Hilbertian spaces. The isomorphisms (5.4.4) and (5.4.5) are also true in cohomology.

Theorem 1.3.2 follows from the analog of (5.1.2) for $J(\mathcal{F})$ and (5.4.1)–(5.4.3).

5.5 Short Exact Sequence of Conormal Reduced Cohomology

The goal of this section is to prove Theorem 1.3.3; i.e., the exactness of (5.1.7). Some remarks will indicate how to modify the argument to get also the exactness of (5.1.6). To begin with, we choose appropriate partial extension maps.

5.5.1 Compatibility of the Maps E_m with $d_{\mathcal{F}}$

For $m\in\mathbb{R}$, take $s\in\mathbb{R}$ such that $s=0$ if $m\geq 0$, and $m>s\in\mathbb{Z}^-$ if $m<0$. For fixed $0<\epsilon<1$, using the tubular neighborhood $T:=T_\epsilon$ of M^0 in M (Sect. 4.3.1), consider the continuous inclusions of $I_{\mathrm{c}}^{(s)}(\mathcal{F}|_T)\subset I^{(s)}(\mathcal{F})$ and $J_{\mathrm{c}}^m(\mathcal{F}|_T)\subset J^m(\mathcal{F})$, using the extension by zero. Let $E_{m,T}:J_{\mathrm{c}}^m(\mathcal{F}|_T)\to I_{\mathrm{c}}^{(s)}(\mathcal{F}|_T)$ be the continuous linear partial extension map constructed in the proofs of the compactly supported versions of Proposition 2.5.1 and Corollary 2.6.4 with $\Lambda\mathcal{F}|_T$ (see Remarks 2.5.3 and 2.5.4). By the Reeb's local stability, the following result follows from its case for suspension foliations, which will be proved in Sect. 5.6.3 (Corollary 5.6.6).

Proposition 5.5.1 $E_{m,T}d_{\mathcal{F}}=d_{\mathcal{F}}E_{m,T}$.

Let $\{\lambda,\mu\}$ be a smooth partition of unity of $\mathbb{R}$ subordinated to the open cover $\{(-\epsilon,\epsilon),\mathbb{R}^\times\}$, which induces the smooth partition of unity $\{\lambda(\rho),\mu(\rho)\}$ of M subordinated to the open cover $\{T,M^1\}$, where $\lambda(\rho)$ (resp., $\mu(\rho)$) is extended by 0 (resp., 1) to the whole of M. According to Remarks 2.5.2 and 2.5.3, take the continuous linear partial extension map $E_m:J^m(\mathcal{F})\to I^{(s)}(\mathcal{F})$ defined by

$$E_m\alpha=E_{m,T}(\lambda(\rho)\,\alpha)+\mu(\rho)\,\alpha\ . \tag{5.5.1}$$

Corollary 5.5.2 $E_m d_{\mathcal{F}} = d_{\mathcal{F}} E_m$.

Proof By the version of Corollary 2.5.6 for $E_{m,T}$ (Corollary 2.6.12), and since $d_{\mathcal{F}}\lambda(\rho) = -d_{\mathcal{F}}\mu(\rho)$ is supported in M^1, we get, for $\alpha \in J^m(\mathcal{F})$,

$$\begin{aligned} E_m d_{\mathcal{F}}\alpha &= E_{m,T}(\lambda(\rho)\, d_{\mathcal{F}}\alpha) + \mu(\rho)\, d_{\mathcal{F}}\alpha \\ &= E_{m,T} d_{\mathcal{F}}(\lambda(\rho)\,\alpha) - E_{m,T}(d_{\mathcal{F}}\lambda(\rho) \wedge \alpha) \\ &\quad + d_{\mathcal{F}}(\mu(\rho)\,\alpha) - d_{\mathcal{F}}\mu(\rho) \wedge \alpha \wedge \omega \\ &= d_{\mathcal{F}} E_{m,T}(\lambda(\rho)\,\alpha) - d_{\mathcal{F}}\lambda(\rho) \wedge \alpha \\ &\quad + d_{\mathcal{F}}(\mu(\rho)\,\alpha) - d_{\mathcal{F}}\mu(\rho) \wedge \alpha \\ &= d_{\mathcal{F}} E_{m,T}(\lambda(\rho)\,\alpha) + d_{\mathcal{F}}(\mu(\rho)\,\alpha) = d_{\mathcal{F}} E_m \alpha\ . \end{aligned}$$

□

5.5.2 *The Maps F_m*

For $s \in \mathbb{R}$ and $m < s - n/2 - 1$, we can consider $R : I^{(s)}(\mathcal{F}) \to J^m(\mathcal{F})$ by the analog of (2.5.37) for $J(\mathcal{F})$ (Sect. 2.6.7). Taking $s' = 0$ if $m \geq 0$, and $m > s' \in \mathbb{Z}^-$ if $m < 0$, let $E_m : J^m(\mathcal{F}) \to I^{(s')}(\mathcal{F})$ be defined like in Sect. 5.5.1. We can also consider

$$E_m = E_m\, j_{s,m} : J^{(s)}(\mathcal{F}) \to I^{(s')}(\mathcal{F})\ . \tag{5.5.2}$$

Then define the continuous linear map

$$F_m := 1 - E_m R : I^{(s)}(\mathcal{F}) \to K^{(s')}(\mathcal{F})\ .$$

Note that

$$E_m R_s + \iota_{s'} F_m = j_{s,s'} : I^{(s)}(\mathcal{F}) \to I^{(s')}(\mathcal{F})\ , \tag{5.5.3}$$

$$F_m \iota_s = j_{s,s'} : K^{(s)}(\mathcal{F}) \to K^{(s')}(\mathcal{F})\ . \tag{5.5.4}$$

Moreover, by Corollary 5.5.2,

$$F_m d_{\mathcal{F}} = d_{\mathcal{F}} F_m\ . \tag{5.5.5}$$

Take smaller numbers, $s_1 < s$, $m_1 < m$ and $s'_1 < s'$, satisfying the same inequalities as s, m and s'. Then, with (5.5.2), the version of Proposition 2.5.5 with $\Lambda\mathcal{F}$ (see Remark 2.5.8) gives

$$j_{s',s'_1} E_m = E_{m_1} j_{s,s_1}\ . \tag{5.5.6}$$

Then, using the definition of F_m, we also get

$$j_{s',s'_1} F_m = F_{m_1} j_{s,s_1} . \tag{5.5.7}$$

Remark 5.5.3 According to Remark 6.1.1, we can also define

$$F_m : I^{(s)}(\mathcal{F}; \Omega M) \to K^{(s')}(\mathcal{F}; \Omega M)$$

satisfying similar properties, using $d^{\mathrm{t}}_{\mathcal{F}}$.

5.5.3 *The Equality* $\ker \bar{R}_* = \operatorname{im} \bar{\iota}_*$

We already know that $\ker \bar{R}_* \supset \operatorname{im} \bar{\iota}_*$. To prove that $\ker \bar{R}_* \subset \operatorname{im} \bar{\iota}_*$, take any class $\overline{[\alpha]} \in \ker \bar{R}_*$ in $\bar{H}^\bullet I(\mathcal{F})$. Hence there is some net $\varphi_l \in J(\mathcal{F})$ such that $R\alpha = \lim_l d_{\mathcal{F}}\varphi_l$ in $J(\mathcal{F})$. We can assume $\varphi_l \in C^\infty_{\mathrm{c}}(M^1; \Lambda\mathcal{F})$ by the density of $C^\infty_{\mathrm{c}}(M^1; \Lambda\mathcal{F})$ in $J(\mathcal{F})$ (Sect. 2.6.7).

Using that $J(\mathcal{F})$ is compactly retractive (Sect. 2.6.7) and arguing like in Sect. 5.2.1, we get that $\{R\alpha, d_{\mathcal{F}}\beta_l\}_l$ is contained in some step $J^{(s)}(\mathcal{F})$, and $R\alpha = \lim_l d_{\mathcal{F}}\varphi_l$ in $J^{(s)}(\mathcal{F})$. Moreover, we can assume $d_{\mathcal{F}}\phi_l$ is a sequence because $J^{(s)}(\mathcal{F})$ is a Fréchet space.

Consider the notation of Sect. 5.5.2. We have $R\alpha = \lim_l d_{\mathcal{F}}\varphi_l$ in $J^m(\mathcal{F})$ by the version of (2.5.36) for $J(\mathcal{F})$ (Sect. 2.6.7). We have $\beta := F_m\alpha \in ZJ^{(s')}(\mathcal{F}) \subset ZJ(\mathcal{F})$ by (5.5.5), obtaining a class $\overline{[\beta]} \in \bar{H}^\bullet J(\mathcal{F})$.

Since the sequence φ_l is in $C^\infty_{\mathrm{c}}(M^1; \Lambda\mathcal{F})$, it is also in $J^m(\mathcal{F})$ and in $I^{(s')}(\mathcal{F})$, and we have $E_m\varphi_l = \varphi_l$ by the version of Corollary 2.5.6 with $J^m(\mathcal{F})$. Hence, by Corollary 5.5.2,

$$\beta = \alpha - E_m R\alpha = \alpha - \lim_l E_m d_{\mathcal{F}}\varphi_l = \alpha - \lim_l d_{\mathcal{F}} E_m\varphi_l = \alpha - \lim_l d_{\mathcal{F}}\varphi_l$$

in $I^{(s')}(\mathcal{F})$, and therefore also in $I(\mathcal{F})$. This shows that $\bar{\iota}_*(\overline{[\beta]}) = \overline{[\alpha]}$ in $\bar{H}^\bullet I(\mathcal{F})$.

Remark 5.5.4 Using cohomology instead of reduced cohomology and a single element φ instead of a net φ_l, the analogous argument gives $\ker R_* \subset \operatorname{im} \iota_*$, obtaining $\ker R_* = \operatorname{im} \iota_*$.

5.5.4 *Injectivity of* $\bar{\iota}_*$

Take any $\overline{[\alpha]} \in H^\bullet K(\mathcal{F})$ with $\bar{\iota}_*(\overline{[\alpha]}) = 0$ in $\bar{H}^\bullet I(\mathcal{F})$. Since $I(\mathcal{F})$ is compactly retractive, $C^\infty(M; \Lambda\mathcal{F})$ is dense in $I(\mathcal{F})$ and every $I^{(s)}(\mathcal{F})$ is a Fréchet space

(Sect. 2.2.2), we get as above that there is some s and a sequence φ_l in $C^\infty(M; \Lambda\mathcal{F})$ such that $\alpha \in K^{(s)}(\mathcal{F})$ and $\alpha = \lim_l d_\mathcal{F}\varphi_l$ in $I^{(s)}(\mathcal{F})$.

Consider again the notation of Sect. 5.5.2. By (5.5.4),

$$\alpha = F_m\alpha = \lim_l F_m d_\mathcal{F}\varphi_l = \lim_l d_\mathcal{F} F_m \varphi_l$$

in $K^{(s)}(\mathcal{F})$, and therefore in $K(\mathcal{F})$. So $\alpha \in \bar{B}K(\mathcal{F}) = BK(\mathcal{F})$ by (5.1.3), and therefore $[\alpha] = 0$ in $H^\bullet K(\mathcal{F})$.

Remark 5.5.5 Like in Remark 5.5.6, we also get the injectivity of ι_*.

5.5.5 *Surjectivity of $\bar{R}_*$*

For any class $\overline{[\alpha]} \in \bar{H}^\bullet J(\mathcal{F})$, the representative α is in some step $J^{(s)}(\mathcal{F})$, and therefore it is also in $ZJ^{(s)}(\mathcal{F})$. With the notation of Sect. 5.5.2, $\beta := E_m\alpha \in ZI^{(s')}(\mathcal{F}) \subset ZI(\mathcal{F})$ by Corollary 5.5.2, and we have $R\beta = \alpha$. This shows that $\overline{[\alpha]} = \bar{R}(\overline{[\beta]})$.

Remark 5.5.6 Using cohomology instead of reduced cohomology, the analogous argument gives the surjectivity of R_*.

5.6 Computations in the Case of a Suspension Foliation

Consider the notation of Sects. 4.2.1 to 4.2.5, where the case of a suspension foliation with a simple foliated flow was considered.

5.6.1 *Description of $(K(\mathcal{F}), d_\mathcal{F})$*

For every $m \in \mathbb{N}_0$ and $s < -1/2$, consider the injection defined by (2.6.38) for the vector bundle $\Lambda\mathcal{F}$,

$$C^\infty(L; \Lambda \otimes \Omega^{-1}NL) \to K^{(s-m)}(M, L; \Lambda\mathcal{F}), \quad \alpha \mapsto \partial_\rho^m \delta_L^\alpha . \tag{5.6.1}$$

Proposition 5.6.1 *Via* (5.6.1), *the operator* $d_\mathcal{F}$ *on* $K^{(s-m)}(M, L; \Lambda\mathcal{F})$ *corresponds to the operator* $d_L - m\eta\wedge$ *on* $C^\infty(L; \Lambda \otimes \Omega^{-1}NL)$.

Proof Consider first the case $m = 0$. According to (3.2.20), for some degree v, take $\alpha \in C^\infty(L; \Lambda^v \otimes \Omega^{-1}NL)$ and $\beta \in C^\infty_c(M; \Lambda^{1,n-1-v})$. We can write $\alpha = \alpha_0 \otimes |\omega|^{-1}$ and $\beta = \beta_0 \wedge \omega$ for some $\alpha_0 \in C^\infty(L; \Lambda^v)$ and

$\beta_0 \in C^\infty_c(M; \Lambda^{0,n-1-v})$. By (2.9.5), (2.8.4), (2.9.8) (or (2.9.2) and the Stokes' theorem), (3.2.20) and (3.2.21), and since $d\omega = \eta \wedge \omega$,

$$\begin{aligned}\langle d_{\mathcal{F}}\delta^\alpha_L, \beta\rangle &= -(-1)^v \langle \delta^\alpha_L, d\beta\rangle = -(-1)^v \langle \delta^\alpha_L, (d\beta_0 + (-1)^{n-1-v}\beta_0 \wedge \eta) \wedge \omega\rangle \\ &= -(-1)^v \int_L \alpha_0 \wedge ((d + \eta\wedge)\beta_0)|_L = -(-1)^v \int_L \alpha_0 \wedge (d + \eta\wedge)(\beta_0|_L) \\ &= \int_L (d_L - \eta\wedge)\alpha_0 \wedge \beta_0|_L = \int_L d_L\alpha \wedge \beta|_L = \langle \delta^{d_L\alpha}_L, \beta\rangle \ .\end{aligned}$$

The general case follows from the previous case because $d_{\mathcal{F}}\partial_\rho = \partial_\rho(d_{\mathcal{F}} - \eta\wedge)$ on $C^{-\infty}(M; \Lambda\mathcal{F})$ by (4.2.8) and (4.2.9), and $\eta \wedge \delta^\alpha_L = \delta^{\eta\wedge\alpha}_L$ by (2.8.2). □

Corollary 5.6.2 *Proposition 5.3.1 is true in this case.*

Proof Apply (2.6.39), (2.6.40) and Proposition 5.6.1. □

5.6.2 A Partial Extension Map on $M_\pm$

The notation of Sects. 2.5.8 and 2.5.13, concerning conormal distributions at the boundary, is also used here. By Proposition 2.5.1, there is a continuous linear partial extension map,

$$E_{m,\pm} : \mathcal{A}^m(M_\pm; \Lambda\mathcal{F}_\pm) \to \dot{\mathcal{A}}^{(s)}(M_\pm; \Lambda\mathcal{F}_\pm) \ ,$$

where $s = 0$ if $m \geq 0$, and $m > s \in \mathbb{Z}^-$ if $m < 0$. According to the proof of Proposition 2.5.1 and Remarks 2.5.2 and 2.5.3, in the case $0 > m > s \in \mathbb{Z}^-$, the homomorphism $E_{m,\pm}$ can be given by the composition

$$\mathcal{A}^m(M_\pm; \Lambda\mathcal{F}_\pm) \xrightarrow{J^N} \mathcal{A}^0(M_\pm; \Lambda\mathcal{F}_\pm) \xrightarrow{E_{0,\pm}} \dot{\mathcal{A}}^{(0)}(M_\pm; \Lambda\mathcal{F}_\pm) \xrightarrow{\partial_\rho^N} \dot{\mathcal{A}}^{(s)}(M_\pm; \Lambda\mathcal{F}_\pm) \ ,$$

where $N = -s \in \mathbb{Z}^+$, $E_{0,\pm}$ is a continuous inclusion map, and J is the endomorphism of $C^\infty(M_\pm; \Lambda\mathcal{F}_\pm)$ given by

$$J\alpha(\rho, y) = \int_1^\rho \alpha(\rho_1, y)\, d\rho_1 \ ,$$

using (2.5.32), (4.2.13) and the identity $\Lambda_{(\rho,y)}\mathcal{F} \equiv \Lambda_y L$.

Consider also the endomorphism $\widetilde{J}$ of $C^\infty(\widetilde{M}_\pm; \Lambda\widetilde{\mathcal{F}}_\pm)$ defined like J,

$$\widetilde{J}\tilde{\alpha}(\tilde{\rho}, \tilde{y}) = \int_1^{\tilde{\rho}} \tilde{\alpha}(\tilde{\rho}_1, y)\, d\tilde{\rho}_1 \ ,$$

using (4.2.12) and the identity $\Lambda_{(\rho,\tilde{y})}\mathcal{F}_\pm \equiv \Lambda_{\tilde{y}}\widetilde{L}$. Clearly, $\widetilde{J}$ corresponds to J via

$$\pi^*_{M_\pm} : C^\infty(M_\pm; \Lambda\mathcal{F}_\pm) \to C^\infty(\widetilde{M}_\pm; \Lambda\widetilde{\mathcal{F}}_\pm) \,.$$

Using (4.2.11), we can also write

$$\widetilde{J}\tilde{\alpha}(\tau, \tilde{y}) = e^{F(\tilde{y})} \int_{e^{-F(\tilde{y})}}^{\tau} \tilde{\alpha}(\tau_1, \tilde{y})\, d\tau_1 \,, \tag{5.6.2}$$

with the change of variable $\tilde{\rho}_1 = e^{F(\tilde{y})}\tau_1$, because $\tilde{\rho} = \pm e^{F(\tilde{y})}\tau$.

For fixed $0 < \epsilon < 1$, consider the collar neighborhoods $T_\pm := T_{\pm,\epsilon}$ and $\widetilde{T}_\pm = \widetilde{T}_{\pm,\epsilon} = \pi^{-1}_{M_\pm}(T_\pm)$ of the boundaries in $M_\pm$ and $\widetilde{M}_\pm$. Using (3.2.3), consider $\eta \in C^\infty(M_\pm; \Lambda^1\mathcal{F}_\pm)$ and $\tilde{\eta} \in C^\infty(\widetilde{M}_\pm; \Lambda^1\widetilde{\mathcal{F}}_\pm)$.

Proposition 5.6.3 *$Jd_{\mathcal{F}_\pm} = (d_{\mathcal{F}_\pm} - \eta\wedge)J$ on $C^\infty_c(T_\pm; \Lambda\mathcal{F}_\pm)$.*

Proof Take open subsets $\widetilde{B} \subset \widetilde{L}$ such that $\pi_L : \widetilde{B} \to B := \pi_L(\widetilde{B})$ is a diffeomorphism. Since the open sets of the form $\varpi_\pm^{-1}(B) \equiv [0,\infty)_\rho \times \widetilde{B}_{\tilde{\varpi}}$ cover $M_\pm$ and J preserves the spaces $C^\infty_c(T_\pm \cap \varpi_\pm^{-1}(B); \Lambda\mathcal{F}_\pm)$, it is enough to prove the stated equality on $C^\infty_c(T_\pm \cap \varpi_\pm^{-1}(B); \Lambda\mathcal{F}_\pm)$. In turn, this follows by checking that $\widetilde{J}d_{\widetilde{\mathcal{F}}_\pm} = (d_{\widetilde{\mathcal{F}}_\pm} - \tilde{\eta}\wedge)\widetilde{J}$ on $C^\infty_c(\widetilde{T}_\pm; \Lambda\widetilde{\mathcal{F}}_\pm)$ because

$$\pi_{M_\pm} \equiv \mathrm{id} \times \pi_L : \tilde{\varpi}_\pm^{-1}(\widetilde{B}) \equiv [0,\infty)_{\tilde{\rho}} \times \widetilde{B}_{\tilde{\varpi}} \to \varpi_\pm^{-1}(B) \equiv [0,\infty)_\rho \times B_\varpi$$

is a diffeomorphism. Let $\tilde{\alpha} \in C^\infty_c(\widetilde{T}_\pm; \Lambda\widetilde{\mathcal{F}}_\pm)$ and $(\tau, \tilde{y}) \in [0,\infty) \times \widetilde{L}$ with $\tau < e^{-F(\tilde{y})}\epsilon$, which means that $(\tau, \tilde{y})$ corresponds to an element of $T_\pm$ via (4.2.11); in particular, $\tilde{\alpha}(e^{-F(\tilde{y})}, \tilde{y}) = 0$. So, by (5.6.2) and since $d_{\widetilde{\mathcal{F}}_\pm}\tau = 0$,

$$\begin{aligned}
\widetilde{J}d_{\widetilde{\mathcal{F}}_\pm}\tilde{\alpha}(\tau, \tilde{y}) &= e^{F(\tilde{y})} \int_{e^{-F(\tilde{y})}}^{\tau} d_{\widetilde{\mathcal{F}}_\pm}\tilde{\alpha}(\tau_1, \tilde{y})\, d\tau_1 = e^{F(\tilde{y})} d_{\widetilde{\mathcal{F}}_\pm} \int_{e^{-F(\tilde{y})}}^{\tau} \tilde{\alpha}(\tau_1, \tilde{y})\, d\tau_1 \\
&= d_{\widetilde{\mathcal{F}}_\pm} e^{F(\tilde{y})} \int_{e^{-F(\tilde{y})}}^{\tau} \tilde{\alpha}(\tau_1, \tilde{y})\, d\tau_1 - e^{F(\tilde{y})}\tilde{\eta}(\tilde{y}) \wedge \int_{e^{-F(\tilde{y})}}^{\tau} \tilde{\alpha}(\tau_1, \tilde{y})\, d\tau_1 \\
&= (d_{\widetilde{\mathcal{F}}_\pm} - \tilde{\eta}\wedge)\widetilde{J}\tilde{\alpha}(x, \tilde{y}) \,.
\end{aligned}$$

□

Proposition 5.6.4 *We have $\partial_\rho d_{\mathcal{F}_\pm} = (d_{\mathcal{F}_\pm} + \eta\wedge)\partial_\rho$ on $C^{-\infty}(M_\pm; \Lambda\mathcal{F}_\pm)$.*

Proof Apply (3.2.3), (3.2.8), and (4.2.15). □

Corollary 5.6.5 *For all $m \in \mathbb{R}$, $E_{m,\pm}d_{\mathcal{F}_\pm} = d_{\mathcal{F}_\pm}E_{m,\pm}$ on $\mathcal{A}^m_c(T_\pm; \Lambda\mathcal{F}_\pm)$.*

Proof It is enough to consider the case $m < 0$. Then apply Propositions 5.6.3 and 5.6.4, using the given definition of $E_{m,\pm}$ and the density of $C^\infty_c(T_\pm; \Lambda\mathcal{F}_\pm)$ in $\mathcal{A}^m_c(T_\pm; \Lambda\mathcal{F}_\pm)$ (see Sect. 2.5.10). □

5.6.3 A Partial Extension Map on M

Let us apply the notation of Sect. 2.6 to the suspension foliation (that notation is compatible with the notation of Sects. 3.1.12, 4.1.2, and 4.2). Recall that $\boldsymbol{M} = M_- \sqcup M_+$, $\boldsymbol{\mathcal{F}}$ is the combination of $\mathcal{F}_\pm$, and $\boldsymbol{\pi} : \boldsymbol{M} \to M$ is the combination of $\pi_\pm : M_\pm \to M$. The version of the commutative diagram (2.6.41) for $\Lambda\boldsymbol{\mathcal{F}} \equiv \boldsymbol{\pi}^*\Lambda\mathcal{F}$ is

$$\begin{array}{ccccc} \mathcal{K}(\boldsymbol{M};\Lambda\boldsymbol{\mathcal{F}}) & \xrightarrow{\iota} & \dot{\mathcal{A}}(\boldsymbol{M};\Lambda\boldsymbol{\mathcal{F}}) & \xrightarrow{R} & \mathcal{A}(\boldsymbol{M};\Lambda\boldsymbol{\mathcal{F}}) \\ \downarrow{\scriptstyle \boldsymbol{\pi}_*} & & \downarrow{\scriptstyle \boldsymbol{\pi}_*} & & \downarrow{\scriptstyle \boldsymbol{\pi}_*}\,\cong \\ K(\mathcal{F}) & \xrightarrow{\iota} & I(\mathcal{F}) & \xrightarrow{R} & J(\mathcal{F})\,. \end{array} \tag{5.6.3}$$

Moreover $d_{\boldsymbol{\mathcal{F}}} \in \mathrm{Diff_b}(\boldsymbol{M};\Lambda\boldsymbol{\mathcal{F}})$ is the lift of $d_{\mathcal{F}}$. Hence the operators defined by $d_{\boldsymbol{\mathcal{F}}}$ on the spaces of the top row of (5.6.3) correspond to the operators defined by $d_{\mathcal{F}}$ on the spaces of its bottom row via the homomorphisms $\boldsymbol{\pi}_*$ (Sect. 2.6). According to Sect. 2.5.14, $d_{\boldsymbol{\mathcal{F}}}$ preserves the subspaces $\dot{\mathcal{A}}^{(s)}(\boldsymbol{M};\Lambda\boldsymbol{\mathcal{F}})$ and $\mathcal{A}^m(\boldsymbol{M};\Lambda\boldsymbol{\mathcal{F}})$.

The partial extension maps of Sect. 5.6.2,

$$E_{m,\pm} : \mathcal{A}^m(M_\pm;\Lambda\mathcal{F}_\pm) \to \dot{\mathcal{A}}^{(s)}(M_\pm;\Lambda\mathcal{F}_\pm)\,,$$

can be combined to define a continuous linear partial extension map

$$\boldsymbol{E}_m : \mathcal{A}^m(\boldsymbol{M};\Lambda\boldsymbol{\mathcal{F}}) \to \dot{\mathcal{A}}^{(s)}(\boldsymbol{M};\Lambda\boldsymbol{\mathcal{F}})\,.$$

Then, according to Corollary 2.6.4 and its proof, a continuous linear partial extension map $E_m : J^m(\mathcal{F}) \to I^{(s)}(\mathcal{F})$ is given by the composition

$$J^m(\mathcal{F}) \xrightarrow{\boldsymbol{\pi}_*^{-1}} \mathcal{A}^m(\boldsymbol{M};\Lambda\boldsymbol{\mathcal{F}}) \xrightarrow{\boldsymbol{E}_m} \dot{\mathcal{A}}^{(s)}(\boldsymbol{M};\Lambda\boldsymbol{\mathcal{F}}) \xrightarrow{\boldsymbol{\pi}_*} I^{(s)}(\mathcal{F})\,,$$

which is a continuous inclusion map if $m \geq 0$. Recall that $T \equiv (-\epsilon,\epsilon)_\rho \times L_\varpi$ and

$$\boldsymbol{T} = T_- \sqcup T_+ = \boldsymbol{\pi}^{-1}(T) \equiv [0,\epsilon)_{\boldsymbol{\rho}} \times \partial\boldsymbol{M}_{\boldsymbol{\varpi}}\,.$$

Like in Sect. 5.5.4, consider the restriction $E_{m,T} : J^m_{\mathrm{c}}(T;\Lambda\mathcal{F}) \to I^{(s)}_{\mathrm{c}}(T;\Lambda\mathcal{F})$ of E_m. Suppose $\epsilon < 1$, like in Sect. 5.6.2.

Corollary 5.6.6 *For all $m \in \mathbb{R}$, $E_{m,T}$ satisfies $E_{m,T}d_{\mathcal{F}} = d_{\mathcal{F}}E_{m,T}$.*

Proof Apply Corollary 5.6.5. □

5.7 Functoriality and Leafwise Homotopy Invariance

5.7.1 *Pull-Back of Conormal Leafwise Currents*

Let M' be another closed manifold, and let $\phi : M' \to M$ be a smooth map transverse to $\mathcal{F}$. Then $\mathcal{F}' := \phi^*\mathcal{F}$ is another transversely oriented foliation of codimension one satisfying the conditions (C) and (D) in Sect. 4.1.2 with $M'^0 := \phi^{-1}(M^0)$.

Remark 5.7.1 The results of Sect. 5.7 have direct extensions to the case where M or M' may not be compact, with the condition that M^0 and M'^0 are compact.

According to Sect. 2.2.8, the map (3.2.23) has a continuous extension

$$\phi^* : I(\mathcal{F}) \to I(\mathcal{F}') . \tag{5.7.1}$$

defined as the composition

$$I(\mathcal{F}) \xrightarrow{\phi^*} I(M', M'^0; \phi^*\Lambda\mathcal{F}) \xrightarrow{\phi^*} I(\mathcal{F}') , \tag{5.7.2}$$

like (2.8.19), using (2.2.20) with $E = \Lambda\mathcal{F}$. We can also describe (5.7.1) as the restriction of (2.8.18) to conormal currents of bidegree $(0, \bullet)$, like in (3.2.34). The map (5.7.1) is also a restriction of (3.2.34).

Similarly, the analogs of (2.2.20) with $E = \Lambda\mathcal{F}$ for (2.6.42) and (2.6.43) induce continuous homomorphisms

$$\phi^* : K(\mathcal{F}) \to K(\mathcal{F}') , \tag{5.7.3}$$

$$\phi^* : J(\mathcal{F}) \to J(\mathcal{F}') . \tag{5.7.4}$$

By passing to cohomology and reduced cohomology, we get continuous homomorphisms,

$$\left\{\begin{array}{c} \phi^* : H^\bullet K(\mathcal{F}) \to H^\bullet K(\mathcal{F}') , \\ \phi^* : H^\bullet I(\mathcal{F}) \to H^\bullet I(\mathcal{F}') , \quad \phi^* : \bar{H}^\bullet I(\mathcal{F}) \to \bar{H}^\bullet I(\mathcal{F}') , \\ \phi^* : H^\bullet J(\mathcal{F}) \to H^\bullet J(\mathcal{F}') , \quad \phi^* : \bar{H}^\bullet J(\mathcal{F}) \to \bar{H}^\bullet J(\mathcal{F}') . \end{array}\right. \tag{5.7.5}$$

The assignment of the homomorphisms (5.7.1)–(5.7.5) is functorial.

5.7.2 *Description of $\phi^* : K(\mathcal{F}) \to K(\mathcal{F}')$*

For $\omega' = \phi^*\omega$ and $\eta' = \phi^*\eta$, we have $T\mathcal{F}' = \ker\omega'$ and $d\omega' = \eta' \wedge \omega'$ (the Frobenius integrability condition for $\mathcal{F}'$). Thus

$$\phi^* : C^\infty(M^0; \Lambda) \to C^\infty(M'^0; \Lambda) \tag{5.7.6}$$

is a cochain map for $d_{s\eta}$ and $d_{s\eta'}$ ($s \in \mathbb{R}$) (Sect. 2.9.3). In other words, ϕ induces

$$\phi^* : C^\infty(M^0; \Lambda \otimes \Omega^s NM^0) \to C^\infty(M'^0; \Lambda \otimes \Omega^s NM'^0) , \tag{5.7.7}$$

given by

$$\phi^*(\alpha \otimes |\omega|^s) = \phi^*\alpha \otimes |\omega'|^s , \tag{5.7.8}$$

which is another cochain map for the de Rham differentials defined with the flat bundle structures of $\Omega^s NM^0$ and $\Omega^s NM'^0$.

If ρ is a defining function of M^0 in some open neighborhood T, then $\rho' := \phi^*\rho$ is a defining function of M'^0 in $T' = \phi^{-1}(T)$, and (5.7.7) satisfies

$$\phi^*(\alpha \otimes |d\rho|^s) = \phi^*\alpha \otimes |d\rho'|^s . \tag{5.7.9}$$

Note the compatibility of (5.7.8) and (5.7.9) with (4.2.5). Furthermore, the inverse image of $T := T_\epsilon \equiv (-\epsilon, \epsilon)_\rho \times M^0_\varpi$, for $\epsilon > 0$ small enough, is a tubular neighborhood $T' \equiv (-\epsilon, \epsilon)_{\rho'} \times M'^0_{\varpi'}$ of M'^0 in M', where $\varpi' : T' \to M'^0$ satisfies $\phi\varpi' = \varpi\phi$ as maps $T' \to M^0$. Thus

$$\phi \equiv \mathrm{id} \times \phi : T' \equiv (-\epsilon, \epsilon) \times M'^0 \to T \equiv (-\epsilon, \epsilon) \times M^0 ,$$

which is proper because M'^0 is compact. We can use these tubular neighborhoods to define the operators ∂_ρ and $\partial_{\rho'}$ on $C^\infty_{\mathrm{c}}(T; \Lambda\mathcal{F})$ and $C^\infty_{\mathrm{c}}(T'; \Lambda\mathcal{F}')$ (Sect. 2.6), which are used in the identities of Proposition 5.3.1 for $K(\mathcal{F})$ and $K(\mathcal{F}')$ (Sect. 5.6.1). Clearly,

$$\partial_{\rho'}\phi^* = \phi^*\partial_\rho , \tag{5.7.10}$$

as maps $C^\infty_{\mathrm{c}}(T; \Lambda\mathcal{F}) \to C^\infty_{\mathrm{c}}(T'; \Lambda\mathcal{F}')$,

Proposition 5.7.2 *According to Proposition 5.3.1, the map* (5.7.3) *is given by*

$$\phi^* \equiv \bigoplus_k \phi^* \equiv \bigoplus_k \phi^* ,$$

where the terms of the first direct sum are given by (5.7.6), *and the terms of the second direct sum are given by* (5.7.7), *taking* $s = -k - 1$.

Proof The second identity follows from the first one and (5.7.8). To prove the first identity, by (5.7.10), it is enough to consider the term with $k = 0$.

For $\alpha \in C^\infty(M^0; \Lambda)$, let $u = \alpha \otimes |d\rho|^{-1} \in C^\infty(M^0; \Lambda \otimes \Omega^{-1}NM^0)$. Using the first identity of Proposition 5.3.1 for $k = 0$, we have $u \equiv \delta^u_{M^0} = \varpi^*\alpha \cdot \rho^*\delta_0$ in $K(\mathcal{F})$, using Dirac sections (Sect. 2.2.6). Here, $\rho^*\delta_0 \in K(T, M^0)$ is defined because $\rho : T \to (-\epsilon, \epsilon)$ is transverse to 0. Moreover $u' := \phi^*u = \phi^*\alpha \cdot |d\rho'|^{-1}$

by (5.7.9). As before, $u' \equiv \delta^{u'}_{M'^0} = \varpi'^*\phi^*\alpha \cdot \rho'^*\delta_0$ in $K(\mathcal{F}')$. Take a sequence $f_i \in C^\infty_{\mathrm{c}}(-\epsilon, \epsilon)$ converging to δ_0 in $C^{-\infty}_{\mathrm{c}}(-\epsilon, \epsilon)$. Then

$$\begin{aligned}\phi^*\delta^u_{M^0} &= \phi^*(\varpi^*\alpha \cdot \rho^*\delta_0) = \lim_i \phi^*(\varpi^*\alpha \cdot \rho^* f_i) = \lim_i \phi^*\varpi^*\alpha \cdot \phi^*\rho^* f_i \\ &= \lim_i \varpi'^*\phi^*\alpha \cdot \rho'^* f_i = \varpi'^*\phi^*\alpha \cdot \rho'^*\delta_0 = \delta^{u'}_{M'^0} .\end{aligned}$$

□

Remark 5.7.3 The equality $\partial_{\rho'}\phi^* = \phi^*\partial_\rho$ has a continuous extension as maps $C^{-\infty}_{\mathrm{c}}(T; \Lambda\mathcal{F}) \to C^{-\infty}_{\mathrm{c}}(T'; \Lambda\mathcal{F}')$, and the computations of the above proof also work also with $\alpha \in C^{-\infty}(M^0; \Lambda)$. So we get similar expressions of $\phi^* : C^{-\infty}_{M^0}(M; \Lambda\mathcal{F}) \to C^{-\infty}_{M'^0}(M'; \Lambda\mathcal{F}')$ according to Remark 5.3.4.

5.7.3 Push-Forward of Conormal Leafwise Currents

With the notation and conditions of Sect. 5.7.1, suppose that moreover ϕ is a submersion such that the vertical bundle $\mathcal{V}$ is oriented (Sect. 3.2.15). Thus $\phi : M'^0 \to M^0$ is also a submersion whose vertical bundle is $\mathcal{V}|_{M'^0} \subset TM'^0$, also oriented. Then the case of (3.2.33) on smooth leafwise forms has a continuous extension

$$\phi_* : I_{\mathrm{c/cv}}(\mathcal{F}') \to I_{\mathrm{c/\cdot}}(\mathcal{F}) . \tag{5.7.11}$$

This map can be described as the restriction of the map (2.8.24) to conormal currents of bidegree $(0, \bullet)$, like (3.2.33) in Sect. 3.2.15. We can also describe (5.7.11) as the composition

$$I_{\mathrm{c/cv}}(M', L'; \Lambda\mathcal{F}) \xrightarrow{\pi_{\mathrm{top}}} I_{\mathrm{c/\,cv}}(M', L'; \phi^*\Lambda\mathcal{F} \otimes \Omega_{\,\mathrm{fiber}}) \xrightarrow{\phi_*} I_{\mathrm{c/\cdot}}(M, L; \Lambda\mathcal{F}) ,$$

like in (3.2.35), where ϕ_* is given by (2.2.23) for $E = \Lambda\mathcal{F}$. The map (5.7.11) is also a restriction of the case of (3.2.33) for leafwise currents.

According to Sect. 2.6.15, the map (5.7.11) induces homomorphisms

$$\phi_* : K(\mathcal{F}') \to K(\mathcal{F}) , \tag{5.7.12}$$

$$\phi_* : J_{\mathrm{c/cv}}(\mathcal{F}') \to J_{\mathrm{c/\cdot}}(\mathcal{F}) . \tag{5.7.13}$$

Like in Sect. 5.7.1, we get induced continuous homomorphisms,

$$\left\{\begin{array}{ll} \multicolumn{2}{c}{\phi_* : H^\bullet K(\mathcal{F}') \to H^\bullet K(\mathcal{F}) ,} \\ \phi_* : H^\bullet I_{\mathrm{c}}(\mathcal{F}') \to H^\bullet I_{\mathrm{c}}(\mathcal{F}) , & \phi_* : \bar{H}^\bullet I_{\mathrm{c}}(\mathcal{F}') \to \bar{H}^\bullet I_{\mathrm{c}}(\mathcal{F}) , \\ \phi_* : H^\bullet J_{\mathrm{c}}(\mathcal{F}') \to H^\bullet J_{\mathrm{c}}(\mathcal{F}) , & \phi_* : \bar{H}^\bullet J_{\mathrm{c}}(\mathcal{F}') \to \bar{H}^\bullet J_{\mathrm{c}}(\mathcal{F}) . \end{array}\right. \tag{5.7.14}$$

The assignments of homomorphisms (5.7.11)–(5.7.14) are clearly functorial.

5.7.4 Description of $\phi_* : K(\mathcal{F}') \to K(\mathcal{F})$

For ϕ as above, consider the notation of Sect. 5.7.2. Then

$$\phi_* : C^\infty(M'^0; \Lambda) \to C^\infty(M^0; \Lambda) \tag{5.7.15}$$

is a cochain map for $d_{s\eta}$ and $d_{s\eta'}$ ($s \in \mathbb{R}$) (Sect. 2.9.3). That is, ϕ induces

$$\phi_* : C^\infty(M'^0; \Lambda \otimes \Omega^s NM'^0) \to C^\infty(M^0; \Lambda \otimes \Omega^s NM^0) , \tag{5.7.16}$$

given by

$$\phi_*(\alpha \otimes |\omega'|^s) = \phi_*\alpha \otimes |\omega|^s , \tag{5.7.17}$$

which is another cochain map for the de Rham differentials defined with the flat bundle structures of $\Omega^s NM^0$ and $\Omega^s NM'^0$ induced by the Bott flat $T\mathcal{F}$-partial connection (Sect. 3.1.3). Like in (5.7.9) and (5.7.10), we have

$$\phi_*(\alpha \otimes |d\rho'|^s) = \phi_*\alpha \otimes |d\rho|^s , \tag{5.7.18}$$

$$\partial_\rho \phi_* = \phi_* \partial_{\rho'} , \tag{5.7.19}$$

where (5.7.19) holds as maps $C^\infty_{\mathrm{c}}(T'; \Lambda\mathcal{F}') \to C^\infty_{\mathrm{c}}(T; \Lambda\mathcal{F})$.

Proposition 5.7.4 *According to Proposition 5.3.1, the map* (5.7.12) *is given by*

$$\phi_* \equiv \bigoplus_k \phi_* \equiv \bigoplus_k \phi_* ,$$

where the terms of the first direct sum are given by (5.7.15), *and the terms of the second direct sum are given by* (5.7.16), *taking* $s = -k - 1$.

Proof The second identity follows from the first one and (5.7.17). To prove the first identity, by (5.7.19), it is enough to consider the term with $k = 0$.

For $\beta \in C^\infty(M'^0; \Lambda)$, let $v' = \beta \otimes |d\rho'|^{-1} \in C^\infty(M'^0; \Lambda \otimes \Omega^{-1} NM'^0)$. Like in the proof of Proposition 5.7.2, we have $v' \equiv \delta^{v'}_{M'^0} = \varpi'^*\beta \cdot \rho'^*\delta_0$ in $K(\mathcal{F}')$. Moreover $v := \phi_* v' = \phi_*\beta \cdot |d\rho|^{-1}$ by (5.7.18), with $v \equiv \delta^v_{M^0} = \varpi'^*\phi_*\beta' \cdot \rho^*\delta_0$ in $K(\mathcal{F})$. Take a sequence $f_i \in C^\infty_{\mathrm{c}}(-\epsilon, \epsilon)$ converging to δ_0 in $C^{-\infty}_{\mathrm{c}}(-\epsilon, \epsilon)$. We get

$$\phi_*\delta^{v'}_{M'^0} = \phi_*(\varpi'^*\beta \cdot \rho'^*\delta_0) = \lim_i \phi_*(\varpi'^*\beta \cdot \rho'^* f_i) = \lim_i \phi_*\varpi^*\beta \cdot \rho^* f_i$$

$$= \lim_i \varpi^* \phi_* \beta \cdot \rho^* f_i = \varpi^* \phi^* \beta \cdot \rho^* \delta_0 = \delta^v_{M^0} \, .$$

□

The analog of Remark 5.7.3 for $\phi_* : K(\mathcal{F}') \to K(\mathcal{F})$ is true.

5.7.5 Leafwise Homotopy Invariance

With the notation of Sect. 3.2.16, let $H : (M' \times I, \mathcal{F}' \times I) \to (M, \mathcal{F})$ $(I = [0, 1])$ be a smooth leafwise homotopy such that H_0 is transverse to M^0 and $H_0^{-1}(M^0) = M'^0$. Then, for every $p' \in M'$, the map $H_{t*} : N_{p'}\mathcal{F}' \to N_{H_t(p')}\mathcal{F}$ is the composition of $H_{0*} : N_{p'}\mathcal{F}' \to N_{H_0(p')}\mathcal{F}$ with the parallel transport along the leafwise path $s \in [0, t] \mapsto H_s(p')$. It follows that every H_t is transverse to M^0 and $H_t^{-1}(M^0) = M'^0$. Hence H is transverse to M^0 and $H^{-1}(M^0) = M'^0 \times I$. Then, by (3.2.38) and according to Sects. 5.7.1 and 5.7.3, the corresponding leafwise homotopy operator $\mathsf{h} : C^\infty(M; \Lambda\mathcal{F}) \to C^\infty(M'; \Lambda\mathcal{F}')$ has continuous linear extensions,

$$\mathsf{h} : K(\mathcal{F}) \to K(\mathcal{F}') \, , \quad \mathsf{h} : I(\mathcal{F}) \to I(\mathcal{F}') \, , \quad \mathsf{h} : J(\mathcal{F}) \to J(\mathcal{F}') \, .$$

By continuity and according to Sect. 3.2.16, we have $H_1^* - H_0^* = \mathsf{h} d_{\mathcal{F}} + d_{\mathcal{F}'} \mathsf{h}$ with H_0^* and H_1^* given by (5.7.1), (5.7.3) and (5.7.4). Hence we get the following.

Proposition 5.7.5 *Let $\phi, \psi : (M', \mathcal{F}') \to (M, \mathcal{F})$ be smooth foliated maps transverse to M^0 with $\phi^{-1}(M^0) = \psi^{-1}(M^0) = M'^0$. If ϕ is leafwise homotopic to ψ, then ϕ and ψ induce the same homomorphisms* (5.7.5).

5.8 Action of Foliated Flows on the Conormal Sequence

Let $\phi = \{\phi^t\}$ be a foliated flow with transversely simple preserved leaves on a compact foliated manifold $(M, \mathcal{F})$. The homomorphisms (5.7.5) induced by the maps ϕ^t define actions of $\mathbb{R}$ on $H^\bullet K(\mathcal{F})$, $H^\bullet I(\mathcal{F})$ and $H^\bullet J(\mathcal{F})$, denoted by $\phi^* = \{\phi^{t*}\}$, and actions on $\bar{H}^\bullet I(\mathcal{F})$ and $\bar{H}^\bullet J(\mathcal{F})$, denoted by $\bar{\phi}^* = \{\bar{\phi}^{t*}\}$. By Proposition 5.7.5, they only depend on the flow-leafwise-homotopy class of ϕ (Sect. 3.1.7).

With the notation of Sect. 4.3.3, the foliated flow $\xi = \{\xi^t\}$ has transversely simple preserved leaves and satisfies $\bar{\xi} = \bar{\phi}$ and $\xi^t = \mathrm{id}$ on M^0. By Proposition 3.1.4, there is a flow-leafwise homotopy between ϕ and ξ, and therefore $\phi^* = \xi^*$ on $H^\bullet K(\mathcal{F})$. Consider the tubular neighborhood with defining function, $T_\epsilon \equiv (-\epsilon, \epsilon)_\rho \times M^0_\varpi$, like in Sect. 5.7.2.

Proposition 5.8.1 *According to Corollary 5.3.2 and Remark 5.3.3,*

$$\phi^{t*} \equiv \bigoplus_{k,L} e^{-(k+1)\varkappa_L t} \equiv \bigoplus_{k,L} e^{-(k+1)\varkappa_L t}$$

on $H^\bullet K(\mathcal{F})$, where k runs in $\mathbb{N}_0$ and L in $\pi_0 M^0$.

Proof Since $\xi^{t*}\rho = e^{\varkappa_L t}\rho$ on every $T_{L,\epsilon} \cap \xi^{-t}(T_{L,\epsilon})$ by (4.2.7), it follows from (5.7.9) and Proposition 5.7.2 that

$$\xi^{t*} \equiv \bigoplus_{k,L} e^{-(k+1)\varkappa_L t} \equiv \bigoplus_{k,L} e^{-(k+1)\varkappa_L t}$$

on $K(\mathcal{F})$, according to Proposition 5.3.1 and Remark 5.3.3. Hence $\phi^{t*} = \xi^{t*}$ has the stated expression on $H^\bullet K(\mathcal{F})$. □

Propositions 5.3.1 and 5.8.1, Corollary 5.3.2, and Remark 5.3.3 show Theorem 1.3.1.

Chapter 6
Dual-Conormal Leafwise Reduced Cohomology

6.1 Dual-Conormal Sequence of Leafwise Differential Forms

Assume the conditions of Sect. 5.1 on $(M, \mathcal{F})$. According to Sect. 2.3.4, the LCHS

$$I'(\mathcal{F}) = I'\Lambda^\bullet(\mathcal{F}) := I'(M, M^0; \Lambda\mathcal{F})$$

is a topological complex with $d_\mathcal{F}$. It induces the *dual-conormal leafwise cohomology* and *dual-conormal leafwise reduced cohomology* of $\mathcal{F}$ (or of $(\mathcal{F}, M^0)$). The notation $BI'(\mathcal{F})$, $ZI'(\mathcal{F})$, $\bar{B}I'(\mathcal{F})$, $H^\bullet I'(\mathcal{F})$ and $\bar{H}^\bullet I'(\mathcal{F})$ is used like in Sect. 5.1.

For a leafwise flat vector bundle E, we can also consider the topological complex

$$I'(\mathcal{F}; E) = I'\Lambda^\bullet(\mathcal{F}; E) = I'(M, M^0; \Lambda\mathcal{F} \otimes E)$$

with $d_\mathcal{F}$. The LCHS $I'(\mathcal{F}; E)$ is also defined for an arbitrary vector bundle E, missing the leafwise differential map $d_\mathcal{F}$.

Moreover, the LCHSs

$$I'^{(s)}(\mathcal{F}) = I'^{(s)}\Lambda^\bullet(\mathcal{F}) = I'^{(s)}(M, M^0; \Lambda\mathcal{F}) \quad (s \in \mathbb{R}) .$$

also become topological complexes with $d_\mathcal{F}$. The notation $BI'^{(s)}(\mathcal{F})$, $ZI'^{(s)}(\mathcal{F})$, $\bar{B}I'^{(s)}(\mathcal{F})$, $H^\bullet I'^{(s)}(\mathcal{F})$ and $\bar{H}^\bullet I'^{(s)}(\mathcal{F})$ is used like in Sect. 5.1.

Remark 6.1.1 Although ΩM has no leafwise flat structure in general, we can assume $\mathcal{F}$ is oriented by working locally or passing to the double cover of orientations of $\mathcal{F}$. Then we can apply (3.2.19)–(3.2.21) and the leafwise flat structure of $\Omega N\mathcal{F}$ to define $d_\mathcal{F}$ and $d_\mathcal{F}^{\mathrm{t}}$ on every $I^{(s)}(\mathcal{F}; \Omega M) \equiv I^{(s)}(\mathcal{F}; \Omega N\mathcal{F})$. Since the condition of being in $I^{(s)}(\mathcal{F}; \Omega M)$ is local for elements of $C^{-\infty}(M; \Lambda\mathcal{F} \otimes \Omega)$, this procedure gives the definition of $d_\mathcal{F} = d_\mathcal{F}^{\mathrm{tt}}$.

J. A. Álvarez López et al., *A Trace Formula for Foliated Flows*, Lecture Notes in Mathematics 2387, https://doi.org/10.1007/978-3-032-15413-2_6

For $s' \leq s$ in $\mathbb{R}$, we have the continuous linear restriction maps (Sect. 2.3.1)

$$j'_s : I'(\mathcal{F}) \to I'^{(s)}(\mathcal{F}) , \quad j'_{s,s'} : I'^{(s)}(\mathcal{F}) \to I'^{(s')}(\mathcal{F}) , \tag{6.1.1}$$

where $j'_s = j^{\mathrm{t}}_{-s}$ and $j'_{s,s'} = j^{\mathrm{t}}_{-s',-s}$ for the version of (5.1.1) with ΩM. The induced homomorphisms in cohomology and reduced cohomology are denoted by j'_{s*}, $j'_{s,s'*}$, $\bar{j}'_{s*}$ and $\bar{j}'_{s,s'*}$. The homomorphisms $j'_{s,s'*}$ and $\bar{j}'_{s,s'*}$ form projective spectra, giving rise to projective limits as $s \uparrow +\infty$. Like in (5.1.2), the maps j'_{s*} and $\bar{j}'_{s*}$ induce canonical continuous linear maps,

$$\begin{cases} \tilde{j}'_* := \varprojlim j'_{s*} : H^\bullet I'(\mathcal{F}) \to \widetilde{H}^\bullet I'(\mathcal{F}) := \varprojlim H^\bullet I'^{(s)}(\mathcal{F}) , \\ \hat{j}'_* := \varprojlim \bar{j}'_{s*} : \bar{H}^\bullet I'(\mathcal{F}) \xrightarrow{\cong} \widehat{H}^\bullet I'(\mathcal{F}) := \varprojlim \bar{H}^\bullet I'^{(s)}(\mathcal{F}) , \end{cases} \tag{6.1.2}$$

where the second one is a linear isomorphism (Sect. 6.2). The canonical maps of the inductive limits to the steps are denoted by

$$\tilde{j}_{s*} : \widetilde{H}^\bullet I'(\mathcal{F}) \to H^\bullet I'^{(s)}(\mathcal{F}) , \quad \hat{j}_{s*} : \widehat{H}^\bullet I'(\mathcal{F}) \to \bar{H}^\bullet I'^{(s)}(\mathcal{F}) .$$

Using the above type of notation, the LCHSs $J'(\mathcal{F})$ and $K'(\mathcal{F})$ are also topological complexes with $d_{\mathcal{F}}$ (Sect. 2.7.5), with corresponding spaces $BJ'(\mathcal{F})$, $ZJ'(\mathcal{F})$, $\bar{B}J'(\mathcal{F})$, $H^\bullet J'(\mathcal{F})$ and $\bar{H}^\bullet J'(\mathcal{F})$, and the same for $K'(\mathcal{F})$.

Similarly, we have topological complexes $J'^{(s)}(\mathcal{F})$, $J'^{m}(\mathcal{F})$ and $K'^{(s)}(\mathcal{F})$ ($s, m \in \mathbb{R}$) (Sects. 2.7.4 and 2.6.13), with corresponding spaces $BJ'^{(s)}(\mathcal{F})$, $ZJ'^{(s)}(\mathcal{F})$, $\bar{B}J'^{(s)}(\mathcal{F})$, $H^\bullet J'^{(s)}(\mathcal{F})$ and $\bar{H}^\bullet J'^{(s)}(\mathcal{F})$, and the same for $J'^{m}(\mathcal{F})$ and $K'^{(s)}(\mathcal{F})$. There are obvious versions for $J'^{(s)}(\mathcal{F})$ and $K'^{(s)}(\mathcal{F})$ of the maps (6.1.1) (Sect. 2.7.1), also denoted by j'_s and $j'_{s,s'}$, giving rise to projective spectra in cohomology and reduced cohomology, and the corresponding projective limits. In the case of $J'(\mathcal{F})$, the maps $j'_{s,s'}$ and j'_s are continuous inclusions (Sect. 2.7.1).

There are also continuous inclusion maps (Sect. 2.7.1)

$$\begin{cases} j'_m : J'(\mathcal{F}) \hookrightarrow J'^{m}(\mathcal{F}) , \quad j'_{m,m'} : J'^{m}(\mathcal{F}) \hookrightarrow J'^{m'}(\mathcal{F}) \quad (m' \leq m) , \\ j'_{m,s} : J^{m}(\mathcal{F}) \hookrightarrow J^{(s)}(\mathcal{F}) \quad (m > s + n/2 + 1) , \\ j'_{s,m} : J^{(s)}(\mathcal{F}) \hookrightarrow J^{m}(\mathcal{F}) \quad (s \geq m, 0) , \end{cases} \tag{6.1.3}$$

denoted like in (5.1.1) with some abuse of notation. The homomorphisms induced by the maps $j'_{m,m'}$ in cohomology and reduced cohomology form projective spectra whose inductive limits as $m \uparrow +\infty$ agree with the previous ones for $J(\mathcal{F})$, and the maps j'_m induce a continuous linear isomorphism analogous to (6.1.2).

It will be shown (Corollary 6.3.2) that the canonical projections are TVS-identities,

$$\begin{cases} H^\bullet K'(\mathcal{F}) \equiv \bar{H}^\bullet K'(\mathcal{F}) , \quad H^\bullet K'^{(s)}(\mathcal{F}) \equiv \bar{H}^\bullet K'^{(s)}(\mathcal{F}) , \\ H^\bullet K'(\mathcal{F}) \equiv \varprojlim H^\bullet K'^{(s)}(\mathcal{F}) , \quad \bar{H}^\bullet K'(\mathcal{F}) \equiv \varprojlim \bar{H}^\bullet K'^{(s)}(\mathcal{F}) . \end{cases} \tag{6.1.4}$$

The version of the bottom row of (2.7.8) with $\Lambda\mathcal{F}$ is a short exact sequence of continuous homomorphisms of topological complexes,

$$0 \leftarrow K'(\mathcal{F}) \overset{R'}{\leftarrow} I'(\mathcal{F}) \overset{\iota'}{\leftarrow} J'(\mathcal{F}) \leftarrow 0 , \tag{6.1.5}$$

using the notation $R' = \iota^{\mathrm{t}}$ and $\iota' = R^{\mathrm{t}}$. The exactness of the induced sequences,

$$0 \leftarrow H^\bullet K'(\mathcal{F}) \overset{R'_*}{\longleftarrow} H^\bullet I'(\mathcal{F}) \overset{\iota'_*}{\leftarrow} H^\bullet J'(\mathcal{F}) \leftarrow 0 , \tag{6.1.6}$$

$$0 \leftarrow H^\bullet K'(\mathcal{F}) \overset{\bar{R}'_*}{\longleftarrow} \bar{H}^\bullet I'(\mathcal{F}) \overset{\bar{\iota}'_*}{\leftarrow} \bar{H}^\bullet J'(\mathcal{F}) \leftarrow 0 , \tag{6.1.7}$$

will be proved in Sect. 6.5; in particular, this shows Theorem 1.3.6.

Taking the transpose of the analog of (5.1.8) with ΩM, we get continuous linear maps

$$R'_s : K'^{(s)}(\mathcal{F}) \to I'^{(s)}(\mathcal{F}) , \quad \iota'_s : I'^{(s)}(\mathcal{F}) \to J'^{(s)}(\mathcal{F}) . \tag{6.1.8}$$

Like in Sect. 5.1, the subscript "s" may be also added to the elements of the cohomologies or reduced cohomologies of $K'^{(s)}(\mathcal{F})$, $I'^{(s)}(\mathcal{F})$ or $J'^{(s)}(\mathcal{F})$.

6.2 Projective Limits in Reduced Cohomology

The goal of this section is to prove the linear isomorphism (6.1.2), and its version for $J'(\mathcal{F})$. The case of $K'(\mathcal{F})$ is given by (6.1.4).

To simplify the notation, we write

$$\widetilde{H}^\bullet I'(\mathcal{F}) = \varprojlim H^\bullet I'^{(s)}(\mathcal{F}) , \quad \widehat{H}^\bullet I'(\mathcal{F}) = \varprojlim \bar{H}^\bullet I'^{(s)}(\mathcal{F}) ,$$

as $s \uparrow +\infty$, and the canonical maps of the projective limits to the steps are denoted by

$$\tilde{\jmath}'_s : \widetilde{H}^\bullet I'(\mathcal{F}) \to H^\bullet I'^{(s)}(\mathcal{F}) , \quad \hat{\jmath}'_s : \widehat{H}^\bullet I'(\mathcal{F}) \to \bar{H}^\bullet I'^{(s)}(\mathcal{F}) .$$

The same type of notation is used in the cases of $J'(\mathcal{F})$ and $K'(\mathcal{F})$.

Lemma 6.2.1 *$BI'(\mathcal{F})$ is dense in every $BI'^{(s)}(\mathcal{F})$.*

Proof Use that the image of $J'(\mathcal{F})$ is dense in $J'^{(s)}(\mathcal{F})$ (Sect. 2.3.1) and $d_{\mathcal{F}}$ is continuous on $J'(\mathcal{F})$ and $J'^{(s)}(\mathcal{F})$ (Sect. 2.7.5). □

Recall that $\bar{B}J'(\mathcal{F})$ (resp., $\bar{B}J'^{(s)}(\mathcal{F})$) denotes the closure of $BJ'(\mathcal{F})$ (resp., $BJ'^{(s)}(\mathcal{F})$) in $J'(\mathcal{F})$ (resp., $J'^{(s)}(\mathcal{F})$).

Corollary 6.2.2 *As vector spaces,*

$$\bar{B}J'(\mathcal{F}) = \bigcap_s \bar{B}J'^{(s)}(\mathcal{F}) .$$

Proof By the definition of the projective topology of $\bigcap_s J'^{(s)}(\mathcal{F})$ [Sch71, Section II.5] and using Lemma 6.2.1, we get that $BJ'(\mathcal{F})$ is dense in $\bigcap_s \bar{B}J'^{(s)}(\mathcal{F})$. Moreover, this intersection is closed in $J'(\mathcal{F})$. Then the stated equality is true. □

Lemma 6.2.3 *As vector spaces,*

$$ZJ'(\mathcal{F}) = \bigcap_s ZJ'^{(s)}(\mathcal{F}) .$$

Proof Consider the commutative diagram

$$\begin{array}{ccccccccc} 0 & \longrightarrow & \bigcap_s ZJ'^{(s)}(\mathcal{F}) & \longrightarrow & \bigcap_s J'^{(s)}(\mathcal{F}) & \xrightarrow{d_{\mathcal{F}}} & \bigcap_s BJ'^{(s)}(\mathcal{F}) & & \\ & & \uparrow & & \| & & \uparrow & & \\ 0 & \longrightarrow & ZJ'(\mathcal{F}) & \longrightarrow & J'(\mathcal{F}) & \xrightarrow{d_{\mathcal{F}}} & BJ'(\mathcal{F}) & \longrightarrow & 0 . \end{array}$$

Here, the central vertical equality is the analog (2.5.54), the arrows that are not given by $d_{\mathcal{F}}$ and do not go to 0 denote inclusion maps, and the bottom row is exact. Since the surjective maps $d_{\mathcal{F}} : J'^{(s)}(\mathcal{F}) \to BJ'^{(s)}(\mathcal{F})$ form a homomorphism between projective spectra whose kernel is the projective spectrum consisting of the spaces $ZJ'^{(s)}(\mathcal{F})$, the top row is also exact [Wen03, Proposition 3.1.8]. Thus the left-hand-side vertical arrow is an equality of vector spaces. □

Proposition 6.2.4 *The canonical map $\bar{H}^\bullet J'(\mathcal{F}) \to \widehat{H}^\bullet J'(\mathcal{F})$ is a linear isomorphism.*

Proof Consider the commutative diagram

$$\begin{array}{ccccccccc} 0 & \longrightarrow & \bigcap_s \bar{B}J'^{(s)}(\mathcal{F}) & \longrightarrow & \bigcap_s ZJ'^{(s)}(\mathcal{F}) & \longrightarrow & \widehat{H}^\bullet J'(\mathcal{F}) & \longrightarrow & 0 \\ & & \| & & \| & & \uparrow & & \\ 0 & \longrightarrow & \bar{B}J'(\mathcal{F}) & \longrightarrow & ZJ'(\mathcal{F}) & \longrightarrow & \bar{H}^\bullet J'(\mathcal{F}) & \longrightarrow & 0 . \end{array}$$

Here, Corollary 6.2.2 and 6.2.3 give the vertical equalities of vector spaces, the vertical arrow is canonical, and the other maps are canonical; in particular, the bottom row is exact. Lemma 6.2.1 also shows that every $BJ^{(s)}(\mathcal{F})$ is dense in $BJ^{(s')}(\mathcal{F})$ for $s' < s$. Hence the right derived functor $\varprojlim^1$ satisfies $\varprojlim^1 \bar{B}J^{(s)}(\mathcal{F}) = 0$ as $s \uparrow +\infty$ [Wen03, Theorem 3.2.1], obtaining that the top row is also exact by [Wen03, Corollary 3.1.5]. Then the result follows. □

On the other hand, the kind of arguments that will be given in Sect. 6.5 can be adapted to show the exactness of the sequence

$$0 \leftarrow H^\bullet K'(\mathcal{F}) \xleftarrow{\widehat{R}'_*} \widehat{H}^\bullet I'(\mathcal{F}) \xleftarrow{\widehat{\iota}'_*} \widehat{H}^\bullet J'(\mathcal{F}) \leftarrow 0 , \tag{6.2.1}$$

where $\widehat{R}'_* = \varprojlim \bar{R}'_{s*}$ and $\widehat{\iota}'_* = \varprojlim \bar{\iota}'_{s*}$, using the homomorphisms induced by (6.1.8). This fits into a commutative diagram

$$\begin{array}{ccccccccc} 0 & \longleftarrow & H^\bullet K'(\mathcal{F}) & \longleftarrow & \bar{H}^\bullet I'(\mathcal{F}) & \longleftarrow & \bar{H}^\bullet J'(\mathcal{F}) & \longleftarrow & 0 \\ & & \| & & \downarrow & & \downarrow \cong & & \\ 0 & \longleftarrow & H^\bullet K'(\mathcal{F}) & \longleftarrow & \widehat{H}^\bullet I'(\mathcal{F}) & \longleftarrow & \widehat{H}^\bullet J'(\mathcal{F}) & \longleftarrow & 0 , \end{array}$$

where the top row is the exact sequence (6.1.7), and the vertical arrows are canonical. The last vertical arrow is a linear isomorphism by Proposition 6.2.4. Then the central vertical arrow is also a linear isomorphism by the five lemma.

6.3 Description of $H^\bullet K'(\mathcal{F})$

As explained in Sect. 6.1, there is no loss of generality in assuming $\mathcal{F}$ is oriented, and then we can apply (3.2.19)–(3.2.21) to get $K^{(s)}(\mathcal{F}; \Omega M) \equiv K^{(s)}(\mathcal{F}; \Omega N\mathcal{F})$, where we can consider $d_{\mathcal{F}}$ or $d^{\mathrm{t}}_{\mathcal{F}}$ using the leafwise flat structure of $\Omega N\mathcal{F}$.

Consider the notation of Sect. 5.3. Since $d\omega = \eta \wedge \omega$ and $d_{\mathcal{F}}$ satisfies the derivation rule on products of smooth leafwise currents and smooth leafwise forms (Sect. 3.2.8), it follows that the version of Proposition 5.3.1 with coefficients in $\Omega N\mathcal{F}$ states that

$$\left\{\begin{aligned} K(\mathcal{F}; \Omega N\mathcal{F}) &\equiv \textstyle\bigoplus_k C^\infty(M^0; \Lambda) \equiv \bigoplus_k C^\infty(M^0; \Lambda \otimes \Omega^{-k} N M^0) , \\ d_{\mathcal{F}} &\equiv \textstyle\bigoplus_k d_{-k} \equiv \bigoplus_k d , \end{aligned}\right. \tag{6.3.1}$$

where k runs in $\mathbb{N}_0$. Moreover the subcomplex $K^{(s)}(\mathcal{F}; \Omega N\mathcal{F}) \subset K(\mathcal{F}; \Omega N\mathcal{F})$ corresponds to the finite direct sum with $k < -s - 1/2$. Taking dual spaces and transposing maps, using (2.8.4) and (2.9.8), we get the following consequence.

Corollary 6.3.1 *We have identities of topological complexes,*

$$K'(\mathcal{F}) \equiv \prod_k C^{-\infty}(M^0; \Lambda) \equiv \prod_k C^{-\infty}(M^0; \Lambda \otimes \Omega^k N M^0) ,$$

$$d_{\mathcal{F}} \equiv \prod_k d_k \equiv \prod_k d ,$$

where k runs in $\mathbb{N}_0$. *Moreover the quotient complex* $K'^{(s)}(\mathcal{F})$ *corresponds to the finite direct sum with* $k < s - 1/2$.

Corollary 6.3.2 *We have TVS-identities,*

$$H^\bullet K'(\mathcal{F}) \equiv \prod_k H_k^\bullet(M^0) \equiv \prod_k H^\bullet(M^0, \Omega^k NM^0) ,$$

where k runs in $\mathbb{N}_0$. *Moreover* $H^\bullet K'^{(s)}(\mathcal{F})$ *is the quotient space of* $H^\bullet K'(\mathcal{F})$ *given by the finite product with* $k < s - 1/2$. *In particular,* (6.1.4) *is satisfied.*

Remark 6.3.3 The differential complexes on M^0 used in Corollary 6.3.1 split into direct sums of the same complexes given by leaves $L \subset M^0$. The same applies to their cohomologies, used in Corollary 6.3.2.

Corollary 6.3.4 *There is a canonical TVS-isomorphism,*

$$H^\bullet K'(\mathcal{F}) \equiv H^\bullet(M^0) \oplus H^{n''-\bullet} K(\mathcal{F})' ,$$

Proof Apply Corollaries 5.3.2 and 6.3.2, and (2.9.6). □

6.4 Description of $\bar{H}^\bullet J'(\mathcal{F})$

Like in Sect. 6.4, by (2.7.4) and (2.7.5), for $m \in \mathbb{R}$,

$$J'^m(\mathcal{F}) \cong \rho^m H_{\mathrm{b}}^{-\infty}(\boldsymbol{M}; \Lambda\boldsymbol{\mathcal{F}}) \equiv \rho^{m-\frac{1}{2}} H^{-\infty}(\mathring{\boldsymbol{M}}; \Lambda\mathring{\boldsymbol{\mathcal{F}}}) , \tag{6.4.1}$$

$$J'(\mathcal{F}) \cong \bigcap_m \rho^m H_{\mathrm{b}}^{-\infty}(\boldsymbol{M}; \Lambda\boldsymbol{\mathcal{F}}) = \bigcap_m \rho^m H^{-\infty}(\mathring{\boldsymbol{M}}; \Lambda\mathring{\boldsymbol{\mathcal{F}}}) , \tag{6.4.2}$$

as topological complexes with $d_{\mathcal{F}}$, $d_{\boldsymbol{\mathcal{F}}}$ and $d_{\mathring{\boldsymbol{\mathcal{F}}}}$, using the b-metric $\boldsymbol{g}$ to define $H_{\mathrm{b}}^{-\infty}(\boldsymbol{M}; \Lambda\boldsymbol{\mathcal{F}})$, and using $\boldsymbol{g}|_{\mathring{\boldsymbol{M}}}$ to define $H^{-\infty}(\mathring{\boldsymbol{M}}; \Lambda\mathring{\boldsymbol{\mathcal{F}}})$. The leafwise version of (2.9.4) (Sect. 3.5) also gives isomorphisms of topological complexes,

$$\rho^{-m+\frac{1}{2}} : \left(\rho^{m-\frac{1}{2}} H^{\infty}(\mathring{\boldsymbol{M}}; \Lambda\mathring{\boldsymbol{\mathcal{F}}}), d_{\mathring{\boldsymbol{\mathcal{F}}}}\right) \xrightarrow{\cong} \left(H^{\infty}(\mathring{\boldsymbol{M}}; \Lambda\mathring{\boldsymbol{\mathcal{F}}}), d_{\mathring{\boldsymbol{\mathcal{F}}},m-\frac{1}{2}}\right) . \tag{6.4.3}$$

By (6.4.1) and (6.4.3), and the analog of (3.4.16) for $\Delta_{\mathring{\boldsymbol{\mathcal{F}}},m-\frac{1}{2}}$ in $H^{-\infty}(\mathring{\boldsymbol{M}}; \Lambda\mathring{\boldsymbol{\mathcal{F}}})$ (Sect. 3.5), we get induced TVS-isomorphisms

$$\bar{H}^\bullet J'^m(\mathcal{F}) \cong \bar{H}^\bullet\left(\rho^{m-\frac{1}{2}} H^{-\infty}(\mathring{\boldsymbol{M}}; \Lambda\mathring{\boldsymbol{\mathcal{F}}}), d_{\mathring{\boldsymbol{\mathcal{F}}}}\right) \tag{6.4.4}$$

$$\cong \bar{H}^\bullet\left(H^{-\infty}(\mathring{\boldsymbol{M}}; \Lambda\mathring{\boldsymbol{\mathcal{F}}}), d_{\mathring{\boldsymbol{\mathcal{F}}},m-\frac{1}{2}}\right) \tag{6.4.5}$$

$$\cong \ker \Delta_{\mathring{\boldsymbol{\mathcal{F}}},m-\frac{1}{2}} . \tag{6.4.6}$$

By the analog of (6.1.2) for $J'(\mathcal{F})$ and (6.4.6), the LCHS $\bar{H}^\bullet J(\mathcal{F})$ is a projective limit of a sequence of Hilbertian spaces, and therefore a Fréchet space. The isomorphisms (6.4.4) and (6.4.5) are also true in cohomology.

Theorem 1.3.5 follows from the analog of (6.1.2) for $J'(\mathcal{F})$ and (6.4.1)–(6.4.3).

6.5 Short Exact Sequence of Dual-Conormal Reduced Cohomology

The goal of this section is to prove the exactness of (6.1.7). Some remarks will indicate how to adapt the proof to show also the exactness of (6.1.6) and (6.2.1).

6.5.1 The Maps F'_m

For every $m \in \mathbb{R}$, let

$$F'_m = E^{\mathrm{t}}_{-m} : I'^{(s)}(\mathcal{F}) \to J'^{m}(\mathcal{F}) ,$$

where $s = 0$ if $m \le 0$, and $m < s \in \mathbb{Z}^+$ if $m > 0$, and where

$$E_{-m} : J^{-m}(\mathcal{F}; \Omega M) \to I^{(-s)}(\mathcal{F}; \Omega M)$$

is given by the version of Corollary 5.5.2 with ΩM (see Remark 6.1.1); thus

$$F'_m d_{\mathcal{F}} = d_{\mathcal{F}} F'_m . \tag{6.5.1}$$

Since $s \ge m, 0$, the map $j'_{s,m}$ is defined, and we have

$$F'_m \iota'_s = E^{\mathrm{t}}_{-m} R^{\mathrm{t}}_{-s} = (R_{-s} E_{-m})^{\mathrm{t}} = j^{\mathrm{t}}_{-m,-s} = j'_{s,m} . \tag{6.5.2}$$

6.5.2 The Maps E'_m

For $s \in \mathbb{R}$ and $m > s + n/2 + 1$, let

$$E'_m = F^{\mathrm{t}}_{-m} : K'^{(s')}(\mathcal{F}) \to I'^{(s)}(\mathcal{F}) ,$$

where $s' = 0$ if $m \le 0$, and $m < s' \in \mathbb{Z}^+$ if $m > 0$. Here, we use the map

$$F_{-m} : I^{(-s)}(\mathcal{F}; \Omega M) \to K^{(-s')}(\mathcal{F}; \Omega M)$$

given by the version of Sect. 6.5.1 with coefficients in ΩM (Remark 5.5.3). Consider

$$F'_m = j'_{m,s} F'_m : I'^{(s')}(\mathcal{F}) \to J'^{(s)}(\mathcal{F}) , \tag{6.5.3}$$

which is the transpose of the version of (5.5.2) with coefficients in ΩM,

$$E_{-m} : J^{(-s)}(\mathcal{F}; \Omega M) \to I^{(-s')}(\mathcal{F}; \Omega M) .$$

Then (6.5.2) becomes

$$F'_m \iota'_s = j'_{s,s'} . \tag{6.5.4}$$

Transposing the versions of (5.5.3)–(6.5.1) with coefficients in ΩM, we get

$$\iota'_s F'_m + E'_m R'_{s'} = j'_{s',s} : I'^{(s')}(\mathcal{F}) \to I'^{(s)}(\mathcal{F}) , \tag{6.5.5}$$

$$R'_s E'_m = j'_{s',s} : K'^{(s')}(\mathcal{F}) \to K'^{(s)}(\mathcal{F}) , \tag{6.5.6}$$

$$E'_m d_{\mathcal{F}} = d_{\mathcal{F}} E'_m . \tag{6.5.7}$$

Take greater numbers, $s_1 > s$, $m_1 > m$ and $s'_1 > s'$, satisfying the same inequalities as s, m and s'. Using (6.5.3), the transposition of the versions of (5.5.6) and (5.5.7) with coefficients in ΩM give

$$F'_m j'_{s'_1,s'} = j'_{s_1,s} F'_{m_1} , \tag{6.5.8}$$

$$E'_m j'_{s'_1,s'} = j'_{s_1,s} E'_{m_1} . \tag{6.5.9}$$

6.5.3 *The Equality* $\ker \bar{R}'_* = \operatorname{im} \bar{\iota}'_*$

We already know that $\ker \bar{R}'_* \supset \operatorname{im} \bar{\iota}'_*$. To prove $\ker \bar{R}'_* \subset \operatorname{im} \bar{\iota}'_*$, take any class $\overline{[u]} \in \ker \bar{R}'_*$ in $\bar{H}^\bullet I'(\mathcal{F})$. Thus there is some net v_l in $K'(\mathcal{F})$ such that $R'u = \lim_l d_{\mathcal{F}} v_l$ in $K'(\mathcal{F})$. Write $u_s = j'_s u \in I'^{(s)}(\mathcal{F})$ and $v_{l,s} = j'_s v_l \in K'^{(s)}(\mathcal{F})$. Take s, m and s' satisfying the conditions of Sect. 6.5.2, obtaining $E'_m : K'^{(s')}(\mathcal{F}) \to I'^{(s)}(\mathcal{F})$ and $F'_m : I'^{(s')}(\mathcal{F}) \to J'^{(s)}(\mathcal{F})$. Let $a_s = F'_m u_{s'} \in J'^{(s)}(\mathcal{F})$ and $b_{l,s} = E'_m v_{l,s'} \in I'^{(s)}(\mathcal{F})$. By (6.5.1),

$$d_{\mathcal{F}} a_s = F'_m d_{\mathcal{F}} u_{s'} = 0 .$$

Moreover, by (6.5.5) and (6.5.7),

$$u_s = j'_{s',s} u_{s'} = \iota'_s F'_m u_{s'} + E'_m R'_{s'} u_{s'}$$

$$= \iota'_s a_s + \lim_l E'_m d_{\mathcal{F}} v_{l,s'} = \iota'_s a_s + \lim_l d_{\mathcal{F}} b_{l,s} .$$

Now consider the above notation for greater real numbers s_1, m_1 and s'_1, satisfying the same properties as s, m and s'. By (6.5.8) and (6.5.9),

$$j'_{s_1,s} a_{s_1} = j'_{s_1,s} F'_{m_1} u_{s'_1} = F'_m j'_{s'_1,s'} u_{s'_1} = F'_m u_{s'} = a_s .$$

$$j'_{s_1,s} b_{l,s_1} = j'_{s_1,s} E'_{m_1} v_{l,s'_1} = E'_m j'_{s'_1,s'} v_{l,s'_1} = E'_m v_{l,s'} = b_{l,s} .$$

Therefore, taking $s \uparrow +\infty$, $m \uparrow +\infty$ and $s' \uparrow +\infty$, satisfying the above relations, the elements $a_s \in J'^{(s)}(\mathcal{F})$ and $b_{l,s} \in I'^{(s)}(\mathcal{F})$ define elements $a := (a_s)_s \in ZJ'(\mathcal{F})$ and $b_l := (b_{l,s})_s \in I'(\mathcal{F})$, and we have $u = \iota' a + \lim_l d_{\mathcal{F}} b_l$. Hence $\overline{[u]} = \bar{\iota}_*(\overline{[a]})$.

Remark 6.5.1 A similar argument, taking an element $v \in K'(\mathcal{F})$ instead of a net v_l, shows the equality $\ker R'_* = \operatorname{im} \iota'_*$ in $H^\bullet I'(\mathcal{F})$.

Remark 6.5.2 As before, to prove $\ker \widehat{R}'_* = \operatorname{im} \hat{\iota}'_*$ in (6.2.1), we only have to prove "$\subset$". For any $\hat{u} := (\overline{[u_s]}_s)_s \in \ker \widehat{R}'_*$, there is some $v \in K'(\mathcal{F})$ such that $R'_s u_s = d_{\mathcal{F}} v_s$, where $v_s = j'_s v$. Moreover, $j'_{s',s} u_{s'} = u_s + \lim_l d_{\mathcal{F}} g_{l,s,s'}$ for some net $g_{l,s,s'}$ in $I'^{(s)}(\mathcal{F})$. Take a_s and $b_{s,l}$ as above. The given argument shows that

$$d_{\mathcal{F}} a_s = 0 , \quad u_s + \lim_l d_{\mathcal{F}} g_{l,s',s} = \iota'_s a_s + \lim_l d_{\mathcal{F}} b_{l,s} ,$$

$$j'_{s_1,s} a_{s_1} = a_s + \lim_l d_{\mathcal{F}} F'_m g_{l,s'_1,s'} , \quad j'_{s_1,s} b_{l,s_1} = b_{l,s} .$$

Hence $\hat{a} := (\overline{[a_s]}_s)_s \in \widehat{H}^\bullet J'(\mathcal{F})$ is defined and $\hat{\iota}'_*(\hat{a}) = \hat{u}$.

6.5.4 *Injectivity of* $\bar{\iota}'_*$

Let $\overline{[u]} \in \bar{H}^r J'(\mathcal{F})$ such that $\bar{\iota}'_*(\overline{[u]}) = 0$. This means that there is a net v_l in $I'(\mathcal{F})$ such that $\iota' u = \lim_l d_{\mathcal{F}} v_l$ in $I'(\mathcal{F})$. Write $u_s = j'_s u \in ZK'^{(s)}(\mathcal{F})$ and $v_{l,s} = j'_s v_l \in I'^{(s)}(\mathcal{F})$. With the notation of Sect. 5.5.3, let $b_{l,s} = F'_m v_{l,s'} \in J'^{(s)}(\mathcal{F})$. By (6.5.1) and (6.5.4),

$$u_s = j'_{s',s} u_{s'} = F'_m \iota'_{s'} u_{s'} = \lim_l F'_m d_{\mathcal{F}} v_{l,s'} = \lim_l d_{\mathcal{F}} b_{l,s} .$$

Like in Sect. 6.5.3, it can be shown that, taking $s \uparrow +\infty$, $m \uparrow +\infty$ and $s' \uparrow +\infty$ as above, the elements $b_{l,s} \in J'^{(s)}(\mathcal{F})$ define elements $b_l := (b_{l,s})_s \in J'(\mathcal{F})$, and we have $u = \lim_l d_{\mathcal{F}} b_l$. Thus $\overline{[u]} = 0$ in $\bar{H}^r J'(\mathcal{F})$.

Remark 6.5.3 Like in Remark 6.5.1, we also get the injectivity of ι'_*.

Remark 6.5.4 To prove the injectivity of $\hat{\iota}'_*$, take any $\hat{u} := (\overline{[u_s]_s})_s \in \ker \hat{\iota}'_*$ in (6.2.1). Then there is some net $v_{l,s}$ in every $I'^{(s)}(\mathcal{F})$ such that $\iota'_s u_s = \lim_l d_{\mathcal{F}} v_{l,s}$ in $I'^{(s)}(\mathcal{F})$. Moreover, $j'_{s',s} u_{s'} = u_s + \lim_l d_{\mathcal{F}} g_{l,s,s'}$ for some net $g_{l,s,s'}$ in $J'^{(s)}(\mathcal{F})$. Take $b_{l,s}$ as before. The above argument shows that

$$u_s + \lim_l d_{\mathcal{F}} g_{l,s',s} = \lim_l d_{\mathcal{F}} b_{l,s} .$$

So $\hat{u} = 0$ in $\widehat{H}^\bullet J'(\mathcal{F})$.

6.5.5 *Surjectivity of* $\bar{R}'_*$

Take any $[u] \in H^\bullet K(\mathcal{F})$, and write $u_s = j'_s u \in ZK^{(s)}(\mathcal{F})$. With the notation of Sect. 6.5.2, we have $v_s := E'_m v_{s'} \in ZI^{(s')}(\mathcal{F})$ by (6.5.7), and $R'_s v_s = j'_{s',s} u_{s'} = u_s$ by (6.5.6).

Now consider the above notation for greater real numbers s_1, m_1 and s'_1, satisfying the same properties as s, m and s'. By (6.5.9),

$$j'_{s_1,s} v_{s_1} = j'_{s_1,s} E'_{m_1} v_{s'_1} = E'_m j'_{s'_1,s'} v_{s'_1} = E'_m v_{s'} = v_s .$$

So $v := (v_s)_s \in ZI'(\mathcal{F})$ satisfies $R'v = u$, and therefore $\bar{R}'_*(\overline{[v]}) = \overline{[u]}$.

Remark 6.5.5 Using cohomology instead of reduced cohomology, the analogous argument gives the surjectivity of R'_*.

Remark 6.5.6 To prove the surjectivity of $\widehat{R}'_*$ in (6.2.1), for any $[u] \in H^\bullet K(\mathcal{F})$, define u_s and v_s as above. We also have $R'_s v_s = u_s$ and $j'_{s_1,s} v_{s_1} = v_s$. Thus $\hat{v} := (\overline{[v_s]_s})_s \in \widehat{H}^\bullet I'(\mathcal{F})$ and $\widehat{R}'_* \hat{v} = [u]$.

6.6 Functoriality and Leafwise Homotopy Invariance

6.6.1 *Pull-Back of Dual-Conormal Leafwise Currents*

Consider the notation and conditions of Sect. 5.7.3 (including the conditions of Sect. 5.7.1). According to Sect. 2.3.5, the map (3.2.23) has a continuous extension

$$\phi^* : I'(\mathcal{F}) \to I'(\mathcal{F}') , \tag{6.6.1}$$

defined as the composition

$$I'(\mathcal{F}) \xrightarrow{\phi^*} I'(M', M'^0; \phi^* \Lambda \mathcal{F}) \xrightarrow{\phi^*} I'(\mathcal{F}') ,$$

like (5.7.2), using (2.3.9) with $E = \Lambda\mathcal{F}$. We can also describe (6.6.1) as the restriction of (2.8.24) to dual-conormal currents of bidegree $(0, \bullet)$, like in (3.2.33). The map (6.6.1) is also a restriction of (3.2.33).

Similarly, the analogs of (2.2.20) with $E = \Lambda\mathcal{F}$ for (2.7.9) and (2.7.10) induce continuous homomorphisms

$$\phi^* : K'(\mathcal{F}) \to K'(\mathcal{F}') , \tag{6.6.2}$$

$$\phi^* : J'(\mathcal{F}) \to J'(\mathcal{F}') . \tag{6.6.3}$$

By passing to cohomology and reduced cohomology, we get continuous homomorphisms,

$$\left\{\begin{aligned} &\phi^* : H^\bullet K'(\mathcal{F}) \to H^\bullet K'(\mathcal{F}') ,\\ \phi^* : H^\bullet I'(\mathcal{F}) \to H^\bullet I'(\mathcal{F}') , \quad &\phi^* : \bar{H}^\bullet I'(\mathcal{F}) \to \bar{H}^\bullet I'(\mathcal{F}') ,\\ \phi^* : H^\bullet J'(\mathcal{F}) \to H^\bullet J'(\mathcal{F}') , \quad &\phi^* : \bar{H}^\bullet J'(\mathcal{F}) \to \bar{H}^\bullet J'(\mathcal{F}') . \end{aligned}\right. \tag{6.6.4}$$

The assignment of the homomorphisms (6.6.1)–(6.6.4) is functorial.

6.6.2 Description of $\phi^* : K'(\mathcal{F}) \to K'(\mathcal{F}')$

Consider the notation and conditions of Sect. 5.7.2, and assume also that ϕ is a submersion. By the density of the space of smooth forms in the space of currents, we get from (5.7.6) that

$$\phi^* : C^{-\infty}(M^0; \Lambda) \to C^{-\infty}(M'^0; \Lambda) \tag{6.6.5}$$

is a cochain map for $d_{s\eta}$ and $d_{s\eta'}$ ($s \in \mathbb{R}$), and we get from (5.7.7) that

$$\phi^* : C^{-\infty}(M^0; \Lambda \otimes \Omega^s NM^0) \to C^{-\infty}(M'^0; \Lambda \otimes \Omega^s NM'^0) \tag{6.6.6}$$

is another cochain map for the de Rham differentials defined with the flat bundle structures of $\Omega^s NM^0$ and $\Omega^s NM'^0$.

Proposition 6.6.1 *According to Corollary 6.3.1, the map* (6.6.2) *is given by*

$$\phi^* \equiv \prod_k \phi^* \equiv \prod_k \phi^* ,$$

where the terms of the first direct sum are given by (6.6.5), *and the terms of the second direct sum are given by* (6.6.6), *taking* $s = k$.

Proof Apply Propositions 2.8.1, 2.8.2 and 5.7.4. □

6.6.3 Push-Forward of Dual-Conormal Leafwise Currents

Consider the notation and conditions of Sect. 5.7.3 (containing those of Sect. 5.7.1). Then the case of (3.2.33) on smooth leafwise forms has a continuous extension

$$\phi_* : I'_{\mathrm{c/cv}}(\mathcal{F}') \to I'_{\mathrm{c/\cdot}}(\mathcal{F}) . \tag{6.6.7}$$

This map can be described as the restriction of the map (2.8.33) to dual-conormal currents of bidegree $(0, \bullet)$, like (3.2.33) in Sect. 3.2.15. We can also describe (6.6.7) as the composition

$$I'_{\mathrm{c/cv}}(M', L'; \Lambda\mathcal{F}) \xrightarrow{\pi_{\mathrm{top}}} I'_{\mathrm{c/\,cv}}(M', L'; \phi^*\Lambda\mathcal{F} \otimes \Omega_{\mathrm{\,fiber}}) \xrightarrow{\phi_*} I'_{\mathrm{c/\cdot}}(M, L; \Lambda\mathcal{F}) ,$$

like in (3.2.35), where ϕ_* is given by (2.3.11) with $E = \Lambda\mathcal{F}$. The map (6.6.7) is also a restriction of the case of (3.2.33) for leafwise currents.

According to Sect. 2.7.7, the map (6.6.7) induces homomorphisms

$$\phi_* : K'(\mathcal{F}') \to K'(\mathcal{F}) , \tag{6.6.8}$$

$$\phi_* : J'_{\mathrm{c/cv}}(\mathcal{F}') \to J'_{\mathrm{c/\cdot}}(\mathcal{F}) . \tag{6.6.9}$$

Like in Sect. 6.6.1, we get induced continuous homomorphisms,

$$\left\{\begin{array}{c} \phi_* : H^\bullet K'(\mathcal{F}') \to H^\bullet K'(\mathcal{F}) , \\ \phi_* : H^\bullet I'_{\mathrm{c}}(\mathcal{F}') \to H^\bullet I'_{\mathrm{c}}(\mathcal{F}) , \quad \phi_* : \bar{H}^\bullet I'_{\mathrm{c}}(\mathcal{F}') \to \bar{H}^\bullet I'_{\mathrm{c}}(\mathcal{F}) , \\ \phi_* : H^\bullet J'_{\mathrm{c}}(\mathcal{F}') \to H^\bullet J'_{\mathrm{c}}(\mathcal{F}) , \quad \phi_* : \bar{H}^\bullet J'_{\mathrm{c}}(\mathcal{F}') \to \bar{H}^\bullet J'_{\mathrm{c}}(\mathcal{F}) . \end{array}\right. \tag{6.6.10}$$

The assignments of homomorphisms (6.6.7)–(6.6.10) are clearly functorial.

6.6.4 Leafwise Homotopy Invariance

Consider the notation and conditions of Sect. 5.7.5, and assume that every H_t is a submersion. Like in Sect. 5.7.5, according to Sects. 6.6.1 and 6.6.3, the corresponding leafwise homotopy operator $\mathsf{h} : C^\infty(M; \Lambda\mathcal{F}) \to C^\infty(M'; \Lambda\mathcal{F}')$ has continuous linear extensions,

$$\mathsf{h} : K'(\mathcal{F}) \to K'(\mathcal{F}') , \quad \mathsf{h} : I'(\mathcal{F}) \to I'(\mathcal{F}') , \quad \mathsf{h} : J'(\mathcal{F}) \to J'(\mathcal{F}') .$$

By continuity and according to Sect. 3.2.16, we have $H_1^* - H_0^* = \mathsf{h}d_{\mathcal{F}} + d_{\mathcal{F}'}\mathsf{h}$ with H_0^* and H_1^* given by (6.6.1), (6.6.2), and (6.6.3). Hence we get the following.

Proposition 6.6.2 *Let $\phi, \psi : (M', \mathcal{F}') \to (M, \mathcal{F})$ be smooth foliated maps transverse to M^0 with $\phi^{-1}(M^0) = \psi^{-1}(M^0) = M'^0$. If ϕ is leafwise homotopic to ψ, then ϕ and ψ induce the same homomorphisms* (6.6.4).

6.7 Action of Foliated Flows on the Dual-Conormal Sequence

Consider the notation and conditions of Sect. 5.8.

Proposition 6.7.1 *According to Corollary 6.3.2 and Remark 6.3.3,*

$$\phi^{t*} \equiv \prod_{k,L} e^{k\varkappa_L t} \equiv \prod_{k,L} e^{k\varkappa_L t}$$

on $H^\bullet K(\mathcal{F})$, where k runs in $\mathbb{N}_0$ and L in $\pi_0 M^0$.

Proof Argue like in the proof of Proposition 5.8.1 and its previous observations, using Corollary 6.3.1, Remark 6.3.3 and Proposition 6.6.2. □

Corollaries 6.3.1 and 6.3.2, Remark 6.3.3, and Proposition 6.7.1 show Theorem 1.3.4.

Chapter 7
Contribution from M^1

7.1 Operators on a Suspension Foliation

Consider again the notation of Sect. 4.2, where the case of a weakly simple foliated flow $\phi = \{\phi^t\}$ on a suspension foliated manifold $(M, \mathcal{F})$ was described. Equip $\mathring{M}_\pm$ with $g_\pm$, obtaining that $\mathring{\mathcal{F}}_\pm$ is of bounded geometry (Proposition 4.2.1). We can assume ϕ is of $\mathbb{R}$-local bounded geometry on $\mathring{M}_\pm$ by Proposition 4.2.15 and according to Sect. 2.4.7. Thus, on $\tilde{\mathring{M}}_\pm \equiv (\mathring{M}_\pm, \tilde{g}_\pm)$, $\tilde{\mathring{\mathcal{F}}}_\pm$ is of bounded geometry and $\tilde{\phi}$ is of $\mathbb{R}$-local bounded geometry. Consider the leafwise perturbed operators for $(\tilde{\mathring{M}}_\pm, \tilde{\mathring{\mathcal{F}}}_\pm)$ and $(\mathring{M}_\pm, \mathring{\mathcal{F}}_\pm)$ defined by the leafwise-closed form η_0 and the leafwise-exact form $\tilde{\eta}_0$ (Sect. 3.3.1). For any $\psi \in \mathcal{A}$, $f \in C^\infty_c(\mathbb{R})$ and $z \in \mathbb{C}$, the operator

$$\mathring{P}_\pm = \int_{-\infty}^{+\infty} \phi_z^{t*}\, \psi(D_{\mathring{\mathcal{F}}_\pm, z})\, f(t)\, dt \tag{7.1.1}$$

on $H^{-\infty}(\mathring{M}_\pm; \Lambda\mathring{\mathcal{F}}_\pm)$ is a version of (3.4.17) for ϕ_z^{t*} and $D_{\mathring{\mathcal{F}}_\pm, z}$, and therefore it is smoothing by the corresponding analog of (3.4.18). Let $\mathring{K}_\pm = K_{\mathring{P}_\pm}$.

By (4.2.1) and (4.2.2), for $\gamma \in \Gamma$ and $t \in \mathbb{R}$, the equality $\tilde{\phi}_z^{t*} T_\gamma^* = T_\gamma^* \tilde{\phi}_z^{t*}$ means that, for all $x \in \mathbb{R}$,

$$\tilde{\phi}_{x,z}^{t*} T_\gamma^* = T_\gamma^* \tilde{\phi}_{a_\gamma x, z}^{t*} \tag{7.1.2}$$

on $C^\infty(\widetilde{L}, \Lambda)$ (Sect. 2.9.4). Consider also the notation of Sect. 2.9.9 for the regular covering $\pi = \pi_L : \widetilde{L} \to L$ used in the suspension construction; in particular, recall the notation $\tilde{k}_z$. Recall that $h_\pm(\gamma) = \varkappa^{-1} \ln a_\gamma$ for $\gamma \in \Gamma$, and $D_\pm(x, \tilde{y}) = \varkappa^{-1} \ln |x|$ for $(x, \tilde{y}) \in \tilde{\mathring{M}}_\pm$ (Sect. 4.2.1). Thus, by the version of (3.4.21) for the leafwise perturbed differential complex (Sect. 3.5), and by (4.2.1), (4.2.2),

J. A. Álvarez López et al., *A Trace Formula for Foliated Flows*, Lecture Notes in Mathematics 2387, https://doi.org/10.1007/978-3-032-15413-2_7

and (7.1.2), if $\hat{\psi} \in C^\infty_{\mathrm{c}}(\mathbb{R})$, then

$$\mathring{K}_\pm([x, \tilde{y}], [x', \tilde{y}']) \equiv \sum_{\gamma\in\Gamma} \mathring{K}_{\pm,\gamma}([x, \tilde{y}], [x', \tilde{y}']) \tag{7.1.3}$$

for all $(x, \tilde{y}), (x', \tilde{y}') \in \mathring{\widetilde{M}}_\pm$, where

$$\begin{aligned}\mathring{K}_{\pm,\gamma}&([x, \tilde{y}], [x', \tilde{y}'])\\ &= \frac{1}{|\varkappa|}\tilde{\phi}_{x,z}^{\frac{1}{\varkappa}\ln\frac{x'}{a_\gamma x}*} T_\gamma^*\tilde{k}_z\Big(\gamma\cdot\tilde{\phi}_x^{\frac{1}{\varkappa}\ln\frac{x'}{a_\gamma x}}(\tilde{y}), \tilde{y}'\Big) f\Big(\frac{1}{\varkappa}\ln\frac{x'}{a_\gamma x}\Big)\Big|\frac{dx'}{x'}\Big| .\end{aligned} \tag{7.1.4}$$

According to Sect. 2.5.26, for the boundary-defining function ρ on $M_\pm$ (Sect. 4.2.5), let ρ and ρ' denote its lifts to $(M_\pm)^2$ from the left and right factors, and let $s = \rho/\rho' : (M_\pm)^2 \to [0, \infty]$. We have corresponding smooth functions ρ, ρ' and s on $(M_\pm)^2_{\mathrm{b}}$. Similarly, let η and η' denote the lifts of η from the left and right factors. Using (4.2.13), we get

$$(M_\pm)^2 \equiv [0, \infty)_\rho \times [0, \infty)_{\rho'} \times L^2 , \quad (\mathring{M}_\pm)^2 \equiv (0, \infty)^2 \times L^2 .$$

Then

$$(M_\pm)^2_{\mathrm{b}} \equiv [0, \infty)_\rho \times [0, \infty]_s \times L^2 ,$$

with boundary components $\mathrm{lb} = \{s = 0\}$, $\mathrm{rb} = \{s = \infty\}$ and $\mathrm{ff} = \{\rho = 0\}$. Moreover

$$\Delta_{\mathrm{b}} \equiv \{\, (\rho, 1, y, y) \mid \rho \ge 0,\ y \in L \,\} .$$

With the above identities, the restriction of $\beta_{\mathrm{b}} : (M_\pm)^2_{\mathrm{b}} \to (M_\pm)^2$ to the interior corresponds to the diffeomorphism

$$(0, \infty)^2 \times L^2 \to (0, \infty)^2 \times L^2 , \quad (\rho, s, y, y') \mapsto (\rho, \rho s^{-1}, y, y') .$$

Similar observations apply to $\widetilde{M}_\pm$, using $\widetilde{L}$ instead of L, and using the lifts $\tilde{\rho}$, $\tilde{\rho}'$ and $\tilde{s}$ instead of ρ, ρ' and s. The subscript "$\pm$" will be added to the notation Δ_{b} and $\Delta_{\mathrm{b},0} = \Delta_{\mathrm{b}} \cap \mathrm{ff}$ if needed.

Let $\mathring{\kappa}_\pm$ be the C^∞ section of $\beta_{\mathrm{b}}^*(\Lambda\mathcal{F}_\pm \boxtimes (\Lambda\mathcal{F}_\pm^* \otimes \Omega M_\pm))$ on the interior of $(M_\pm)^2_{\mathrm{b}}$ that corresponds to $\mathring{K}_\pm$ via β_{b}. If $\hat{\psi} \in C^\infty_{\mathrm{c}}(\mathbb{R})$, then, using the changes of variables

$$x = \pm e^{-F(\tilde{y})}\tilde{\rho} , \quad x' = \pm e^{-F(\tilde{y}')}\tilde{\rho}' ,$$

with

$$\ln\frac{x'}{x} = F(\tilde{y}) - F(\tilde{y}') - \ln\tilde{s}\,, \quad \frac{dx'}{x'} = -\tilde{\eta}' + \frac{d\tilde{\rho}'}{\tilde{\rho}'}\,, \quad \frac{d\tilde{s}}{\tilde{s}} = -\frac{d\tilde{\rho}'}{\tilde{\rho}'}\,,$$
$$x' = 0 \Leftrightarrow \tilde{\rho}' = 0 \Leftrightarrow \tilde{s} = \infty\,, \quad x' = \pm\infty \Leftrightarrow \tilde{\rho}' = \infty \Leftrightarrow \tilde{s} = 0\,,$$

it follows from (7.1.3) and (7.1.4) that

$$\mathring{K}_{\pm}(\rho, \rho', [\tilde{y}], [\tilde{y}']) = \sum_{\gamma\in\Gamma} \mathring{K}_{\pm,\gamma}(\rho, \rho', [\tilde{y}], [\tilde{y}'])\,, \tag{7.1.5}$$

$$\mathring{\kappa}_{\pm}(\rho, s, [\tilde{y}], [\tilde{y}']) = \sum_{\gamma\in\Gamma} \mathring{\kappa}_{\pm,\gamma}(\rho, s, [\tilde{y}], [\tilde{y}'])\,, \tag{7.1.6}$$

where

$$\begin{aligned}
&\mathring{K}_{\pm,\gamma}(\rho, \rho', [\tilde{y}], [\tilde{y}']) \\
&= \frac{1}{|\varkappa|}\tilde{\phi}_{\pm e^{-F(\tilde{y})}\rho,z}^{\frac{1}{\varkappa}(F(\tilde{y})-F(\tilde{y}')+\ln\frac{\rho'}{a_\gamma\rho})*} T_\gamma^*\,\tilde{k}_z\Big(\gamma\cdot\tilde{\phi}_{\pm e^{-F(\tilde{y})}\rho}^{\frac{1}{\varkappa}(F(\tilde{y})-F(\tilde{y}')+\ln\frac{\rho'}{a_\gamma\rho})}(\tilde{y}), \tilde{y}'\Big) \\
&\quad\times f\Big(\frac{1}{\varkappa}\Big(F(\tilde{y}) - F(\tilde{y}') + \ln\frac{\rho'}{a_\gamma\rho}\Big)\Big)\Big|\frac{d\rho'}{\rho'}\Big|\,,
\end{aligned} \tag{7.1.7}$$

and

$$\begin{aligned}
&\mathring{\kappa}_{\pm,\gamma}(\rho, s, [\tilde{y}], [\tilde{y}']) \\
&= \frac{1}{|\varkappa|}\tilde{\phi}_{\pm e^{-F(\tilde{y})}\rho,z}^{\frac{1}{\varkappa}(F(\tilde{y})-F(\tilde{y}')-\ln a_\gamma s)*} T_\gamma^*\,\tilde{k}_z\Big(\gamma\cdot\tilde{\phi}_{\pm e^{-F(\tilde{y})}\rho}^{\frac{1}{\varkappa}(F(\tilde{y})-F(\tilde{y}')-\ln a_\gamma s)}(\tilde{y}), \tilde{y}'\Big) \\
&\quad\times f\Big(\frac{1}{\varkappa}\big(F(\tilde{y}) - F(\tilde{y}') - \ln a_\gamma s\big)\Big)\Big|\frac{ds}{s}\Big|\,.
\end{aligned} \tag{7.1.8}$$

Let us look for more general conditions on ψ to get (7.1.6) by using the Fréchet algebra and $\mathbb{C}[z]$-module $\mathcal{A}$ (Sect. 2.9.8). Notice that every $\mathring{\kappa}_{\pm,\gamma}(\rho, s, [\tilde{y}], [\tilde{y}'])$ has a C^∞ extension to $\rho = 0$.

Lemma 7.1.1 *If $\psi \in \mathcal{A}$, then, given any fundamental domain $\boldsymbol{F} \subset \widetilde{L}$, the series in (7.1.6) converges with all covariant derivatives, uniformly on $\rho \geq 0$, $0 < s < \infty$ and $\tilde{y}, \tilde{y}' \in \boldsymbol{F}$. Moreover its sum is $\mathring{\kappa}_{\pm}(\rho, s, [\tilde{y}], [\tilde{y}'])$ for $\rho > 0$.*

Proof Since $\tilde{\phi}$ is of $\mathbb{R}$-local bounded geometry on $\mathring{\widetilde{M}}_{\pm}$ with $\tilde{g}_{\pm}$ and $\operatorname{supp} f$ is compact, we can take $R > 0$ and $\boldsymbol{K} \subset \widetilde{L}^2$ like in the proof of Proposition 4.2.7 with $\operatorname{supp} f \subset I$ for any compact $I \subset \mathbb{R}$. Using (2.9.19) with this $\boldsymbol{K}$, for any

$W > 0$, we get

$$\left|\tilde{k}_z(\gamma \cdot \tilde{\phi}_x^t(\tilde{y}), \tilde{y}')\right| \leq C_1' e^{-\frac{W}{c_1}|\gamma|} \|\psi\|_{\mathcal{A},W,N}$$

for $\gamma \in \Gamma$, $x \in \mathbb{R}^{\pm}$, $t \in \operatorname{supp} f$ and $\tilde{y}, \tilde{y}' \in \boldsymbol{F}$. Using again the $\mathbb{R}$-local bounded geometry of $\tilde{\phi}$ on $\mathring{\widetilde{M}}_{\pm}$ with $\tilde{g}_{\pm}$ and compactness of I, it follows that there is some $C_2 = C_2(z, W) > 0$ such that

$$\left|\mathring{\kappa}_{\pm,\gamma}(\rho, s, [\tilde{y}], [\tilde{y}'])\right| \leq C_2 e^{-\frac{W}{c_1}|\gamma|} \|\psi\|_{\mathcal{A},W,N} \|f\|_{I,C^0} \tag{7.1.9}$$

for $\gamma \in \Gamma$, $\rho \geq 0$, $s > 0$ and $\tilde{y}, \tilde{y}' \in \boldsymbol{F}$. By (2.9.20) and (7.1.9), if $W > c_1 W_0$, then the series in (7.1.6) converges uniformly on $\rho \geq 0$, $s > 0$ and $\tilde{y}, \tilde{y}' \in \boldsymbol{F}$, and the norm of its sum is $\leq C\|\psi\|_{\mathcal{A},W,N}$ for some $C = C(z, W, N) > 0$.

With more generality, by the $\mathbb{R}$-local bounded geometry of $\tilde{\phi}$ on $\mathring{\widetilde{M}}_{\pm}$ and the compactness of I, the higher order derivatives of $\tilde{\phi}_x^t(\tilde{y})$ with respect to x, t and $\tilde{y}$ (in normal coordinates) are also uniformly bounded for $x \in \mathbb{R}^{\pm}$, $t \in I$ and $\tilde{y} \in \widetilde{L}$. Hence, for every $m \in \mathbb{N}_0$, it follows from (2.9.19) that

$$\left|\nabla_{\tilde{y}}^{m_1} \nabla_{\tilde{y}'}^{m_2} \tilde{k}_z(\gamma \cdot \tilde{\phi}_x^t(\tilde{y}), \tilde{y}')\right| \leq C_1' e^{-\frac{W}{c_1}|\gamma|} \|\psi\|_{\mathcal{A},W,N+m}$$

for $\gamma \in \Gamma$, $x \in \mathbb{R}^{\pm}$, $t \in I$, $\tilde{y}, \tilde{y}' \in \boldsymbol{F}$ and $m_1 + m_2 \leq m$. Moreover, since I and $\boldsymbol{F}$ are compact, there is some $c_3 \in \mathbb{R}$ such that, for all $\tilde{y}, \tilde{y}' \in \boldsymbol{F}$,

$$\ln a_\gamma s > c_3 \Rightarrow \varkappa^{-1}\big(F(\tilde{y}) - F(\tilde{y}') - \ln a_\gamma s\big) \notin I\,.$$

Thus we can assume $s^{-1} < e^{-c_3} a_\gamma$, yielding $s^{-1} < e^{c_0|\gamma| - c_3}$ by (4.2.19). Hence there is some $C_3 = C_3(z, W, m) > 0$ such that

$$\begin{aligned}\left|\partial_\rho^{m_1} \partial_s^{m_2} \nabla_{[\tilde{y}]}^{m_3} \nabla_{[\tilde{y}']}^{m_4} \mathring{\kappa}_{\pm,\gamma}(\rho, s, [\tilde{y}], [\tilde{y}'])\right| \\ \leq C_3 e^{(mc_0 - \frac{W}{c_1})|\gamma|} \|\psi\|_{\mathcal{A},W,N+m} \|f\|_{I,C^m}\,,\end{aligned} \tag{7.1.10}$$

for $\gamma \in \Gamma$, $\rho \geq 0$, $s > 0$, $\tilde{y}, \tilde{y}' \in \boldsymbol{F}$ and $m_1 + \cdots + m_4 \leq m$. By (2.9.20) and (7.1.10), if $W > c_1(mc_0 + W_0)$, then the series defined by the covariant derivatives of order $\leq m$ of the terms in (7.1.6) is also convergent, uniformly on $\rho \geq 0$, $s > 0$ and $\tilde{y}, \tilde{y}' \in \boldsymbol{F}$, and the norm of its sum is $\leq C'\|\psi\|_{\mathcal{A},W,N+m} \|f\|_{I,C^m}$ for some $C' = C'(z, W, N, m) > 0$.

We already know that the sum of the series in (7.1.6) is $\mathring{\kappa}_{\pm}(\rho, s, [\tilde{y}], [\tilde{y}'])$ for $\rho > 0$ if $\hat{\psi} \in C_c^\infty(\mathbb{R})$. Then this also holds when $\psi \in \mathcal{A}$, as follows by taking a convergent sequence $\psi_k \to \psi$ in $\mathcal{A}$ with $\widehat{\psi_k} \in C_c^\infty(\mathbb{R})$, and using the above estimates of the sum. □

Remark 7.1.2 Like in Remark 2.9.3, Lemma 7.1.1 is true for any $\psi \in \mathcal{S}$ since Γ is abelian. But $\psi \in \mathcal{A}$ is needed for the estimates (7.1.9) and (7.1.10), which will be used later.

Proposition 7.1.3 *If $\psi \in \mathcal{A}$, then $\mathring{\kappa}_\pm$ has a C^∞ extension $\kappa_\pm$ to $(M_\pm)^2_b$, also given by (7.1.6) and (7.1.8) using C^∞ extensions $\kappa_{\pm,\gamma}$ of the sections $\mathring{\kappa}_{\pm,\gamma}$ to $(M_\pm)^2_b$, which vanishes to all orders at* $\mathrm{lb} \cup \mathrm{rb}$. *Therefore $\kappa_\pm = \kappa_{P_\pm}$ for some $P_\pm \in \Psi_b^{-\infty}(M_\pm; \Lambda\mathcal{F}_\pm)$ induced by $\mathring{P}_\pm$.*

Proof By Lemma 7.1.1, $\mathring{\kappa}_\pm$ extends smoothly to ff ($\rho = 0$ and $0 < s < \infty$).

Take any compact $I \subset \mathbb{R}$ containing $\operatorname{supp} f$. According to (7.1.8), the sum in (7.1.6) can be taken for $\gamma \in \Gamma$ with

$$\varkappa^{-1}(F(\tilde{y}) - F(\tilde{y}') - \ln a_\gamma s) \in I .$$

Then, since $\tilde{y}, \tilde{y}' \in \boldsymbol{F}$, there exists $R > 0$ such that

$$\ln s - R < \ln a_\gamma < \ln s + R .$$

Combining this with (4.2.19), we get

$$c_0^{-1}(\pm \ln s - R) < |\gamma| . \tag{7.1.11}$$

By (2.9.20), (7.1.9), and (7.1.11), for any $W > c_1 W_0$, there is some $C_2' = C_2'(z, W) > 0$ such that, for $\rho \geq 0$, $s > 0$ and $y, y' \in L$,

$$\begin{aligned}
\left|\mathring{\kappa}_\pm(\rho, s, y, y')\right| &< C_2 \sum_{|\gamma| > c_0^{-1}(\pm \ln s - R)} e^{-c_1^{-1} W |\gamma|} \, \|\psi\|_{\mathcal{A}, W, N} \, \|f\|_{I, C^0} \\
&< C_2 e^{-(c_1^{-1} W - W_0) c_0^{-1}(\pm \ln s - R)} \sum_{\gamma \in \Gamma} e^{-W_0 |\gamma|} \, \|\psi\|_{\mathcal{A}, W, N} \, \|f\|_{I, C^0} \\
&< C_2' s^{\mp (c_1^{-1} W - W_0) c_0^{-1}} \, \|\psi\|_{\mathcal{A}, W, N} \, \|f\|_{I, C^0} .
\end{aligned} \tag{7.1.12}$$

Using (7.1.10) and (7.1.11), we similarly get that, for $m \in \mathbb{N}_0$, if $W > c_1(m c_0 + W_0)$, then there is some $C_3' = C_3'(z, W, m) > 0$ such that

$$\begin{aligned}
&\left|\partial_\rho^{m_1} \partial_s^{m_2} \nabla_y^{m_3} \nabla_{y'}^{m_4} \mathring{\kappa}_\pm(\rho, s, y, y')\right| \\
&\quad < C_3' s^{\mp (c_1^{-1} W - m c_0 - W_0) c_0^{-1}} \, \|\psi\|_{\mathcal{A}, W, N+m} \, \|f\|_{I, C^m}
\end{aligned} \tag{7.1.13}$$

for $\rho \geq 0$, $s > 0$, $y, y' \in L$ and $m_1 + m_2 + m_3 + m_4 \leq m$. Since W is arbitrarily large, it follows that $\mathring{\kappa}_\pm$ also extends smoothly to $\mathrm{lb} \cup \mathrm{rb}$ ($s = 0, \infty$), where it vanishes to all orders. □

Notation 7.1.4 The subscripts "ψ", "f" or "z" may be added to the notation $\mathring{P}_\pm$, $\mathring{K}_\pm$, $\mathring{K}_{\pm,\gamma}$, $\mathring{\kappa}_\pm$, $\kappa_\pm$ and $P_\pm$ if needed.

Proposition 7.1.5 *The bilinear map*

$$\mathcal{A} \times C^\infty_c(\mathbb{R}) \to C^\infty\big((M_\pm)^2_b; \beta_b^*(\Lambda\mathcal{F}_\pm \boxtimes (\Lambda\mathcal{F}^*_\pm \otimes \Omega M_\pm))\big) , \quad (\psi, f) \mapsto \kappa_{\pm,\psi,f} ,$$

is continuous.

Proof This is an additional consequence of (7.1.12) and (7.1.13). □

Recall the notation $\phi_L = \{\phi^t_L\} = \{\phi^t_0\}$ on $M^0 \equiv L$ and $\tilde{\phi}_{\tilde{L}} = \{\tilde{\phi}^t_{\tilde{L}}\} = \{\tilde{\phi}^t_0\}$ on $\widetilde{M}^0 \equiv \widetilde{L}$ (Sect. 4.2.2), and the trivialization ν of ${}_+N\partial M_\pm$ (Sect. 4.2.5). Recall also that the indicial family is defined in Sect. 2.5.28.

Proposition 7.1.6 *We have*

$$I_{\nu_\pm}(P_{\pm,z}, \lambda) \equiv \int_{-\infty}^{+\infty} \phi^{t*}_{L,z+i\lambda}\, \psi(D_{L,z+i\lambda})\, e^{i\lambda\varkappa t} f(t)\, dt .$$

Proof By (2.5.63), it is enough to show that the Schwartz kernel of the smoothing operator

$$\int_{-\infty}^{+\infty} \phi^{t*}_{L,z+i\lambda}\, \psi(D_{L,z+i\lambda})\, e^{i\lambda\varkappa t} f(t)\, dt$$

on $C^{-\infty}(L; \Lambda)$ is given by

$$\int_0^\infty s^{-i\lambda} \kappa_{\pm,z}(0, s, y, y')\, \frac{ds}{s} ,$$

at every $(y, y') \in L^2$. By Lemma 7.1.1 and Proposition 7.1.3, for all $\tilde{y}, \tilde{y}' \in \widetilde{L}$,

$$\begin{aligned}
&\int_0^\infty s^{-i\lambda} \kappa_{\pm,z}(0, s, [\tilde{y}], [\tilde{y}'])\, \frac{ds}{s}\\
&= \frac{1}{|\varkappa|} \sum_{\gamma\in\Gamma} \int_0^\infty s^{-i\lambda}\, \tilde{\phi}_{0,z}^{\frac{1}{\varkappa}(F(\tilde{y})-F(\tilde{y}')-\ln a_\gamma s)*}\, T^*_\gamma\\
&\quad \circ \tilde{k}_z\Big(\gamma \cdot \tilde{\phi}_0^{\frac{1}{\varkappa}(F(\tilde{y})-F(\tilde{y}')-\ln a_\gamma s)}(\tilde{y}), \tilde{y}'\Big)\\
&\quad \times f\Big(\frac{1}{\varkappa}\Big(F(\tilde{y}) - F(\tilde{y}') - \ln a_\gamma s\Big)\Big)\frac{ds}{s}\\
&= \sum_{\gamma\in\Gamma} e^{i\lambda(F(\tilde{y}')-F(\tilde{y})-\ln a_\gamma)} \int_{-\infty}^{+\infty} \tilde{\phi}^{t*}_{0,z} T^*_\gamma\, \tilde{k}_z(\gamma \cdot \tilde{\phi}^t_0(\tilde{y}), \tilde{y}') e^{i\lambda\varkappa t} f(t)\, dt ,
\end{aligned}$$

where we have used the change of variable

$$t = \varkappa^{-1}(F(\tilde{y}) - F(\tilde{y}') - \ln s + \ln a_\gamma)\ ,$$

with

$$s = e^{F(\tilde{y})-F(\tilde{y}')+\ln a_\gamma - \varkappa t}\ , \quad dt = -\frac{ds}{\varkappa s}\ ,$$

$$s = 0 \Leftrightarrow t = \operatorname{sign}(\varkappa)\infty\ , \quad s = \infty \Leftrightarrow t = -\operatorname{sign}(\varkappa)\infty\ .$$

By Proposition 2.9.2,

$$k_{z+i\lambda}([\tilde{y}], [\tilde{y}']) = \sum_{\gamma\in\Gamma} T_\gamma^* \tilde{k}_{z+i\lambda}(\gamma\cdot\tilde{y}, \tilde{y}')\ .$$

Moreover, by (2.9.4),

$$\tilde{k}_{z+i\lambda}(\tilde{y}, \tilde{y}') = e^{i\lambda(F(\tilde{y}')-F(\tilde{y}))}\tilde{k}_z(\tilde{y}, \tilde{y}')\ .$$

So, by (4.2.2) and (4.2.4),

$$\begin{aligned}
&\sum_{\gamma\in\Gamma} e^{i\lambda(F(\tilde{y}')-F(\tilde{y})-\ln a_\gamma)}\tilde{\phi}_{0,z}^{t*} T_\gamma^*\, \tilde{k}_z(\gamma\cdot\tilde{\phi}_0^t(\tilde{y}), \tilde{y}') \\
&\quad = \sum_{\gamma\in\Gamma} e^{i\lambda(F(\tilde{y}')-F(\gamma\cdot\tilde{y}))}\tilde{\phi}_{0,z}^{t*} T_\gamma^*\, \tilde{k}_z(\gamma\cdot\tilde{\phi}_0^t(\tilde{y}), \tilde{y}') \\
&\quad = \sum_{\gamma\in\Gamma} e^{i\lambda(\tilde{\phi}_0^{t*}F-F)(\gamma\cdot\tilde{y})}\tilde{\phi}_{0,z}^{t*} e^{i\lambda(F(\tilde{y}')-F(\gamma\cdot\tilde{\phi}_0^t(\tilde{y})))} T_\gamma^*\, \tilde{k}_z(\gamma\cdot\tilde{\phi}_0^t(\tilde{y}), \tilde{y}') \\
&\quad = \tilde{\phi}_{0,z+i\lambda}^{t*} \sum_{\gamma\in\Gamma} T_\gamma^* \tilde{k}_{z+i\lambda}(\gamma\cdot\tilde{\phi}_0^t(\tilde{y}), \tilde{y}') \equiv \phi_{0,z+i\lambda}^{t*}\, k_{z+i\lambda}(\phi_0^t([\tilde{y}]), [\tilde{y}'])\ .
\end{aligned}$$

□

Notation 7.1.7 In Notation 7.1.4, we may also add the subscript "u" if we use a family of functions $\psi_u \in \mathcal{A}$ depending on a parameter u. This also applies to k_z and $\tilde{k}_z$.

The identity element of Γ is denoted by e.

Proposition 7.1.8 *If $\psi_u(x) = e^{-ux^2}$, then $(\kappa_{\pm,u} - \kappa_{\pm,e,u})|_{\Delta_{b,\pm}} \to 0$ as $u \downarrow 0$ in the C^∞ topology.*

Proof For $\gamma \in \Gamma$ and $p = [x, \tilde{y}] = [a_\gamma x, \gamma\cdot\tilde{y}] \in \mathring{M}_\pm$, by (4.2.1) and (4.2.2),

$$\phi^{-h_\pm(\gamma)}(p) = \left[x, \phi_{a_\gamma x}^{-h_\pm(\gamma)}(\gamma\cdot\tilde{y})\right] = \left[x, \gamma\cdot\phi_x^{-h_\pm(\gamma)}(\tilde{y})\right]\ .$$

Thus, using that π_M defines an isometric diffeomorphism of $\{x\} \times \widetilde{L} \equiv \widetilde{L}$ to L_p, it follows from Corollary 4.2.8 that there is some $c_3 > 0$, independent of p and γ, such that, if $h_\pm(\gamma) \in \operatorname{supp} f$, then

$$d_{\widetilde{L}}\big(\gamma \cdot \phi_x^{-h_\pm(\gamma)}(\tilde{y}), \tilde{y}\big) = d_{\mathcal{F}}\big(\phi^{-h_\pm(\gamma)}(p), p\big) \geq c_3|\gamma| \,.$$

Therefore, by (2.9.14) and since ϕ is of $\mathbb{R}$-local bounded geometry, for $m_1, m_2 \in \mathbb{N}_0$, $0 < u < u_0$, $\gamma \in \Gamma$, $\rho > 0$ and $\tilde{y} \in \widetilde{L}$, we get

$$\big|\partial_\rho^{m_1} \nabla_{\tilde{y}}^{m_2} \tilde{k}_{u,z}\big(\gamma \cdot \phi_{\pm e^{-F(\tilde{y})}\rho}^{-h_\pm(\gamma)}(\tilde{y}), \tilde{y}\big)\big| \leq C_1 u^{-(n-1+m_1+m_2)/2} e^{-C_2 c_3^2 |\gamma|^2/u} \,,$$

where $\tilde{k}_{u,z}$ is the Schwartz kernel of $\psi_u(D_{\widetilde{L},z}) = e^{-u\Delta_{\widetilde{L},z}}$. Using again the $\mathbb{R}$-local bounded geometry of ϕ and the compactness of $\operatorname{supp} f$, it follows that there is some $C_3 > 0$ such that

$$|\partial_\rho^{m_1} \nabla_y^{m_2} \kappa_{\gamma,u}(\rho, 1, y, y)| \leq C_3 u^{-(n-1+m_1+m_2)/2} e^{-C_2 c_3^2 |\gamma|^2/u}$$

for $m_1, m_2 \in \mathbb{N}_0$, $0 < u < u_0$, $\gamma \in \Gamma$, $\rho > 0$ and $y \in L$. So

$$|\partial_\rho^{m_1} \nabla_y^{m_2} (\kappa_{\pm,u} - \kappa_{\pm,e,u})(\rho, 1, y, y)| \leq C_3 u^{-(n-1+m_1+m_2)/2} \sum_{\gamma \in \Gamma \setminus \{e\}} e^{-C_2 c_3^2 |\gamma|^2/u} \,,$$

which converges to zero as $u \downarrow 0$. □

Corollary 7.1.9 *If $\psi_u(x) = e^{-ux^2}$ and $f(0) = 0$, then $\kappa_{\pm,u}|_{\Delta_{b,\pm}} \to 0$ as $u \downarrow 0$ in the C^∞ topology.*

Recall the notation $e(\mathcal{F}_\pm, g_{\mathcal{F}_\pm})$ if $n - 1$ is even (Sects. 3.2.7 and 4.2.6), and also the notation $C^{0,\infty}_{\varpi\pm}(M_\pm; {}^b\Omega)$ (Sect. 3.1.8).

Corollary 7.1.10 *If $\psi_u(x) = e^{-ux^2}$, then*

$$\lim_{u \downarrow 0} \operatorname{str}(\kappa_{\pm,u}|_{\Delta_{b,\pm}}) \equiv \begin{cases} f(0)\, e(\mathcal{F}_\pm, g_{\mathcal{F}_\pm})\, |\omega_\pm| & \text{if } n-1 \text{ is even} \\ 0 & \text{if } n-1 \text{ is odd} \end{cases}$$

in $C^{0,\infty}_{\varpi\pm}(M_\pm; {}^b\Omega)$, using the identity $\Delta_{b,\pm} \equiv M_\pm$.

Proof By Propositions 7.1.3 and 7.1.8, and (7.1.5), (7.1.7) and (4.2.17), for all $\rho > 0$ and $y = [\tilde{y}] \in L$ with $\tilde{y} \in \widetilde{L}$,

$$\begin{aligned} &\lim_{u \downarrow 0} \operatorname{str} \mathring{\kappa}_{\pm,u}(\rho, 1, y, y) \\ &\qquad = \lim_{u \downarrow 0} \operatorname{str} \mathring{\kappa}_{\pm,e,u}(\rho, 1, y, y) = \lim_{u \downarrow 0} \operatorname{str} \mathring{K}_{\pm,e,u}(\rho, \rho, y, y) \\ &\qquad = \frac{f(0)}{|\varkappa|} \left|\frac{d\rho}{\rho}\right| \lim_{u \downarrow 0} \operatorname{str} \tilde{k}_{z,u}(\tilde{y}, \tilde{y}) = f(0)\, |\omega_\pm|(y) \lim_{u \downarrow 0} \operatorname{str} \tilde{k}_{z,u}(\tilde{y}, \tilde{y}) \,. \end{aligned}$$

But, by Theorem 2.9.4,

$$\lim_{u\downarrow 0} \operatorname{str} \tilde{k}_{z,u}(\tilde{y}, \tilde{y}) = e(\widetilde{L}, g_{\widetilde{L}})(\tilde{y}) = e(L, g_L)(y) \equiv e(\mathcal{F}_\pm, g_{\mathcal{F}_\pm})(\rho, y)$$

if $n - 1$ is even, and

$$\lim_{u\downarrow 0} \operatorname{str} \tilde{k}_{z,u}(\tilde{y}, \tilde{y}) = 0$$

if $n - 1$ is odd. □

7.2 Operators on the Components M_l^1

Consider the notation of Sects. 4.3.3 and 4.3.4; in particular, consider the boundary-defining function $\rho = \rho_l$ on every M_l and the trivialization $\nu = \nu_l$ of ${}_+N\partial M_l$. According to Sect. 2.5.26, consider also the lifts of ρ to M_l^2 from the left and right factors, ρ and ρ', and the function $s = s_l = \rho/\rho' : M_l^2 \to [0, \infty]$, as well as the corresponding functions ρ, ρ' and s on $(M_l)^2_{\mathrm{b}}$. Equip $\mathring{M}_l$ with the Riemannian metric $g_{\mathrm{b},l}$, so that $\mathring{\mathcal{F}}_l$ becomes a Riemannian foliation of bounded geometry (Sect. 3.4.1). Consider the leafwise perturbed operators for $(\mathring{M}_l, \mathring{\mathcal{F}}_l)$ defined by the leafwise-closed form η_0, which agrees with η on the collar neighborhood of the boundary we have fixed. For any $\psi \in \mathcal{A}$, $f \in C_c^\infty(\mathbb{R})$, $z \in \mathbb{C}$ and every index l, the operator

$$\mathring{P}_l = \int_{-\infty}^{+\infty} \phi_{l,z}^{t*}\, \psi(D_{\mathring{\mathcal{F}}_{l,z}}) f(t)\, dt$$

on $H^{-\infty}(\mathring{M}_l; \Lambda\mathring{\mathcal{F}}_l)$ is a twisted version of (3.4.17), which is smoothing by the appropriate analog of (3.4.18) (Sect. 3.5). Let

$$\mathring{K}_l = K_{\mathring{P}_l} \in C^\infty(\mathring{M}_l^2; \Lambda\mathring{\mathcal{F}}_l \boxtimes (\Lambda\mathring{\mathcal{F}}_l^* \otimes \Omega\mathring{M}_l)) .$$

Lemma 7.2.1 *For any compact $I \subset \mathbb{R}$ containing* $\operatorname{supp} f$*, and for all $k, m \in \mathbb{N}_0$ and $a \in \mathbb{R}$, there are some $C', C'' > 0$ and $N \in \mathbb{N}_0$, depending only on I, k, m and a, such that*

$$\|\mathring{P}_l\|_{\rho^a H^k, \rho^a H^m} \le C'' \|\psi\|_{\mathcal{A},C',N} \|f\|_{I,C^N} .$$

Proof By (2.9.4) and (4.3.1), $D_{\mathring{\mathcal{F}}_{l,z}} = \rho^a D_{\mathring{\mathcal{F}}_{l,z+a,z-a}} \rho^{-a}$ (see Sect. 3.3.3). So the result follows from the analog of (3.4.20) for $D_{\mathring{\mathcal{F}}_{l,z+a,z-a}}$ (Sect. 3.5). □

Proposition 7.2.2 *The kernel $\mathring{K}_l$ has a C^∞ extension to $M_l^2 \setminus (\partial M_l)^2$ that vanishes to all orders on $(\partial M_l \times \mathring{M}_l) \cup (\mathring{M}_l \times \partial M_l)$.*

Proof We will use the arguments from the proof of [Hör83, Theorem 5.2.6].

For any $q \in \mathring{M}_l$ and $\alpha \in \Lambda_q \mathcal{F}_l \otimes \Omega_q^{-1} \mathring{M}_l$, we have $\delta_q^\alpha \in H^k(\mathring{M}_l; \Lambda \mathring{\mathcal{F}}_l)$ for any $k < -n/2$, and $\|\delta_q^\alpha\|_k \leq C_k |\alpha|$, where $C_k > 0$ is independent of q and α (Sect. 2.2.6). Therefore, by the definition of weighted Sobolev spaces and the properties of Dirac sections at submanifolds (Sects. 2.1.12 and 2.2.6), for all $a \in \mathbb{R}$, we have $\delta_q^\alpha \in \rho^a H^k(\mathring{M}_l; \Lambda \mathring{\mathcal{F}}_l)$ and

$$\|\delta_q^\alpha\|_{\rho^a H^k} \leq C_k \rho(q)^{-a} |\alpha| .$$

Moreover, for any $\alpha \in C^\infty(\mathring{M}_l; \Lambda \mathring{\mathcal{F}}_l \otimes \Omega^{-1} \mathring{M}_l)$, the map

$$\mathring{M}_l \to \rho^a H^k(\mathring{M}_l; \Lambda \mathring{\mathcal{F}}_l) , \quad q \mapsto \delta_q^{\alpha(q)} ,$$

is continuous by the continuity of (2.2.13).

Fix any compact $I \subset \mathbb{R}$ containing $\operatorname{supp} f$. By Lemma 7.2.1, we have $\mathring{P}_l \delta_q^\alpha \in \rho^a H^m(\mathring{M}_l; \Lambda \mathring{\mathcal{F}}_l)$ for any $m \in \mathbb{N}_0$, and

$$\left\| \mathring{P}_l \delta_q^\alpha \right\|_{\rho^a H^m} \leq C'_m \rho(q)^{-a} \|\psi\|_{\mathcal{A}, C', N} \|f\|_{I, C^N} |\alpha|$$

for $q \in \mathring{M}_l$ and $\alpha \in \Lambda_q \mathring{\mathcal{F}}_l \otimes \Omega_q^{-1} \mathring{M}_l$, where $C'_m > 0$ is independent of a, q and α. Moreover, for any $\alpha \in C^\infty(\mathring{M}_l; \Lambda \mathring{\mathcal{F}}_l \otimes \Omega^{-1} \mathring{M}_l)$, the map

$$\mathring{M}_l \to \rho^a H^m(\mathring{M}_l; \Lambda \mathring{\mathcal{F}}_l) , \quad q \mapsto \mathring{P}_l \delta_q^{\alpha(q)} ,$$

is continuous. On the other hand, by (2.2.14), for all $q \in \mathring{M}_l$ and $\alpha \in \Lambda_q \mathring{\mathcal{F}}_l \otimes \Omega_q^{-1} \mathring{M}_l$,

$$\mathring{K}_l(\cdot, q)(\alpha) = \mathring{P}_l \delta_q^\alpha \in C^\infty(\mathring{M}_l; \Lambda \mathring{\mathcal{F}}_l) .$$

It follows that the map

$$\mathring{M}_l \to \rho^a H^m(\mathring{M}_l; \Lambda \mathring{\mathcal{F}}_l \otimes \Omega_q \mathring{M}_l) , \quad q \mapsto \mathring{K}_l(\cdot, q) ,$$

is continuous for any $a \in \mathbb{R}$ and $m \in \mathbb{N}_0$, with

$$\left\| \mathring{K}_l(\cdot, q) \right\|_{\rho^a H^m} \leq C'_m \rho(q)^{-a} \|\psi\|_{\mathcal{A}, C', N} \|f\|_{I, C^N}$$

for all $q \in \mathring{M}_l$. Using the Sobolev embedding theorem, we conclude that $\mathring{K}_l$ is continuous on $\mathring{M}_l^2$, and

$$\left| \mathring{K}_l(p, q) \right| \leq C \left(\frac{\rho(p)}{\rho(q)} \right)^a \|\psi\|_{\mathcal{A}, C', N}, \|f\|_{I, C^N}$$

$$= C\, s(p,q)^a\, \|\psi\|_{\mathcal{A},C',N}\, \|f\|_{I,C^N}\,,$$

for all $a \in \mathbb{R}$ and $p, q \in \mathring{M}_l$, where $C, C' > 0$ and $N \in \mathbb{N}_0$ are independent of a, p and q. So $\mathring{K}_l$ extends to a continuous section on $M_l^2 \setminus (\partial M_l)^2$, which vanishes on $(\partial M_l \times \mathring{M}_l) \cup (\mathring{M}_l \times \partial M_l)$.

For any $D_1, D_2 \in \mathrm{Diff}_{\mathrm{b}}^k(M_l; \Lambda\mathcal{F}_l)$, applying the above arguments to the operator $D_1 \mathring{P}_l D_2$ and using (2.1.28), it follows that, for all $a \in \mathbb{R}$ and $p, q \in \mathring{M}_l$,

$$\left|D_{1,p}\, D_{2,q}^{\mathrm{t}}\, \mathring{K}_l(p,q)\right| \le C\, s(p,q)^a\, \|\psi\|_{\mathcal{A},C',N}\, \|f\|_{I,C^N}\,, \tag{7.2.1}$$

where $C, C' > 0$ and $N \in \mathbb{N}_0$ are independent of a, p and q. □

Let $\mathring{\kappa}_l$ be the C^∞ section of $\beta_{\mathrm{b}}^*(\Lambda\mathcal{F}_l \boxtimes (\Lambda\mathcal{F}_l^* \otimes \Omega M_l))$ on the interior of $(M_l)_{\mathrm{b}}^2$ that corresponds to $\mathring{K}_l$ via β_{b}^*, using the notation of Sect. 2.5.26. Fix $0 < \epsilon < \epsilon_0$ like in Sect. 4.3.3, and consider the notation of Sects. 4.3.2 and 4.3.3. Let L be a boundary component of M_l, which can be identified with a leaf of $\mathcal{F}$ in M^0. For $0 < \sigma \le \epsilon_0$, via the identity $\mathring{T}_{L,l,\sigma} \equiv \mathring{T}'_{L,l,\sigma}$, the sections in $C_{\mathrm{c}}^\infty(\mathring{M}_l; \Lambda\mathring{\mathcal{F}}_l)$ and $C_{\mathrm{c}}^\infty(\mathring{M}_l; \Lambda\mathring{\mathcal{F}}_l \otimes \Omega)$ supported in $\mathring{T}_{L,l,\sigma}$ can be identified with sections in $C_{\mathrm{c}}^\infty(\mathring{M}'_l; \Lambda\mathring{\mathcal{F}}'_l)$ and $C_{\mathrm{c}}^\infty(\mathring{M}'_l; \Lambda\mathring{\mathcal{F}}'_l \otimes \Omega)$ supported in $\mathring{T}'_{L,l,\sigma}$. Similarly, according to Sect. 7.1,

$$\beta_{\mathrm{b}}^{-1}(\mathring{T}_{L,l,\sigma}^2) \equiv \beta_{\mathrm{b}}^{-1}(\mathring{T}_{L,l,\sigma}^{\prime 2}) \equiv \{\, (\rho, s, y, y') \in (0,\infty)^2 \times L^2 \mid \rho, \rho s^{-1} < \sigma \,\}\,.$$

The operator (7.1.1), studied in Sect. 7.1, is now expressed as

$$\mathring{P}'_{L,l} = \int_{-\infty}^{+\infty} \phi_{L,l,z}^{\prime t*}\, \psi(D_{\mathring{\mathcal{F}}'_{L,l,z}}) f(t)\, dt\,.$$

Let $\mathring{K}'_{L,l} = K_{\mathring{P}'_{L,l}}$, with lift $\mathring{\kappa}'_{L,l}$ to the interior of $(M'_{L,l})_{\mathrm{b}}^2$, and let $\kappa'_{L,l} = \kappa_{P'_{L,l}}$ denote the extension of $\mathring{\kappa}'_{L,l}$ to $(M'_{L,l})_{\mathrm{b}}^2$ given by Proposition 7.1.3.

The subscripts of Notations 7.1.4 and 7.1.7 may also be used with $\mathring{P}_l$, $\mathring{K}_l$, $\mathring{\kappa}_l$, $\mathring{P}'_{L,l}$, $\mathring{K}'_{L,l}$, $\mathring{\kappa}'_{L,l}$ and $\kappa'_{L,l}$.

Proposition 7.2.3 *Given $\psi \in \mathcal{A}$ and $u > 0$, take $\psi_u \in \mathcal{A}$ defined by $\psi_u(x) = \psi(ux)$, and consider the restrictions of $\mathring{\kappa}_{l,u}$ and $\mathring{\kappa}'_{L,l,u}$ to $\beta_b^{-1}(\mathring{T}_{L,l,\epsilon}^2) \equiv \beta_b^{-1}(\mathring{T}_{L,l,\epsilon}^{\prime 2})$. There is some $0 < \epsilon' < \epsilon$ such that, for any $R > 0$, $m, N \in \mathbb{N}_0$ and $a \in \mathbb{R}$, there exist $\widehat{C}, W > 0$ and $N' \in \mathbb{N}_0$ so that, for $m_1 + m_2 + m_3 + m_4 \le m$, $0 < u \le 1$ and $(\rho, s, y, y') \in \beta_b^{-1}(\mathring{T}_{L,l,\epsilon'}^2)$,*

$$\left|\partial_\rho^{m_1} \partial_s^{m_2} \nabla_y^{m_3} \nabla_{y'}^{m_4} (\mathring{\kappa}_{l,u} - \mathring{\kappa}'_{L,l,u})(\rho, s, y, y')\right| \le \widehat{C} e^{-\frac{R}{u}} \rho^N s^a\, \|\psi\|_{\mathcal{A},W,N'}\, \|f\|_{I,C^{N'}}\,.$$

Proof Take $C > 0$ and $c \geq 1$ like in Lemma 4.2.4 and Corollary 4.2.5, and take $c' > 0$ like in Proposition 4.2.9, for the suspension foliation $\mathcal{F}'_{L,l}$ on $M'_{L,l}$ and any compact $I \subset \mathbb{R}$ containing $\operatorname{supp} f$.

Claim 7.2.4 For $\alpha, \beta \in C^\infty_{\mathrm{c}}(\mathring{M}_{L,l}; \Lambda\mathring{\mathcal{F}}_{L,l})$, $\alpha', \beta' \in C^\infty_{\mathrm{c}}(\mathring{M}'_{L,l}; \Lambda\mathring{\mathcal{F}}'_{L,l})$ and $\xi \in \mathbb{R}$, let

$$\alpha(\xi) = e^{i\xi D_{\mathring{\mathcal{F}}_{L,l},z}}\alpha\;, \qquad \beta(\xi) = e^{i\xi D_{\mathring{\mathcal{F}}_{L,l},z}}\beta\;,$$

$$\alpha'(\xi) = e^{i\xi D_{\mathring{\mathcal{F}}'_{L,l},z}}\alpha'\;, \qquad \beta'(\xi) = e^{i\xi D_{\mathring{\mathcal{F}}'_{L,l},z}}\beta'\;.$$

The following properties hold for $0 < \sigma, \tau < \epsilon$:

(i) If α and α' are supported in $\mathring{T}_{L,l,\sigma} \equiv \mathring{T}'_{L,l,\sigma}$ and agree there, then $\alpha(\xi)$ and $\alpha'(\xi)$ are supported in $\mathring{T}_{L,l,\epsilon} \equiv \mathring{T}'_{L,l,\epsilon}$ and agree there for $|\xi| < C \ln \frac{\epsilon}{c\sigma}$.
(ii) If β and β' agree on $\mathring{T}_{L,l,\epsilon} \equiv \mathring{T}'_{L,l,\epsilon}$, then $\beta(\xi) \equiv \beta'(\xi)$ on $\mathring{T}_{L,l,\tau} \equiv \mathring{T}'_{L,l,\tau}$ for $|\xi| < C \ln \frac{\epsilon}{c\tau}$.

This is a consequence of Corollary 4.2.5 and the leafwise twisted version of (3.4.9) applied to the equation $\partial_\xi \mu(\xi) = i D_{\mathring{\mathcal{F}}_l,z}\mu(\xi)$ on $\mathring{T}_{L,l,\epsilon} \equiv \mathring{T}'_{L,l,\epsilon}$, where $\mu(\xi) = \alpha(\xi) \equiv \alpha'(\xi)$ in (i), and $\mu(\xi) = \beta(\xi) \equiv \beta'(\xi)$ in (ii).

Claim 7.2.5 Let α, α', $\alpha(\xi)$ and $\alpha'(\xi)$ be defined like in Claim 7.2.4, and let $0 < \sigma < \epsilon$ and $0 < \tau < \epsilon, \epsilon/c'$. If α and α' are supported in $\mathring{T}_{L,l,\sigma} \equiv \mathring{T}'_{L,l,\sigma}$ and agree there, then $\phi^{t*}\alpha(\xi) \equiv \phi'^{t*}\alpha'(\xi)$ on $\mathring{T}_{L,l,\tau} \equiv \mathring{T}'_{L,l,\tau}$ for any $t \in I$ and $|\xi| < C(\ln \frac{\epsilon}{c\sigma} + \ln \frac{\epsilon}{cc'\tau})$.

By Claim 7.2.4 (i), if $\xi < C \ln \frac{\epsilon}{c\sigma}$, then $\alpha(\xi)$ and $\alpha'(\xi)$ are supported in $\mathring{T}_{L,l,\epsilon} \equiv \mathring{T}'_{L,l,\epsilon}$ and agree there. Thus, by Claim 7.2.4 (ii), if $|\zeta| < C \ln \frac{\epsilon}{cc'\tau}$, then $\alpha(\xi + \zeta) \equiv \alpha'(\xi + \zeta)$ on $\mathring{T}_{L,l,c'\tau} \equiv \mathring{T}'_{L,l,c'\tau}$. Hence $\phi^{t*}\alpha(\xi + \zeta) \equiv \phi'^{t*}\alpha'(\xi + \zeta)$ on $\mathring{T}_{L,l,\tau} \equiv \mathring{T}'_{L,l,\tau}$ for all $t \in I$ since $\phi^t(\mathring{T}_{L,l,\tau}) \subset \mathring{T}_{L,l,c'\tau}$ by Proposition 4.2.9. This shows Claim 7.2.5.

Take any $\mu \in C^\infty(\mathbb{R})$ such that $0 \leq \mu \leq 1$, $\operatorname{supp}\mu \subset (-\infty, 0]$, and $\mu = 1$ on $(-\infty, -\ln 2]$. For $0 < \sigma < \epsilon$, let $\chi_\sigma = \mu(\ln\rho - \ln\sigma) \in C^\infty_{\mathrm{ub}}(\mathring{M}_l)$. We have $\chi_\sigma \geq 0$, $\operatorname{supp}\chi_\sigma \subset \mathring{T}_{L,l,\sigma}$, and $\chi_\sigma = 1$ on $\mathring{T}_{L,l,\sigma/2}$. Moreover $\chi_\sigma \in C^\infty_{\mathrm{ub}}(\mathring{M}_l)$ and $\|\chi_\sigma\|_{C^m_{\mathrm{ub}}}$ is independent of σ for $m \in \mathbb{N}_0$ because $d(\ln\rho) \in C^\infty_{\mathrm{ub}}(\mathring{M}_l; T^*\mathring{M}_l)$ (Sect. 4.3.3). Let also $0 < \tau < \epsilon, \epsilon/c'$ and define χ_τ as above. Then the operator $\chi_\tau(\mathring{P}_{l,u} - \mathring{P}'_{L,l,u})\chi_\sigma$ is well defined on $H^{-\infty}(\mathring{M}_l; \Lambda\mathring{\mathcal{F}}_l)$ via the identity $\mathring{T}_{L,l,\epsilon} \equiv \mathring{T}'_{L,l,\epsilon}$.

Let $\alpha \in C^\infty_{\mathrm{c}}(\mathring{M}_l; \Lambda\mathring{\mathcal{F}}_l)$ and $\beta \in C^\infty_{\mathrm{c}}(\mathring{M}_l; \Lambda\mathring{\mathcal{F}}_l^* \otimes \Omega)$. By Claim 7.2.5 and the version of (3.4.11) for $\xi u D_{\mathcal{F}_l,z}$ and $\xi u D_{\mathcal{F}'_{L,l},z}$ instead of tD_0 (Sect. 3.5),

$$\left\langle \chi_\tau(\mathring{P}_{l,u} - \mathring{P}'_{L,l,u})\chi_\sigma\alpha, \beta\right\rangle = \frac{1}{2\pi}\int_{|\xi| > \frac{C}{u}\ln\frac{\epsilon^2}{c^2c'\sigma\tau}} \int_{-\infty}^{+\infty} \hat{\psi}(\xi) A_{l,z,u}(t,\xi) f(t)\, d\xi\, dt\;,$$

where

$$A_{l,z,u}(t,\xi) = \left\langle \left(\phi_{l,z}^{t*} e^{i\xi u D_{\mathring{\mathcal{F}}_{l,z}}} - \phi_{L,l,z}'^{t*} e^{i\xi u D_{\mathring{\mathcal{F}}'_{L,l,z}}} \right) \chi_\sigma \alpha, \chi_\tau \beta \right\rangle .$$

Then, by the version of (3.4.10) for $\xi u D_{\mathcal{F}_{l,z}}$ and $\xi u D_{\mathcal{F}'_{L,l,z}}$ instead of tD_0 (Sect. 3.5), since ϕ_l and $\phi'_{L,l}$ are of $\mathbb{R}$-local bounded geometry, and using that $\|\chi_\sigma\|_{C^k_{\mathrm{ub}}}$ and $\|\chi_\tau\|_{C^k_{\mathrm{ub}}}$ are finite and independent of σ and τ for all $k \in \mathbb{N}_0$, we get that, for all $m \in \mathbb{R}$,

$$\begin{aligned} |A_{l,z,u}(t,\xi)| &\le \left\| \left(\phi_{l,z}^{t*} e^{i\xi u D_{\mathring{\mathcal{F}}_{l,z}}} - \phi_{L,l,z}'^{t*} e^{i\xi u D_{\mathring{\mathcal{F}}'_{L,l,z}}} \right) \chi_\sigma \alpha \right\|_m \|\chi_\tau \beta\|_{-m} \\ &\le C' e^{C_m|\xi|} \|\chi_\sigma \alpha\|_m \|\chi_\tau \beta\|_{-m} \le C'' e^{C_m|\xi|} \|\alpha\|_m \|\beta\|_{-m} , \end{aligned}$$

for some $C_m, C', C'' > 0$ independent of α, β, σ, τ, and $u \in (0,1]$. So, for all $W > 0$,

$$\begin{aligned} &\left| \left\langle \chi_\tau (\mathring{P}_{l,u} - \mathring{P}'_{L,l,u}) \chi_\sigma \alpha, \beta \right\rangle \right| \\ &\quad \le \frac{1}{2\pi} \int_{|\xi| > \frac{C}{u}(\ln \frac{\epsilon}{c\sigma} + \ln \frac{\epsilon}{cc'\tau})} \int_{-\infty}^{+\infty} \left| \hat{\psi}(\xi) \right| |A_{l,z,u}(t,\xi)| \, |f(t)| \, d\xi \, dt \\ &\quad \le C'' \|\alpha\|_m \|\beta\|_{-m} \|f\|_{L^1} \int_{|\xi| > \frac{C}{u}(\ln \frac{\epsilon}{c\sigma} + \ln \frac{\epsilon}{cc'\tau})} e^{C_m|\xi|} \left| \hat{\psi}(\xi) \right| d\xi \\ &\quad \le C'' \|\alpha\|_m \|\beta\|_{-m} \|f\|_{L^1} e^{-\frac{CW}{u}(\ln \frac{\epsilon}{c\sigma} + \ln \frac{\epsilon}{cc'\tau})} \int_{-\infty}^{\infty} e^{(W+C_m)|\xi|} \left| \hat{\psi}(\xi) \right| d\xi , \end{aligned}$$

for some $C_m, C', C'' > 0$ independent of α, β, σ, τ, and $u \in (0,1]$. Now, assume

$$\sigma < \frac{\epsilon}{ce} , \quad \tau < \frac{\epsilon}{cc'e} . \tag{7.2.2}$$

Thus $\ln \frac{\epsilon}{c\sigma}, \ln \frac{\epsilon}{cc'\tau} > 1$, obtaining

$$\begin{aligned} e^{-\frac{CW}{u}(\ln \frac{\epsilon}{c\sigma} + \ln \frac{\epsilon}{cc'\tau})} &\le e^{-\frac{CW}{u}(1 + \frac{1}{2}(\ln \frac{\epsilon}{c\sigma} + \ln \frac{\epsilon}{cc'\tau}))} \le e^{-\frac{CW}{u}} e^{-\frac{CW}{2}(\ln \frac{\epsilon}{c\sigma} + \ln \frac{\epsilon}{cc'\tau})} \\ &= e^{-\frac{CW}{u}} e^{-\frac{CW}{2}(\ln \frac{\epsilon}{c} + \ln \frac{\epsilon}{cc'})} (\sigma\tau)^{\frac{CW}{2}} \le e^{-\frac{CW}{u}} (\sigma\tau)^{\frac{CW}{2}} . \end{aligned}$$

Hence

$$\begin{aligned} &\left| \left\langle \chi_\tau (\mathring{P}_{l,u} - \mathring{P}'_{L,l,u}) \chi_\sigma \alpha, \beta \right\rangle \right| \\ &\qquad \le C''' e^{-\frac{CW}{u}} (\sigma\tau)^{\frac{CW}{2}} \|\psi\|_{\mathcal{A},W+C_m,0} \|f\|_{I,C^0} \|\alpha\|_m \|\beta\|_{-m} , \end{aligned}$$

for some $C''' > 0$ independent of α, β, σ, τ, and $u \in (0, 1]$, but involving the length of I. Thus, for any $R > 0$, $N \in \mathbb{N}_0$ and $m \in \mathbb{R}$, there are some $\widehat{C}$, $W > 0$, such that, for all σ and τ as in (7.2.2), and every $u \in (0, 1]$,

$$\left\| \chi_\tau(\mathring{P}_{I,u} - \mathring{P}'_{L,I,u})\chi_\sigma \right\|_m \le \widehat{C}e^{-\frac{R}{u}}\sigma^N\tau^N \|\psi\|_{\mathcal{A},W,0} \|f\|_{I,C^0} .$$

Using the arguments of the proof of (7.2.1), we similarly get that, for any $R > 0$, $N \in \mathbb{N}_0$ and $m, m' \in \mathbb{R}$, there are $\widehat{C}$, $W > 0$ and $N' \in \mathbb{N}_0$ such that, for all σ and τ as in (7.2.2), and every $u \in (0, 1]$,

$$\left\| \chi_\tau(\mathring{P}_{I,u} - \mathring{P}'_{L,I,u})\chi_\sigma \right\|_{m,m'} \le \widehat{C}e^{-\frac{R}{u}}\sigma^N\tau^N \|\psi\|_{\mathcal{A},W,N'} \|f\|_{I,C^{N'}} .$$

Moreover, for any $a \in \mathbb{R}$, replacing α with $\rho^{-a}\alpha$ and β with $\rho^a\beta$ in the above argument, we also get

$$\begin{aligned}\left\langle \chi_\tau(\mathring{P}_{I,u} - \mathring{P}'_{L,I,u})\chi_\sigma\rho^{-a}\alpha, \rho^a\beta \right\rangle \\ = \frac{1}{2\pi}\int_{|\xi|u>C(\ln\frac{\epsilon}{c\sigma}+\ln\frac{\epsilon}{cc'\tau})}\int_{-\infty}^{+\infty} \hat{\psi}(\xi)B_{I,z}(t,\xi,a)f(t)\,d\xi\,dt ,\end{aligned}$$

where

$$\begin{aligned}B_{I,z}(t,\xi,a) &= \left\langle \left(\phi_{I,z}^{t*}e^{i\xi uD_{\mathring{\mathcal{F}}_{I,z}}} - \phi_{L,I,z}^{\prime t*}e^{i\xi uD_{\mathring{\mathcal{F}}'_{L,I,z}}}\right)\chi_\sigma\rho^{-a}\alpha, \chi_\tau\rho^a\beta \right\rangle \\ &= \left\langle \left(\rho^a\phi_{I,z}^{t*}e^{i\xi uD_{\mathring{\mathcal{F}}_{I,z}}}\rho^{-a} - \rho^a\phi_{L,I,z}^{\prime t*}e^{i\xi uD_{\mathring{\mathcal{F}}'_{L,I,z}}}\rho^{-a}\right)\chi_\sigma\alpha, \chi_\tau\beta \right\rangle \\ &= \left\langle \left(\phi_{I,z-a}^{t*}e^{i\xi uD_{\mathring{\mathcal{F}}_{I,z-a,z+a}}} - \phi_{L,I,z-a}^{\prime t*}e^{i\xi uD_{\mathring{\mathcal{F}}'_{L,I,z-a,z+a}}}\right)\chi_\sigma\alpha, \chi_\tau\beta \right\rangle .\end{aligned}$$

Then, proceeding as above, we obtain

$$\left\| \chi_\tau(\mathring{P}_{I,u} - \mathring{P}'_{L,I,u})\chi_\sigma \right\|_{\rho^aH^m,\rho^aH^{m'}} \le \widehat{C}e^{-\frac{R}{u}}\sigma^N\tau^N \|\psi\|_{\mathcal{A},W,N'} \|f\|_{I,C^{N'}} ,$$

for some $\widehat{C}$, $W > 0$ and $N' \in \mathbb{N}_0$, depending only on R, N, m, m' and a. Using the Sobolev embedding theorem as in Proposition 7.2.2, it follows that, for any $a \in \mathbb{R}$, $R > 0$ and $N \in \mathbb{N}_0$, there are some $\widehat{C}$, $W > 0$ and $N' \in \mathbb{N}_0$ such that

$$\begin{aligned}\left| \chi_\tau(p)(\mathring{K}_{I,u}(p,q) - \mathring{K}'_{L,I,u}(p,q))\chi_\sigma(q) \right| \\ \le \widehat{C}e^{-\frac{R}{u}}\left(\frac{\rho(p)}{\rho(q)}\right)^a\sigma^N\tau^N \|\psi\|_{\mathcal{A},W,N'} \|f\|_{I,C^{N'}} ,\end{aligned}$$

for all $p, q \in \mathring{T}_{L,l,\epsilon}$ and $u \in (0, 1]$, and every σ and τ as in (7.2.2). Put

$$\epsilon' = \min\left(\frac{\epsilon}{4ce}, \frac{\epsilon}{4cc'e}\right) .$$

For $p, q \in \mathring{T}_{L,l,\epsilon'}$, we set $\tau = 3\rho(p)$ and $\sigma = 3\rho(q)$. It is clear that σ and τ satisfy (7.2.2) and $\chi_\tau(p) = \chi_\sigma(q) = 1$ (since $\rho(p) < \tau/2, \rho(q) < \sigma/2$). Therefore, by the above estimate, we get

$$\begin{aligned}\left|(\mathring{K}_{l,u} - \mathring{K}'_{L,l,u})(p, q)\right| &\le 9^N \widehat{C} e^{-\frac{R}{u}} s(p, q)^a \rho(p)^N \rho(q)^N \, \|\psi\|_{A,W,N'} \, \|f\|_{l,C^{N'}} \\ &= 9^N \widehat{C} e^{-\frac{R}{u}} s(p, q)^{a-N} \rho(p)^{a+N} \, \|\psi\|_{A,W,N'} \, \|f\|_{l,C^{N'}} ,\end{aligned}$$

for all $p, q \in \mathring{T}_{L,l,\epsilon'}$ and $u \in (0, 1]$.

For any $k \in \mathbb{N}_0$, taking arbitrary operators $D_1, D_2 \in \mathrm{Diff}^k_{\mathrm{b}}(M_l; \Lambda\mathcal{F}_l)$ and $D'_1, D'_2 \in \mathrm{Diff}^k_{\mathrm{b}}(M'_{L,l}; \Lambda\mathcal{F}'_{L,l})$ with $D_i \equiv D'_i$ on $\mathring{T}_{L,l,\epsilon} \equiv \mathring{T}'_{L,l,\epsilon}$ $(i = 1, 2)$, and applying the above arguments to the operators $D_1 \mathring{P}_{l,u} D_2$ and $D'_1 \mathring{P}'_{l,u} D'_2$, we obtain that, for all $a \in \mathbb{R}$ and $N \in \mathbb{N}_0$, there are some $\widehat{C}, W > 0$ and $N' \in \mathbb{N}_0$ such that, for all $p, q \in \mathring{T}_{L,l,\epsilon'}$ and $u \in (0, 1]$,

$$\left|D_{1,p} D^{\mathrm{t}}_{2,q} (\mathring{K}_{l,u} - \mathring{K}'_{L,l,u})(p, q)\right| \le \widehat{C} e^{-\frac{R}{u}} s(p, q)^a \rho(p)^N \, \|\psi\|_{A,W,N'} \, \|f\|_{l,C^{N'}} . \tag{7.2.3}$$

Consider the vector bundle $S = \Lambda\mathcal{F}_l \boxtimes (\Lambda\mathcal{F}^*_l \otimes \Omega M_l)$ over M_l^2. Recall that $\mathrm{Diff}^k_{\mathrm{b}}((M_l)^2_{\mathrm{b}}; \beta^*_{\mathrm{b}} S)$ is $C^\infty((M_l)^2_{\mathrm{b}})$-spanned by the lift of $\mathrm{Diff}^k_{\mathrm{b}}(M_l^2; S)$, and $\mathrm{Diff}^k_{\mathrm{b}}(M_l^2; S)$ is $C^\infty(M_l^2)$-spanned by the lift of $\mathrm{Diff}^k_{\mathrm{b}}(M_l; \Lambda\mathcal{F}_l)$ from the left-factor projection and the lift of $\mathrm{Diff}^k_{\mathrm{b}}(M_l; \Lambda\mathcal{F}^*_l \otimes \Omega)$ from right-factor projection (Sect. 2.5.26). Then it follows from (7.2.3) that, for all $A \in \mathrm{Diff}^k_{\mathrm{b}}((M_l)^2_{\mathrm{b}}; \beta^*_{\mathrm{b}} S)$, $a \in \mathbb{R}$ and $N \in \mathbb{N}_0$, there are some $\widehat{C}, W > 0$ and $N' \in \mathbb{N}_0$ such that, on $\beta_{\mathrm{b}}^{-1}(\mathring{T}^2_{L,l,\epsilon'})$,

$$\left|A(\mathring{\kappa}_{l,u} - \mathring{\kappa}'_{L,l,u})\right| \le \widehat{C} e^{-\frac{R}{u}} s^a \rho^N \, \|\psi\|_{A,W,N'} \, \|f\|_{l,C^{N'}} .$$

Since a and N are arbitrary, this indeed holds with $A \in \mathrm{Diff}^k((M_l)^2_{\mathrm{b}}; \beta^*_{\mathrm{b}} S)$, after possibly increasing $\widehat{C}$, obtaining the stated inequality. □

Proposition 7.2.3 means that $\mathring{\kappa}_{l,u} - \mathring{\kappa}'_{L,l,u}$ has a $\dot{C}^\infty$ extension on the open subset $(T_{L,l,\epsilon})^2_{\mathrm{b}} \subset (M_l)^2_{\mathrm{b}}$ over $T_{\epsilon'} \equiv T'_{\epsilon'}$.

Recall that $\phi_L = \{\phi^t_L\}$ denotes the restriction of ϕ_l, or of ϕ, to any boundary leaf L of $\mathcal{F}_l$.

Corollary 7.2.6 *The section $\mathring{\kappa}_l$ has a C^∞ extension κ_l to $(M_l)^2_{\mathrm{b}}$, which vanishes to all orders at* lb ∪ rb, *and therefore $\mathring{P}_l$ defines an operator $P_l \in \Psi_b^{-\infty}(M_l; \Lambda\mathcal{F}_l)$.*

Moreover

$$I_{\nu_l}(P_l,\lambda)\equiv\bigoplus_L\int_{-\infty}^{+\infty}\phi^{l*}_{L,z+i\lambda}\,\psi(D_{L,z+i\lambda})\,e^{i\lambda\varkappa_L t}\,f(t)\,dt$$

$$\in\Psi^{-\infty}(\partial M_l;\Lambda)\equiv\bigoplus_L\Psi^{-\infty}(L;\Lambda)\;,$$

where L runs in $\pi_0(\partial M_l)$.

Proof This follows from Propositions 7.1.3, 7.1.6, 7.2.2, and 7.2.3. □

The subscripts of Notations 7.1.4 and 7.1.7 may also be used with P_l and κ_l. If needed, the subscript "l" is also added to the notation of the b-diagonal Δ_{b} of $(M_l)^2_{\mathrm{b}}$, and to $\Delta_{\mathrm{b},0}=\Delta_{\mathrm{b}}\cap\mathrm{ff}$.

Corollary 7.2.7 *The bilinear map*

$$\mathcal{A}\times C^\infty_c(\mathbb{R})\to C^\infty\big((M_l)^2_b;\,\beta_b^*(\Lambda\mathcal{F}_l\boxtimes(\Lambda\mathcal{F}_l^*\otimes\Omega M_l))\big)\;,\quad(\psi,f)\mapsto\kappa_{l,\psi,f}\;,$$

is continuous.

Proof Apply (7.2.1) and Propositions 7.1.5 and 7.2.3. □

Corollary 7.2.8 *If* $\psi_u(x)=e^{-ux^2}$ $(u>0)$ *and* $f(0)=0$, *then there is some* $0<\epsilon'<\epsilon$ *such that* $\kappa_{l,u}\to0$ *on* $\Delta_{b,l}\cap\beta_b^{-1}(T^2_{l,\epsilon'})\equiv T_{l,\epsilon'}$, *in the* C^∞ *topology, as* $u\downarrow0$.

Proof This is a consequence of Corollary 7.1.9 and Proposition 7.2.3. □

Corollary 7.2.9 *If* $\psi_u(x)=e^{-ux^2}$, *then there is some* $0<\epsilon'<\epsilon$ *such that*

$$\lim_{u\downarrow0}\mathrm{str}(\kappa_{l,u}|_{\Delta_{b,l}})\equiv\begin{cases}f(0)\,e(\mathcal{F}_l,g_{\mathcal{F}_l})\,|\omega_{b,l}| & \text{if } n-1 \text{ is even}\\ 0 & \text{if } n-1 \text{ is odd}\end{cases}$$

in $C^{0,\infty}_{\varpi_l}(T_{l,\epsilon'};{}^b\Omega)$, *using the identity* $\Delta_{b,l}\cap\beta_b^{-1}(T^2_{l,\epsilon'})\equiv T_{l,\epsilon'}$.

Proof This is a consequence of Corollary 7.1.10 and Proposition 7.2.3. □

Proposition 7.2.10 *We have*

$$d_{\mathcal{F}_l,z}\in\mathrm{Diff}^1_b(M_l;\Lambda\mathcal{F}_l)\;,\quad I_{\nu_l}(d_{\mathcal{F}_l,z},\lambda)=d_{\partial M_l,z+i\lambda}\;.$$

Proof By (3.2.28), $d_{\mathcal{F}_l,z}\in\mathrm{Diff}^1(\mathcal{F}_l;\Lambda\mathcal{F}_l)\subset\mathrm{Diff}^1_{\mathrm{b}}(M_l;\Lambda\mathcal{F}_l)$. By (2.5.62) and (4.3.1),

$$I_{\nu_l}(d_{\mathcal{F}_l,z},\lambda)=(\rho^{-i\lambda}d_{\mathcal{F}_l,z}\rho^{i\lambda})_\partial=(d_{\mathcal{F}_l,z}+i\lambda\rho^{-1}d_{\mathcal{F}_l}\rho\wedge)_\partial=d_{\partial M_l,z+i\lambda}\;.$$

Alternatively, we can use (2.5.64) and (3.2.28) to describe $I_{\nu_l}(d_{\mathcal{F}_l,z},\lambda)$. □

Recall that $\boldsymbol{M} \equiv \bigsqcup_l M_l$ and $\boldsymbol{\nu}$ is the combination of the sections ν_l (Sect. 4.3.3). This boldface notation of Sects. 4.2.2, 4.3.2, and 4.3.3 allows to simplify the notation of direct sums of section spaces, cohomologies and operators defined on the manifolds M_l. For instance, we get the operators

$$\boldsymbol{P} \equiv \bigoplus_l P_l \in \Psi_{\mathrm{b}}^{-\infty}(\boldsymbol{M}; \Lambda\boldsymbol{\mathcal{F}}) \equiv \bigoplus_l \Psi_{\mathrm{b}}^{-\infty}(M_l; \Lambda\mathcal{F}_l) ,$$

$$d_{\boldsymbol{\mathcal{F}},z} \equiv \bigoplus_l d_{\mathcal{F}_l,z} \in \Psi_{\mathrm{b}}^{1}(\boldsymbol{M}; \Lambda\boldsymbol{\mathcal{F}}) \equiv \bigoplus_l \Psi_{\mathrm{b}}^{1}(M_l; \Lambda\mathcal{F}_l) ,$$

whose indicial operators are

$$I_{\boldsymbol{\nu}}(\boldsymbol{P}, \lambda) \equiv \bigoplus_l I_{\nu_l}(P_l, \lambda) \in \Psi^{-\infty}(\partial\boldsymbol{M}; \Lambda) \equiv \bigoplus_L \Psi^{-\infty}(L; \Lambda) ,$$

$$I_{\boldsymbol{\nu}}(d_{\boldsymbol{\mathcal{F}},z}, \lambda) \equiv \bigoplus_l I_{\nu_l}(d_{\mathcal{F}_l,z}, \lambda) \in \Psi^{1}(\partial\boldsymbol{M}; \Lambda) \equiv \bigoplus_L \Psi^{1}(L; \Lambda) ,$$

where L runs in $\pi_0(\partial\boldsymbol{M}) \equiv \pi_0 M^0 \sqcup \pi_0 M^0$. On the other hand, according to Proposition 7.2.10, $I_{\boldsymbol{\nu}}(d_{\boldsymbol{\mathcal{F}},z}, \lambda) = d_{\partial\boldsymbol{M},z+i\lambda}$. Let also $\kappa = \kappa_{\boldsymbol{P}}$ on $\boldsymbol{M}_{\mathrm{b}}^2 \equiv \bigsqcup_l (M_l)_{\mathrm{b}}^2$, which is the combination of the sections κ_l. In $\boldsymbol{M}_{\mathrm{b}}^2$, we have $\Delta_{\mathrm{b}} \equiv \bigsqcup_l \Delta_{\mathrm{b},l}$ and $\Delta_{\mathrm{b},0} \equiv \bigsqcup_l \Delta_{\mathrm{b},0,l}$. The subscripts of Notations 7.1.4 and 7.1.7 may also be used with $\boldsymbol{P}$ and κ.

When $z = \mu \in \mathbb{R}$ and $\psi(x) = \psi_u(x) = e^{-ux^2}$ $(u > 0)$, the above $\boldsymbol{P} = \boldsymbol{P}_u$ is the operator $\boldsymbol{P}_{\mu,u,f}$ of Sect. 1.3.4. Thus Theorem 1.3.7 is a consequence of Corollaries 2.5.10, 7.2.6, and 7.2.7.

7.3 The Limit of ${}^{\mathrm{b}}\mathrm{Str}(\boldsymbol{P}_u)$ as $u \downarrow 0$

With the notation of Sect. 4.1.1, let $\mathcal{C} = \mathcal{C}(\phi)$, $\mathcal{P} = \mathcal{P}(\phi)$, $\mathcal{C}_l = \mathcal{C}(\phi_l)$ and $\mathcal{P}_l = \mathcal{P}(\phi_l)$. For any leafwise density $\alpha \in C^\infty(M_l; \Omega\mathcal{F}_l)$, we can consider $\alpha\,|\omega_{\mathrm{b},l}| \in C^\infty(M_l; {}^{\mathrm{b}}\Omega)$. In particular, if $n-1$ is even, the leafwise Euler density $e(\mathcal{F}_l, g_{\mathcal{F}_l}) \in C^\infty(M_l; \Omega\mathcal{F}_l)$ (Sect. 3.2.7) gives rise to the b-density $e(\mathcal{F}_l, g_{\mathcal{F}_l})\,|\omega_{\mathrm{b},l}| \in C^\infty(M_l; {}^{\mathrm{b}}\Omega)$, whose b-integral,

$${}^{\mathrm{b}}\chi_{|\omega_{\mathrm{b},l}|}(\mathcal{F}_l) = {}^{\nu_l}\!\!\int_{M_l} e(\mathcal{F}_l, g_{\mathcal{F}_l})\,|\omega_{\mathrm{b},l}| ,$$

can be called the *b-Connes* $|\omega_{\mathrm{b},l}|$*-Euler characteristic* of $\mathcal{F}_l$. This is a b-normalized version of the Connes $|\omega_{\mathrm{b},l}|$-Euler characteristic, where $|\omega_{\mathrm{b},l}|$ is considered as an invariant transverse measure of $\mathcal{F}_l^1$. The usual Connes $|\omega_{\mathrm{b},l}|$-Euler characteristic is not defined because M_l^1 is not compact. If $n-1$ is odd, let ${}^{\mathrm{b}}\chi_{|\omega_{\mathrm{b},l}|}(\mathcal{F}_l) = 0$.

Recall the operator $\boldsymbol{P}$ defined in Sect. 7.2.

Theorem 7.3.1 *If* $\psi_u(x) = e^{-ux^2}$ $(u > 0)$, *then*

$$\lim_{u\downarrow 0} {}^b\mathrm{Str}(\boldsymbol{P}_u) = \sum_l {}^b\chi_{|\omega_{b,l}|}(\mathcal{F}_l) \cdot f(0) + \sum_{c\in\mathcal{C}} \ell(c) \sum_{k\in\mathbb{Z}^\times} \epsilon_c(k) \cdot f(k\ell(c)) \,.$$

To prove this theorem, we consider every $P_{l,u}$, separately. Recall that $\mathring{\kappa}_{l,u,z}$ corresponds to $\mathring{k}_{l,z,u}$ via the restriction of $\beta_{\mathrm{b}} : (M_l)^2_{\mathrm{b}} \to M_l^2$ to the interiors. Thus we are going to study the asymptotic behaviour of $\mathring{k}_{l,z,u}$ as $u \downarrow 0$. The identities $\mathring{M}_l \equiv M_l^1$, $\mathring{\mathcal{F}}_l \equiv \mathcal{F}_l^1$ and $\mathring{T}_{l,\epsilon'} \equiv T^1_{l,\epsilon'}$ $(0 < \epsilon' \le \epsilon)$ will be used without further comment. With the notation of Sects. 4.3.4 and 4.3.5, and adapting the notation of Sect. 3.4.7, let $\mathfrak{G}_l = \operatorname{Hol}\mathcal{F}_l^1$ and $\widetilde{\mathfrak{G}}_l = \operatorname{Hol}\widetilde{\mathcal{F}}_l^1$, with source and target projections, $\boldsymbol{s}, \boldsymbol{r} : \mathfrak{G}_l \to M_l^1$ and $\boldsymbol{s}, \boldsymbol{r} : \widetilde{\mathfrak{G}}_l \to \widetilde{M}_l^1$. The pairs $(\boldsymbol{r}, \boldsymbol{s})$ define identities $\mathfrak{G}_l \equiv \mathcal{R}_l := \mathcal{R}_{\mathcal{F}_l^1}$ and $\widetilde{\mathfrak{G}}_l \equiv \widetilde{\mathcal{R}}_l := \mathcal{R}_{\widetilde{\mathcal{F}}_l^1}$. Let $\Delta_l \subset \mathcal{R}_l$ denote the diagonal. Consider also the vector bundles

$$S_l = \boldsymbol{s}^*\Lambda\mathcal{F}_l^1 \otimes \boldsymbol{r}^*(\Lambda\mathcal{F}_l^{1*} \otimes \Omega\mathcal{F}_l^1) \,, \quad \widetilde{S}_l = \boldsymbol{s}^*\Lambda\widetilde{\mathcal{F}}_l^1 \otimes \boldsymbol{r}^*\big(\Lambda\widetilde{\mathcal{F}}_l^{1*} \otimes \Omega\widetilde{\mathcal{F}}_l^1\big) \,,$$

over $\mathfrak{G}_l$ and $\widetilde{\mathfrak{G}}_l$, and the leafwise Schwartz kernel $\tilde{k}_{l,z,u}$ defined by the Schwartz kernels of the operators $e^{-u\Delta_{\widetilde{L}',z}}$ on the leaves $\widetilde{L}'$ of $\widetilde{\mathcal{F}}_l^1$, for $z \in \mathbb{C}$ (Sect. 3.4.7). By (3.4.21) and since $\tilde{\omega}_{\mathrm{b},l} = D_l^*dx$ (Sect. 4.3.4), for $\tilde{p} \in \widetilde{M}_l^1$ and $p = [\tilde{p}] \in M_l^1$,

$$\mathring{k}_{l,z,u}(p,p) \equiv \sum_{\gamma\in\Gamma_l} \tilde{\phi}_{l,z}^{-h_l(\gamma)*}\, T_\gamma^*\, \tilde{k}_{l,z,u}\big(T_\gamma \tilde{\phi}_l^{-h_l(\gamma)}(\tilde{p}), \tilde{p}\big)\, f(-h_l(\gamma))\, |\omega_{\mathrm{b},l}|(p) \,, \tag{7.3.1}$$

using that $\widetilde{S}_{(\gamma\cdot\tilde{p},\tilde{p})} \equiv S_{(p,p)}$. This defines a convergent series in $C^\infty_{\mathrm{ub}}(\Delta_l; S_l)$.

Any leaf of $\widetilde{\mathcal{F}}_l^1$ is of the form $\widetilde{L}' = \{x\} \times L_l \equiv L_l$ for some $x \in \mathbb{R}$. Then the restriction of $\tilde{g}_{\mathrm{b},l}$ to $\widetilde{L}'$ is identified with a metric $\tilde{g}_{l,x}$ on L_l, $\Delta_{\widetilde{L}',z}$ is identified with the twisted Laplacian $\Delta_{l,x,z}$ on $(L_l, \tilde{g}_{l,x})$ defined by the restriction of $\tilde{\eta}_0$, and $\tilde{k}_{l,z,u}$ on $\widetilde{L}'^2$ is identified with the Schwartz kernel $\tilde{k}_{l,x,u,z}$ of $e^{-u\Delta_{l,x,z}}$, defined on L_l^2.

Theorem 7.3.1 follows from the following result.

Proposition 7.3.2 *Let* $I \subset \mathbb{R}$ *be a compact interval with* $\operatorname{supp} f \subset I$. *Then the following properties hold:*

(i) *If* $I \subset \mathbb{R}^\times$ *and* $I \cap \mathcal{P}_l = \emptyset$, *then*

$$\lim_{u\downarrow 0} {}^b\mathrm{Str}(P_{l,u}) = 0 \,.$$

(ii) *If* $I \subset \mathbb{R}^\times$ *and* $I \cap \mathcal{P}_l = \{t_0\}$, *then*

$$\lim_{u\downarrow 0} {}^b\mathrm{Str}(P_{l,u}) = f(t_0) \sum_{c\in\mathcal{C}_{l,t_0}} \ell(c)\, \epsilon_c(t_0/\ell(c)) \,,$$

where $\mathcal{C}_{l,t_0}$ *consists of the orbits* $c \in \mathcal{C}_l$ *with period* t_0.

(iii) *If* $0 \in I$ *and* $I \cap \mathcal{P}_l = \emptyset$, *then*

$$\lim_{u \downarrow 0} {}^{b}\mathrm{Str}(P_{l,u}) = f(0)\, {}^{b}\chi_{|\omega_{b,l}|}(\mathcal{F}_l) .$$

Proof Choose some $0 < \epsilon' < \epsilon$ satisfying the statements of Corollaries 7.2.8 and 7.2.9, and take some $0 < \epsilon'' < \epsilon'$. Take some $C_3 \geq 1$ satisfying (4.3.5). Since $\tilde{\phi}_l$ is of $\mathbb{R}$-local bounded geometry (Sect. 2.4.7), there is some $R \geq 0$ such that $\tilde{d}_l(\tilde{\phi}_l^t(\tilde{p}), \tilde{p}) \leq R$ for all $\tilde{p} \in \widetilde{M}_l^1$ and $t \in I$. So, by (4.3.5) and the triangle inequality, for all $\tilde{p} \in \widetilde{M}_{l,\epsilon''}^1$ and $\gamma \in \Gamma_l$ with $-h_l(\gamma) \in I$, we get

$$C_3^{-1}|\gamma| - R \leq \tilde{d}_l\big(\gamma \cdot \tilde{\phi}_l^{-h_l(\gamma)}(\tilde{p}), \tilde{p}\big) \leq C_3|\gamma| + R , \tag{7.3.2}$$

using also that $\gamma \cdot \tilde{\phi}_l^{-h_l(\gamma)}(\tilde{p}) = \tilde{\phi}_l^{-h_l(\gamma)}(\gamma \cdot \tilde{p})$ with $\gamma \cdot \tilde{p} \in \widetilde{M}_{l,\epsilon''}^1$.

By the $\mathbb{R}$-local bounded geometry of $\tilde{\phi}_l$ and the compactness of I, there are $C_4, C_5 > 0$ such that, for all $t \in I$,

$$\big|\tilde{\phi}_{l,z}^{t*}\big| \leq C_4 , \quad |f(t)| \leq C_5 . \tag{7.3.3}$$

Assume $I \subset \mathbb{R}^{\times}$ and $I \cap \mathcal{P}_l = \emptyset$ to prove (i). Thus

$$\{\, (p, \phi_l^t(p)) \mid p \in M_{l,\epsilon''}^1,\ t \in I \,\}$$

is a compact subset of $(M_l^1)^2 \setminus \Delta_l$. By Lemma 3.1.1 (ii), there is some $C_6 > 0$ such that $d_{\mathcal{F}_l^1}(\phi_l^t(p), p) \geq C_6$ for all $p \in M_{l,\epsilon''}^1$ and $t \in I$. So, for all $\tilde{p} \in \widetilde{M}_{l,\epsilon''}^1$ and $\gamma \in \Gamma_l$ with $-h_l(\gamma) \in I$,

$$d_{\widetilde{\mathcal{F}}_l^1}\big(\gamma \cdot \tilde{\phi}_l^{-h_l(\gamma)}(\tilde{p}), \tilde{p}\big) \geq C_6 . \tag{7.3.4}$$

Take some $C_7 > 0$ such that, for all $\gamma \in \Gamma_l$ with $-h_l(\gamma) \in I$,

$$C_7|\gamma| \leq \begin{cases} C_3^{-1}|\gamma| - R & \text{if } |\gamma| > C_3 R \\ C_6 & \text{if } |\gamma| \leq C_3 R . \end{cases}$$

Since $\tilde{d}_l \leq d_{\widetilde{\mathcal{F}}_l^1}$ (Sect. 3.1.6), it follows from (7.3.2) and (7.3.4) that, for all $\tilde{p} \in \widetilde{M}_{l,\epsilon''}^1$ and $\gamma \in \Gamma_l$ with $-h_l(\gamma) \in I$,

$$d_{\widetilde{\mathcal{F}}_l^1}\big(\gamma \cdot \tilde{\phi}_l^{-h_l(\gamma)}(\tilde{p}), \tilde{p}\big) \geq C_7|\gamma| . \tag{7.3.5}$$

By (2.9.14) and (7.3.5), and since the leaves of $\widetilde{\mathcal{F}}_l^1$ are of equi-bounded geometry, there are $C_1, C_2, u_0 > 0$ such that, for all $0 < u \leq u_0$, $\tilde{p} \in \widetilde{M}_{l,\epsilon''}^1$ and $\gamma \in \Gamma_l$ with

$-h_l(\gamma) \in I$,

$$\left|\tilde{k}_{l,z,u}(\gamma \cdot \tilde{\phi}_l^{-h_l(\gamma)}(\tilde{p}), \tilde{p})\right| \le C_1 u^{(n-1)/2} e^{-C_2 C_7^2 |\gamma|^2/u} . \tag{7.3.6}$$

Hence, by (7.3.1) and (7.3.3), for all $0 < u \le u_0$ and $p \in M^1_{l,\epsilon''}$,

$$\left|\mathring{k}_{l,z,u}(p,p)\right| \le C_4 C_5 C_1 u^{(n-1)/2} \sum_{\gamma \in \Gamma_l} e^{-C_2 C_7^2 |\gamma|^2/u} , \tag{7.3.7}$$

which converges to zero as $u \downarrow 0$ because Γ_l is of polynomial growth. Since $M^1_{l,\epsilon''}$ is compact, we get

$$\lim_{u \downarrow 0} \int_{p \in M^1_{l,\epsilon''}} \operatorname{str} \mathring{k}_{l,z,u}(p,p) = 0 ,$$

and therefore (i) follows by Corollaries 7.2.8 and 2.5.10.

Now assume $I \subset \mathbb{R}^\times$ and $I \cap \mathcal{P} = \{t_0\}$ to prove (ii), and let $\mathcal{C}_{l,t_0} = \{c_1, \dots, c_m\}$. Then the following properties hold (Sect. 4.3.4):

(N) There is a unique $\gamma_0 \in \Gamma_l$ such that $t_0 = -h_l(\gamma_0)$.
(O) We have $k_j := t_0/\ell(c_j) \in \mathbb{Z}$ $(j = 1, \dots, m)$.
(P) There is some $y_j \in L_l$ such that $\pi_l : \mathbb{R} \times \{y_j\} \to c_j$ is a C^∞ covering map with fundamental domain $[0, \ell(c_j)] \times \{y_j\}$.
(Q) For all $\tilde{p} \in \mathbb{R} \times \{y_j\}$, we have $\gamma_0 \cdot \tilde{\phi}_l^{t_0}(\tilde{p}) = \tilde{p}$.
(R) For all $x \in \mathbb{R}$, every y_j is a simple fixed point of the diffeomorphism $T_{\gamma_0}\tilde{\phi}^{t_0}_{l,x}$ of L_l with $\epsilon_{y_j}(T_{\gamma_0}\tilde{\phi}^{t_0}_{l,x}) = \epsilon_{c_j}(k_j, \phi) = \epsilon_{c_j}(k_j)$.

In particular, there are no other fixed points of $T_{\gamma_0}\tilde{\phi}^{t_0}_{l,x_j}$ in some open neighborhood W_j of y_j in L_l. Then $\pi_l([0, \ell(c_j)] \times W_j)$ is a neighborhood of c_j, whose interior is denoted by V_j, which does not intersect other closed orbits with period in I. Note that $\pi_l : (0, \ell(c_j)) \times W_j \to V_j$ is a C^∞ embedding and $V_j \setminus \pi_l((0, \ell(c_j)) \times W_j) = \pi_l(\{0\} \times W_j)$ is of measure zero. For every $p \in V_j$, let $\tilde{p}$ be the unique point in $[0, \ell(c_j)) \times W_j$ with $\pi_l(\tilde{p}) = p$. We have

$$\begin{aligned}
&\int_{V_j} \operatorname{str}\left(\tilde{\phi}^{t_0 *}_{l,z}\, T^*_{\gamma_0}\, \tilde{k}_{l,z,u}(T_{\gamma_0}\tilde{\phi}^{t_0}_l(\tilde{p}), \tilde{p})\right) f(t_0)\, |\omega_{\mathrm{b},l}|(p) \\
&\quad = \int_{[0,\ell(c_j)] \times W_j} \operatorname{str}\left(\tilde{\phi}^{t_0 *}_{l,z}\, T^*_{\gamma_0}\, \tilde{k}_{l,z,u}(T_{\gamma_0}\tilde{\phi}^{t_0}_l(\tilde{p}), \tilde{p})\right) f(t_0)\, |\tilde{\omega}_{\mathrm{b},l}|(\tilde{p}) \\
&\quad = f(t_0) \int_0^{\ell(c_j)} \int_{W_j} \operatorname{str}\left(\tilde{\phi}^{t_0 *}_{l,z}\, T^*_{\gamma_0}\, \tilde{k}_{l,z,u}((x, T_{\gamma_0}\tilde{\phi}^{t_0}_{l,x}(y)), (x,y)\right) |dx| \\
&\quad = f(t_0) \int_0^{\ell(c_j)} \int_{W_j} \operatorname{str}\left(\tilde{\phi}^{t_0 *}_{l,x,z}\, T^*_{\gamma_0}\, \tilde{k}_{l,x,u,z}(T_{\gamma_0}\tilde{\phi}^{t_0}_{l,x}(y), y\right) |dx| .
\end{aligned}$$

But, by Proposition 2.9.6,

$$\lim_{u\downarrow 0}\int_{W_j}\operatorname{str}\left(\tilde{\phi}_{l,x,z}^{t_0*}\,T_{\gamma_0}^{*}\,\tilde{k}_{l,x,u,z}\big(T_{\gamma_0}\tilde{\phi}_{l,x}^{t_0}(y),y\big)\right)=\epsilon_{y_j}\big(T_{\gamma_0}\tilde{\phi}_{l,x}^{t_0}\big)=\epsilon_{c_j}(k_j)\,.$$

So

$$\lim_{u\downarrow 0}\int_{V_j}\operatorname{str}\left(\tilde{\phi}_{l,z}^{t_0*}\,T_{\gamma_0}^{*}\,\tilde{k}_{l,z,u}\big(T_{\gamma_0}\tilde{\phi}_{l}^{t_0}(\tilde{p}),\tilde{p}\big)\right)f(t_0)\,|\omega_{\mathrm{b},l}|(p)$$
$$=f(t_0)\ell(c_j)\epsilon_{c_j}(k_j)\,. \qquad (7.3.8)$$

By (i), we can assume the length of I is as small as desired. By the $\mathbb{R}$-local bounded geometry of $\tilde{\phi}_l$, if the length of I is small enough, there is some $0<r<C_3^{-1}/2$ such that $d_l(\phi_l^t(\tilde{p}),\phi_l^s(\tilde{p}))\le r$ for all $\tilde{p}\in\widetilde{M}_l^1$ and $t,s\in I$. So, by (4.3.5), (N) and (Q), for all $p\in c_j$ and $\gamma\in\Gamma_l\setminus\{\gamma_0\}$ with $-h_l(\gamma)\in I$,

$$\begin{aligned}
&d_l\big(\gamma\cdot\tilde{\phi}_l^{-h_l(\gamma)}(\tilde{p}),\tilde{p}\big)\\
&\quad\ge d_l\big(\gamma\cdot\tilde{\phi}_l^{-h_l(\gamma)}(\tilde{p}),\gamma_0\cdot\tilde{\phi}_l^{-h_l(\gamma)}(\tilde{p})\big)\\
&\qquad-d_l\big(\gamma_0\cdot\tilde{\phi}_l^{-h_l(\gamma)}(\tilde{p}),\gamma_0\cdot\tilde{\phi}_l^{-h_l(\gamma_0)}(\tilde{p})\big)-d_l\big(\gamma_0\cdot\tilde{\phi}_l^{-h_l(\gamma_0)}(\tilde{p}),\tilde{p}\big)\\
&\quad=d_l\big(\tilde{\phi}_l^{-h_l(\gamma)}(\tilde{p}),\gamma^{-1}\gamma_0\cdot\tilde{\phi}_l^{-h_l(\gamma)}(\tilde{p})\big)-d_l\big(\tilde{\phi}_l^{-h_l(\gamma)}(\tilde{p}),\tilde{\phi}_l^{-h_l(\gamma_0)}(\tilde{p})\big)\\
&\quad\ge C_3^{-1}|\gamma^{-1}\gamma_0|-r\,.
\end{aligned}$$

Thus, by continuity, the neighborhood W_j of every y_j can be chosen so small that, for all $p\in V_j$ and $\gamma\in\Gamma_l\setminus\{\gamma_0\}$ with $-h_l(\gamma)\in I$,

$$d_l\big(\gamma\cdot\tilde{\phi}_l^{-h_l(\gamma)}(\tilde{p}),\tilde{p}\big)\ge C_3^{-1}|\gamma^{-1}\gamma_0|-2r\ge C_3^{-1}-2r>0\,.$$

Hence, by (2.9.14) and since the leaves of $\widetilde{\mathcal{F}}_l^1$ are of equi-bounded geometry, there are $C_1,C_2,u_0>0$ such that, for all $0<u\le u_0$, $p\in V_j$ and $\gamma\in\Gamma_l\setminus\{\gamma_0\}$ with $-h_l(\gamma)\in I$,

$$\left|\tilde{k}_{l,z,u}\big(\gamma\cdot\tilde{\phi}_l^{-h_l(\gamma)}(\tilde{p}),\tilde{p}\big)\right|\le C_1u^{(n-1)/2}e^{-C_2(C_3^{-1}|\gamma^{-1}\gamma_0|-2r)^2/u}\,.$$

Then, by (7.3.3), for all $0<u\le u_0$ and $p\in V_j$,

$$\left|\sum_{\gamma\in\Gamma_l\setminus\{\gamma_0\}}\tilde{\phi}_{l,z}^{-h_l(\gamma)*}\,T_\gamma^{*}\,\tilde{k}_{l,z,u}\big(T_\gamma\tilde{\phi}_l^{-h_l(\gamma)}(\tilde{p}),\tilde{p}\big)\,f(-h_l(\gamma))\,|\omega_{\mathrm{b},l}|(p)\right|$$
$$\le C_4C_5C_1u^{(n-1)/2}\sum_{\gamma\in\Gamma_l\setminus\{\gamma_0\}}e^{-C_2(C_3^{-1}|\gamma^{-1}\gamma_0|-2r)^2/u}\,,$$

which converges to zero as $u \downarrow 0$ because Γ_l is of polynomial growth. So, by (7.3.1) and (7.3.8),

$$\lim_{u\downarrow 0}\int_{p\in V_j} \operatorname{str}\mathring{k}_{l,z,u}(p,p) = f(t_0)\ell(c_j)\epsilon_{c_j}(k_j)\ . \tag{7.3.9}$$

On the other hand, since ϕ has no closed orbits in $T^1_{l,\epsilon}$ (Sect. 4.3.3), we can assume $V_j \subset M^1_{l,\epsilon''}$. Let $\widetilde{V}_j = \pi_l^{-1}(V_j) \subset \widetilde{M}^1_{l,\epsilon''}$. If $p \in M^1_{l,\epsilon''} \setminus (V_1 \cup \cdots \cup V_m)$ and $t \in I$, then $\phi^t(p) \neq p$. Hence, like in the proof of (i), there are $C_7, C_1, C_2, u_0 > 0$ such that (7.3.5) and (7.3.6) hold for all $0 < u \le u_0$, $\tilde{p} \in \widetilde{M}^1_{l,\epsilon''} \setminus (\widetilde{V}_1 \cup \cdots \cup \widetilde{V}_m)$ and $\gamma \in \Gamma_l$ with $-h_l(\gamma) \in I$. Thus (7.3.7) holds for all p in the compact space $M^1_{l,\epsilon''} \setminus (V_1 \cup \cdots \cup V_m)$, yielding

$$\lim_{u\downarrow 0}\int_{p\in M^1_{l,\epsilon''}\setminus(V_1\cup\cdots\cup V_m)} \operatorname{str}\mathring{k}_{l,z,u}(p,p) = 0\ .$$

So (ii) is true by (7.3.9) and Corollaries 2.5.10 and 7.2.8.

Finally, assume $0 \in I$ and $I \cap \mathcal{P}_l = \emptyset$ to prove (iii). By (i), we can suppose again that the length of I is as small as desired. By (7.3.2), there are finitely many elements $\gamma \in \Gamma_l$ such that $-h_l(\gamma) \in I$ and, for all $\tilde{p} \in \widetilde{M}^1_l$,

$$d_{\widetilde{\mathcal{F}}^1_l}\big(\gamma \cdot \tilde{\phi}_l^{-h_l(\gamma)}(\tilde{p}), \tilde{p}\big) > 1\ . \tag{7.3.10}$$

Thus, if I is small enough, we can assume (7.3.10) is true for all $\tilde{p} \in \widetilde{M}^1_l$ and $\gamma \in \Gamma_l \setminus \{e\}$ with $-h_l(\gamma) \in I$. Then, like in the proof of (i), there are $C_7, C_1, C_2, u_0 > 0$ such that (7.3.5) and (7.3.6) hold for all $0 < u \le u_0$, $\tilde{p} \in \widetilde{M}^1_{l,\epsilon''}$ and $\gamma \in \Gamma_l \setminus \{e\}$ with $-h_l(\gamma) \in I$. Hence, by (7.3.3), for all $0 < u \le u_0$ and $p \in M^1_{l,\epsilon''}$,

$$\begin{aligned}\bigg|\sum_{\gamma\in\Gamma_l\setminus\{e\}} \tilde{\phi}_{l,z}^{-h_l(\gamma)*}\, T_\gamma^*\, \tilde{k}_{l,z,u}\big(T_\gamma\tilde{\phi}_l^{-h_l(\gamma)}(\tilde{p}), \tilde{p}\big)\, f(-h_l(\gamma))\, |\omega_{\mathrm{b},l}|(p)\bigg| \\ \le C_4C_5C_1u^{(n-1)/2}\sum_{\gamma\in\Gamma_l} e^{-C_2C_7^2|\gamma|^2/u}\ ,\end{aligned}$$

which converges to zero as $u \downarrow 0$ because Γ_l is of polynomial growth. On the other hand, by (2.9.22) and Theorem 2.9.4,

$$\lim_{u\downarrow 0}\operatorname{str}\tilde{k}_{l,z,u}(\tilde{p},\tilde{p}) = \begin{cases} e(\widetilde{\mathcal{F}}_l, g_{\widetilde{\mathcal{F}}_l})(\tilde{p}) \equiv e(\mathcal{F}_l, g_{\mathcal{F}_l})([\tilde{p}]) & \text{if } n-1 \text{ is even} \\ 0 & \text{if } n-1 \text{ is odd}\ , \end{cases}$$

uniformly on $\tilde{p} \in \widetilde{M}^1_{l,\epsilon''}$. So, by (7.3.1),

$$\lim_{u \downarrow 0} \operatorname{str} \kappa_{l,u} = f(0)\, e(\mathcal{F}_l, g_{\mathcal{F}_l})\, |\omega_{\mathrm{b},l}|$$

uniformly on $\Delta_{\mathrm{b},l} \cap \beta_{\mathrm{b}}^{-1}(M^2_{l,\epsilon''}) \equiv M_{l,\epsilon''}$. Therefore (iii) follows using Corollaries 2.5.10 and 7.2.9 and Remark 2.5.11. □

Remark 7.3.3 The simpler argument given in [ÁLK02, ÁLK08] for the case of Theorem 7.3.1 with no preserved leaves cannot be applied here because now ${}^{\mathrm{b}}\mathrm{Str}(\boldsymbol{P}_u)$ depends on u.

Theorem 1.3.8 is a restatement of Theorem 7.3.1.

7.4 The Limit of ${}^{\mathrm{b}}\mathrm{Str}(P_{\mu,u})$ as $u \uparrow +\infty$ and $\mu \to \pm\infty$

7.4.1 An Expression of ${}^{\mathrm{b}}\mathrm{Tr}([d_{\mathcal{F},\mu}, P_\mu \mathsf{w}])$

From now on, we will only consider $\boldsymbol{P} = \boldsymbol{P}_z$ for $z = \mu \in \mathbb{R}$, written $\boldsymbol{P}_\mu$, with the possible additional subscripts "ψ", "u" or "f". We keep the notation $z = \mu + i\lambda$ for any other $\lambda \in \mathbb{R}$ ($i = \sqrt{-1}$). In the following, L runs in $\pi_0 M^0$. Recall that $\eta_0 = \eta$ around M^0. For $\psi \in \mathcal{A}$, $\mu \in \mathbb{R}$ and $f \in C^\infty_{\mathrm{c}}(\mathbb{R})$, let

$$\begin{aligned} S_{L,\mu} &= -\frac{1}{2\pi} \int_{-\infty}^{+\infty} \eta_L \wedge \psi(D_{L,z}) \hat{f}(-\varkappa_L \lambda)\, d\lambda \\ &= -\frac{1}{2\pi\, |\varkappa_L|} \int_{-\infty}^{+\infty} \eta_L \wedge \psi(D_{L,z}) \widehat{f_L}(\lambda)\, d\lambda\,, \end{aligned} \tag{7.4.1}$$

where $f_L(\lambda) = f(-\lambda/\varkappa_L)$. Again, we may also add the subscripts "ψ" or "f" to the notation $S_{L,\mu}$ if needed. Recall also that w denotes the degree involution. Observe that $\mathsf{w} d_{\mathcal{F},z} = -d_{\mathcal{F},z}\mathsf{w}$ and $I_\nu(\boldsymbol{P}_\mu \mathsf{w}, \lambda) = I_\nu(\boldsymbol{P}_\mu, \lambda)\, \mathsf{w}$ by (2.5.63).

Lemma 7.4.1 *We have*

$${}^{b}\mathrm{Tr}([d_{\mathcal{F},\mu}, \boldsymbol{P}_\mu \mathsf{w}]) = 2 \sum_L \mathrm{Str}(S_{L,\mu})\,.$$

Proof By the version of (2.5.65) with a b-differential operator and a b-pseudodifferential operator of order $-\infty$, Corollary 7.2.6 and Proposition 7.2.10,

$${}^{\mathrm{b}}\mathrm{Tr}([d_{\mathcal{F},\mu}, \boldsymbol{P}_\mu \mathsf{w}]) = -\frac{1}{2\pi i} \int_{-\infty}^{+\infty} \mathrm{Tr}(\partial_\lambda I_\nu(d_{\mathcal{F},\mu}, \lambda)\, I_\nu(\boldsymbol{P}_\mu, \lambda)\mathsf{w})\, d\lambda$$

$$= -\frac{1}{\pi}\sum_L \int_{-\infty}^{+\infty}\int_{-\infty}^{+\infty} \mathrm{Tr}(\eta_L \wedge \psi(D_{L,z})\mathsf{w})\, e^{i\lambda \varkappa_L t} f(t)\, dt\, d\lambda$$

$$= 2\sum_L \mathrm{Str}(S_{L,\mu}) .$$

□

7.4.2 Variation of ${}^b\mathrm{Str}(\boldsymbol{P}_{\mu,u})$ with Respect to u

For any $\psi \in \mathcal{A}$ and $u > 0$, let $\psi_u \in \mathcal{A}$ be defined by $\psi_u(x) = \psi(\sqrt{u}x)$, and consider the corresponding operator $\boldsymbol{P}_{\mu,u}$. Recall that $\boldsymbol{P}_{\mu,u}$ is the operator $\boldsymbol{P}_{\mu,u,f}$ of Sect. 1.3.4 if $\psi(x) = e^{-x^2}$.

Proposition 7.4.2 *If $\psi \in \mathcal{A}$ is even, then*

$$\frac{d}{du}{}^b\mathrm{Str}(\boldsymbol{P}_{\mu,u}) = -\frac{1}{\sqrt{u}}\sum_L \mathrm{Str}(S_{L,(\psi')_u,\mu}) .$$

Proof This result follows like in the heat equation proof of the usual Lefschetz trace formula [AB67, Gil95, Roe98], but the stated derivative does not vanish because the b-trace of commutators may not be zero. To simplify the arguments, consider the change of variables $v = \sqrt{u}$, and let $\psi^v(x) = \psi_u(x) = \psi(vx)$ and $\boldsymbol{P}^v_\mu = \boldsymbol{P}_{\psi^v,\mu,f}$. By Lemma 7.4.1 and since ψ' is odd,

$$\begin{aligned}
&{}^b\mathrm{Str}\left(\int_{-\infty}^{+\infty} \phi_\mu^{t*} d_{\mathcal{F},\mu}\, \psi'(vD_{\mathcal{F},\mu})\, f(t)\, dt\right)\\
&= {}^b\mathrm{Tr}\left(\int_{-\infty}^{+\infty} \phi_\mu^{t*} d_{\mathcal{F},\mu}\, \psi'(vD_{\mathcal{F},\mu})\,\mathsf{w}\, f(t)\, dt\right)\\
&= {}^b\mathrm{Tr}\left(\int_{-\infty}^{+\infty} \phi_\mu^{t*} \psi'(vD_{\mathcal{F},\mu})\,\mathsf{w}\, d_{\mathcal{F},\mu} f(t)\, dt\right) + 2\sum_L \mathrm{Str}(S_{L,(\psi')^v,\mu})\\
&= -{}^b\mathrm{Tr}\left(\int_{-\infty}^{+\infty} \phi_\mu^{t*} \delta_{\mathcal{F},\mu}\, \psi'(vD_{\mathcal{F},\mu})\,\mathsf{w}\, f(t)\, dt\right) + 2\sum_L \mathrm{Str}(S_{L,(\psi')^v,\mu})\\
&= -{}^b\mathrm{Str}\left(\int_{-\infty}^{+\infty} \phi_\mu^{t*} \delta_{\mathcal{F},\mu}\, \psi'(vD_{\mathcal{F},\mu})\, f(t)\, dt\right) + 2\sum_L \mathrm{Str}(S_{L,(\psi')^v,\mu}) .
\end{aligned}$$

So

$$\begin{aligned}
\frac{d}{dv} {}^{\mathrm{b}}\mathrm{Str}\, \boldsymbol{P}_{\mu}^{v} &= {}^{\mathrm{b}}\mathrm{Str}\left(\int_{-\infty}^{+\infty} \phi_{\mu}^{t*} D_{\mathcal{F},\mu} \psi'(v D_{\mathcal{F},\mu}) f(t)\, dt\right) \\
&= {}^{\mathrm{b}}\mathrm{Str}\left(\int_{-\infty}^{+\infty} \phi_{\mu}^{t*} d_{\mathcal{F},\mu} \psi'(v D_{\mathcal{F},\mu}) f(t)\, dt\right) \\
&\quad + {}^{\mathrm{b}}\mathrm{Str}\left(\int_{-\infty}^{+\infty} \phi_{\mu}^{t*} \delta_{\mathcal{F},\mu} \psi'(v D_{\mathcal{F},\mu}) f(t)\, dt\right) \\
&= 2 \sum_{L} \mathrm{Str}(S_{L,(\psi')^{v},\mu}) \,.
\end{aligned}$$

Now apply the chain rule. □

7.4.3 *The Limit of* ${}^{b}\mathrm{Str}(P_{\mu,u})$ *as* $u \uparrow +\infty$ *and* $\mu \to \pm\infty$

Now take $\psi(x) = e^{-x^2}$. Hence $\psi_u(x) = e^{-ux^2}$ and $(\psi')_u(x) = -2\sqrt{u}xe^{-ux^2}$. Thus, by (7.4.1),

$$\mathrm{Str}(S_{L,(\psi')_u,\mu}) = -\frac{1}{\pi|\varkappa_L|}\int_{-\infty}^{+\infty} \mathrm{Str}\left(\eta_L \wedge \delta_z e^{-u\Delta_{L,z}}\right)\widehat{f_L}(\lambda)\, d\lambda \,. \tag{7.4.2}$$

Theorem 7.4.3 *For all $\tau \gg 0$, we can choose every η_L and g_L ($L \in \pi_0 M^0$) so that*

$$\lim_{\mu\uparrow+\infty}\left(\lim_{u_1\uparrow+\infty} {}^{b}\mathrm{Str}(\boldsymbol{P}_{\mu,u_1}) - \lim_{u_0\downarrow 0} {}^{b}\mathrm{Str}(\boldsymbol{P}_{\mu,u_0})\right) = \tau f(0) \,.$$

If $n-1$ is even, this is true for all $\tau \in \mathbb{R}$ and as $\mu \to \pm\infty$.

Proof By Theorem 2.9.7, if $\tau \gg 0$, we can choose every η_L and g_L so that (2.9.25) defines a tempered distribution $Z_{L,\mu} := Z(L, g_L, \eta_L) \in \mathcal{S}'$ for $|\mu| \gg 0$, and $Z_{L,\mu} \to \tau\delta_0$ in $\mathcal{S}'$ as $\mu \to \infty$. If $n-1 = \dim L$ is even, then this is true for all $\tau \in \mathbb{R}$ and as $\mu \to \pm\infty$. Then the result follows because, by (2.9.25), (7.4.2), and Proposition 7.4.2,

$$\begin{aligned}
\lim_{u_1\uparrow+\infty} {}^{\mathrm{b}}\mathrm{Str}(\boldsymbol{P}_{\mu,u_1}) - \lim_{u_0\downarrow 0} {}^{\mathrm{b}}\mathrm{Str}(\boldsymbol{P}_{\mu,u_0}) &= -\frac{1}{\sqrt{u}}\sum_{L}\int_{0}^{\infty} \mathrm{Str}(S_{L,(\psi')_u,\mu})\, du \\
&= -\frac{1}{\pi}\sum_{L}\frac{1}{|\varkappa_L|}\langle Z_{L,\mu}, f_L\rangle \,.
\end{aligned}$$

□

Theorem 7.4.3 gives Theorem 1.3.9 by taking $\tau = 0$ when $n - 1$ is even.

Corollary 7.4.4 *For all $\tau \gg 0$, we can choose every η_L and g_L ($L \in \pi_0 M^0$) so that*

$$\lim_{\mu\uparrow+\infty} \lim_{u\uparrow+\infty} {}^b\mathrm{Str}(\boldsymbol{P}_{\mu,u}) = \left({}^b\chi_{|\omega_b|}(\mathring{\mathcal{F}}) + \tau\right) f(0) + \sum_{c\in\mathcal{C}} \ell(c) \sum_{k\in\mathbb{Z}^\times} \epsilon_c(k)\, f(k\ell(c)) \,.$$

If n is even, this is true for all $\tau \in \mathbb{R}$ and as $\mu \to \pm\infty$.

Proof Apply Theorems 7.3.1 and 7.4.3. □

Like in [ÁLK02, ÁLKL20], by (5.4.6) and (6.4.6), the distributions

$$f \mapsto \lim_{u\uparrow+\infty} {}^{\mathrm{b}}\mathrm{Str}\left(\boldsymbol{P}_{m+\frac{1}{2},u,f}\right), \quad f \mapsto \lim_{u\uparrow+\infty} {}^{\mathrm{b}}\mathrm{Str}\left(\boldsymbol{P}_{m-\frac{1}{2},u,f}\right)$$

can be considered as distributional supertraces of the actions $\boldsymbol{\phi}^*$ of $\mathbb{R}$ on $\bar{H}^\bullet J^m(\mathcal{F})$ and $\bar{H}^\bullet J'^m(\mathcal{F})$. So, by the analogs of (5.1.2) and (6.1.2) for $J(\mathcal{F})$ and $J'(\mathcal{F})$, and using (5.4.6) and (6.4.6), the distributions

$$f \mapsto \lim_{m\downarrow-\infty} \lim_{u\uparrow+\infty} {}^{\mathrm{b}}\mathrm{Str}\left(\boldsymbol{P}_{m+\frac{1}{2},u,f}\right), \quad f \mapsto \lim_{m\downarrow\infty} \lim_{u\uparrow+\infty} {}^{\mathrm{b}}\mathrm{Str}\left(\boldsymbol{P}_{m-\frac{1}{2},u,f}\right)$$

can be considered as distributional supertraces of the actions $\boldsymbol{\phi}^*$ of $\mathbb{R}$ on $\bar{H}^\bullet J(\mathcal{F})$ and $\bar{H}^\bullet J'(\mathcal{F})$, as indicated in Sect. 1.3.6.

References

[AB67] M.F. Atiyah, R. Bott, A Lefschetz fixed point formula for elliptic complexes. I. Ann. Math. (2) **86**, 374–407 (1967). MR 0212836

[Ada75] R.A. Adams, *Sobolev Spaces*, Pure and Applied Mathematics, vol. 65 (Academic Press, New York, 1975). MR 0450957

[ÁL89] J.A. Álvarez López, A finiteness theorem for the spectral sequence of a Riemannian foliation. Illinois J. Math. **33**(1), 79–92 (1989). MR 974012

[ÁLG21] J.A. Álvarez López, P. Gilkey, The local index density of the perturbed de Rham complex. Czechoslovak Math. J. **71**(3), 901–932 (2021). MR 4295254

[ÁLK01] J.A. Álvarez López, Y.A. Kordyukov, Long time behavior of leafwise heat flow for Riemannian foliations. Compos. Math. **125**(2), 129–153 (2001). https://doi.org/10.1023/A:1002492700960. MR 1815391

[ÁLK02] J.A. Álvarez López, Y.A. Kordyukov, *Distributional Betti Numbers of Transitive Foliations of Codimension One*, Foliations: Geometry and Dynamics. Proceedings of the Euroworkshop, Warsaw, Poland, May 29–June 9, 2000 (Singapore) (World Scientific Publishers, Singapore, 2002), pp. 159–183. MR 1882768

[ÁLK08] J.A. Álvarez López, Y.A. Kordyukov, Lefschetz distribution of Lie foliations, in *C^*-algebras and Elliptic Theory II (Basel)*, Trends Math. (Birkhäuser, 2008), pp. 1–40. https://doi.org/10.1007/978-3-7643-8604-7_1. MR 2408134

[ÁLKL14] J.A. Álvarez López, Y.A. Kordyukov, E. Leichtnam, Riemannian foliations of bounded geometry. Math. Nachr. **287**(14–15), 1589–1608 (2014). MR 3266125

[ÁLKL20] J.A. Álvarez López, Y.A. Kordyukov, E. Leichtnam, Analysis on Riemannian foliations of bounded geometry. Münster J. Math. **13**, 221–265 (2020), arXiv:1905.12912

[ÁLKL22] J.A. Álvarez López, Y.A. Kordyukov, E. Leichtnam, Simple foliated flows. Tohoku Math. J. (2) **74**(1), 53–81 (2022). MR 4374665

[ÁLKL24] J.A. Álvarez López, Y.A. Kordyukov, E. Leichtnam, The topology of the space of conormal distributions. J. Pseudo-Differ. Oper. Appl. **15–47**, 1–68 (2024)

[ÁLKL25] J.A. Álvarez López, Y.A. Kordyukov, E. Leichtnam, Zeta invariants of Morse forms. J. Inst. Math. Jussieu **24**(2), 411–480 (2025)

[ÁLT91] J.A. Álvarez López, P. Tondeur, Hodge decomposition along the leaves of a Riemannian foliation. J. Funct. Anal. **99**(2), 443–458 (1991). MR 1121621

[Bar81] K. Barner, On A. Weil's explicit formula. J. Reine Angew. Math. **323**, 139–152 (1981). MR 611448

[BE91] I. Buttig, J. Eichhorn, The heat kernel for p-forms on manifolds of bounded geometry. Acta Sci. Math. (Szeged) **55**(1–2), 33–51 (1991). MR 1124942

J. A. Álvarez López et al., *A Trace Formula for Foliated Flows*, Lecture Notes in Mathematics 2387, https://doi.org/10.1007/978-3-032-15413-2

[BF97] M. Braverman, M. Farber, Novikov type inequalities for differential forms with non-isolated zeros. Math. Proc. Camb. Philos. Soc. **122**, 357–375 (1997). MR 1458239

[BGV04] N. Berline, E. Getzler, M. Vergne, *Heat Kernels and Dirac Operators*, Grundlehren Text Editions (Springer, Berlin, 2004), Corrected reprint of the 1992 original. MR 2273508

[Bou14] H. Bourlès, On the closed graph theorem and the open mapping theorem (2014), arXiv:1411.5500 [math.FA]

[BT82] R. Bott, L.W. Tu, *Differential Forms in Algebraic Topology*, Graduate Texts in Mathematics, vol. 82 (Springer, New York, 1982). MR 658304

[BZ92] J.-M. Bismut, W. Zhang, An extension of a theorem by Cheeger and Müller. Astérisque **205**, 235 pp. (1992), with an appendix by F. Laudenbach. MR 1185803

[CC00] A. Candel, L. Conlon, *Foliations. I*, Graduate Studies in Mathematics, vol. 23 (American Mathematical Society, Providence, RI, 2000). MR 1732868

[CC03] A. Candel, L. Conlon, *Foliations. II*, Graduate Studies in Mathematics, vol. 60 (American Mathematical Society, Providence, RI, 2003). MR 1994394

[CGT82] J. Cheeger, M. Gromov, M. Taylor, Finite propagation speed, kernel estimates for functions of the Laplace operator, and the geometry of complete Riemannian manifolds. J. Differ. Geom. **17**(1), 15–53 (1982). MR 658471

[Che73] P.R. Chernoff, Essential self-adjointness of powers of generators of hyperbolic equations. J. Funct. Anal. **12**, 401–414 (1973). MR 0369890

[CLN85] C. Camacho, A. Lins Neto, *Geometric Theory of Foliations* (Birkhäuser, Boston, 1985), Translated from the Portuguese by Sue E. Goodman. http://dx.doi.org/10.1007/978-1-4612-5292-4. MR 824240

[Con79] A. Connes, *Sur la théorie non commutative de l'intégration*, Algèbres d'opérateurs (Sém., Les Plans-sur-Bex, 1978) (Berlin), Lecture Notes in Math., vol. 725 (Springer, Berlin, 1979), pp. 19–143. MR 548112

[Con82] A. Connes, *A Survey of Foliations and Operator Algebras*, Operator Algebras and Applications, Kingston, 1980, Proc. Sympos. Pure Math., vol. 38-1 (1982), pp. 521–628

[Den98] C. Deninger, Some analogies between number theory and dynamical systems on foliated spaces. Doc. Math. **Extra Vol. I**, 163–186 (1998), Proceedings of the International Congress of Mathematicians, Vol. I (Berlin, 1998). MR 1648030

[Den01] C. Deninger, *Number theory and dynamical systems on foliated spaces*, Jahresber. Deutsch. Math.-Verein. **103** (2001), no. 3, 79–100. MR 1873325

[Den02] C. Deninger, *On the nature of the "explicit formulas" in analytic number theory—a simple example*, Number Theoretic Methods (Iizuka, 2001), Dev. Math., vol. 8 (Kluwer Acad. Publ., Dordrecht, 2002), pp. 97–118. https://doi.org/10.1007/978-1-4757-3675-5_7. MR 1974137

[Den05] C. Deninger, Arithmetic geometry and analysis on foliated spaces, arXiv:math/0505354 (2005)

[Den08] C. Deninger, *Analogies Between Analysis on Foliated Spaces and Arithmetic Geometry*, Groups and Analysis, London Math. Soc. Lecture Note Ser., vol. 354 (Cambridge Univ. Press, Cambridge, 2008), pp. 174–190. https://doi.org/10.1017/CBO9780511721410.010. MR 2528467

[Den23] C. Deninger, Primes, knots and periodic orbits, arXiv:2301.11643 (2023)

[Den26] C. Deninger, *Dynamical systems for arithmetic schemes*, Indag. Math. (N.S.) **37** (2026), no. 1, 25–136.

[dR84] G. de Rham, *Differentiable Manifolds*, Grundlehren der mathematischen Wissenschaften, vol. 266 (Springer, Berlin, 1984), Forms, currents, harmonic forms, Translated from the French by F. R. Smith, With an introduction by S. S. Chern. MR 760450

[DS01] C. Deninger, W. Singhof, A counterexample to smooth leafwise Hodge decomposition for general foliations and to a type of dynamical trace formulas. Ann. Inst. Fourier (Grenoble) **51**(1), 209–219 (2001). MR 1821074

[DS02] C. Deninger, W. Singhof, Real polarizable Hodge structures arising from foliations. Ann. Global Anal. Geom. **21**(4), 377–399 (2002). MR 1910458

[Edw65] R.E. Edwards, *Functional Analysis. Theory and Applications* (Holt, Rinehart and Winston, New York, 1965). MR 0221256

[Eic91] J. Eichhorn, The boundedness of connection coefficients and their derivatives. Math. Nachr. **152**, 145–158 (1991). MR 1121230

[EMT77] D.B.A. Epstein, K.C. Millett, D. Tischler, Leaves without holonomy. J. Lond. Math. Soc. (2) **16**(3), 548–552 (1977). https://doi.org/10.1112/jlms/s2-16.3.548. MR 0464259

[Far95] M. Farber, Singularities of the analytic torsion. J. Differ. Geom. **41**(3), 528–572 (1995). MR 1338482

[Far04] M. Farber, *Topology of Closed One-forms*, Mathematical Surveys and Monographs, vol. 108 (Amer. Math. Soc., Providence, RI, 2004). MR 2034601

[Fed71] E. Fedida, Sur les feuilletages de Lie, C. R. Acad. Sci. Paris Sér. A **272**, 999–1001 (1971). MR 0285025

[Fed73] E. Fedida, *Feuilletages de Lie, feuilletages du plan*, Lecture Notes in Math., vol. 352 (Springer, Berlin, 1973), pp. 183–195

[Gil95] P.B. Gilkey, *Invariance Theory, the Heat Equation, and the Atiyah-Singer Index Theorem*, Studies in Advanced Mathematics, second edn. (CRC Press, Boca Raton, FL, 1995). MR 1396308

[God91] C. Godbillon, *Feuilletages: études géométriques*, Progress in Math., vol. 98 (Birkhäuser Verlag, Boston, 1991). MR 1120547

[GS77] V. Guillemin, S. Sternberg, *Geometric Asymptotics*, Math. Surveys, vol. 14 (American Mathematical Society, Providence, RI, 1977). MR 0516965

[Gui77] V. Guillemin, Lectures on spectral theory of elliptic operators. Duke Math. J. **44**(3), 485–517 (1977). http://projecteuclid.org/euclid.dmj/1077312384. MR 0448452

[Hae62] A. Haefliger, Variétés feuilletées. Ann. Scuola Norm. Sup. Pisa (3) **16**, 367–397 (1962). MR 0189060

[Hae80] A. Haefliger, Some remarks on foliations with minimal leaves. J. Differ. Geom. **15**(2), 269–384 (1980). MR 614370

[Hec72] G. Hector, Sur les feuilletages presque sans holonomie, C. R. Acad. Sci. Paris Sér. A **274**, 1703–1706 (1972). MR 0303550

[Hec77] G. Hector, *Feuilletages en cylindres*, Geometry and Topology (Proc. III Latin Amer. School of Math., Inst. Mat. Pura Aplicada CNPq, Rio de Janeiro, 1976), Lecture Notes in Math., vol. 597 (Springer, Berlin, 1977), pp. 252–270. MR 0451260

[Hec78] G. Hector, *Croissance des feuilletages presque sans holonomie*, Differential Topology, Foliations and Gelfand-Fuks Cohomology (Proc. Sympos., Pontifícia Univ. Católica, Rio de Janeiro, 1976) (Berlin), Lecture Notes in Math., vol. 652 (Springer, Berlin, 1978), pp. 141–182. MR 505659

[HH81] G. Hector, U. Hirsch, *Introduction to the Geometry of Foliations. Part A: Foliations on Compact Surfaces, Fundamentals for Arbitrary Codimension, and Holonomy*, Aspects of Mathematics, vol. E1 (Friedr. Vieweg & Sohn, Braunschweig, 1981). MR 639738

[HH83] G. Hector, U. Hirsch, *Introduction to the Geometry of Foliations. Part B*, Aspects of Mathematics, vol. E3 (Friedr. Vieweg & Sohn, Braunschweig, 1983)

[Hör65] L. Hörmander, Pseudo-differential operators. Commun. Pure Appl. Math. **18**, 501–517 (1965). MR 0180740

[Hor66] J. Horváth, *Topological Vector Spaces and Distributions*, vol. I (Addison-Wesley Publishing Co., Reading, MA, 1966). MR 0205028

[Hör71] L. Hörmander, Fourier integral operators. I. Acta Math. **127**(1–2), 79–183 (1971). MR 388463

[Hör83] L. Hörmander, *The Analysis of Linear Partial Differential Operators. I. Distribution Theory and Fourier Analysis*, Grundlehren der Mathematischen Wissenschaften, vol. 256 (Springer, Berlin, 1983). MR 717035

[Hör85] L. Hörmander, *The Analysis of Linear Partial Differential Operators. III. Pseudodifferential Operators*, Grundlehren der Mathematischen Wissenschaften, vol. 274 (Springer, Berlin, 1985)

[Kim17] J. Kim, On the leafwise cohomology and dynamical zeta functions for fiber bundles over the circle, arXiv:1712.04181 (2017)

[KN65] J.J. Kohn, L. Nirenberg, An algebra of pseudo-differential operators. Commun. Pure Appl. Math. **18**, 269–305 (1965). MR 0176362

[Kom67] H. Komatsu, Projective and injective limits of weakly compact sequences of locally convex spaces. J. Math. Soc. Japan **19**, 366–383 (1967). MR 217557

[Kop06] F. Kopei, A remark on a relation between foliations and number theory, in *Foliations 2005. Proceedings of the international conference, University of Łodź, Łodź, Poland, June 13–24, 2005 (Hackensack, NJ)*, ed. by P. Walczak et al. (World Scientific Publishers, Singapore, 2006), pp. 245–249. MR 2284785

[Kop11] F. Kopei, A foliated analogue of one- and two-dimensional Arakelov theory. Abh. Math. Semin. Univ. Hambg. **81**, 141–189 (2011). MR 2836630

[Köt69] G. Köthe, *Topological Vector Spaces. I*, Die Grundlehren der mathematischen Wissenschaften, vol. 159 (Springer, Berlin, 1969), translated from the German by D.J.H. Garling. MR 0248498

[KP15] Y.A. Kordyukov, V.A. Pavlenko, On Lefschetz formulas for flows on foliated manifolds. Ufa Math. J. **7**(2), 71–101 (2015). MR 3430750

[Lei08] E. Leichtnam, On the analogy between arithmetic geometry and foliated spaces. Rend. Mat. Appl. (7) **28**(2), 163–188 (2008). MR 2463936

[Lei14] E. Leichtnam, On the analogy between L-functions and Atiyah-Bott-Lefschetz trace formulas for foliated spaces. Rend. Mat. Appl. (7) **35**(1–2), 1–34 (2014). MR 3241361

[Mel81] R.B. Melrose, Transformation of boundary problems. Acta Math. **147**(3–4), 149–236 (1981). MR 639039

[Mel93] R.B. Melrose, *The Atiyah-Patodi-Singer Index Theorem*, Research Notes in Mathematics, vol. 4 (A.K. Peters, Wellesley, MA, 1993). MR 1348401

[Mel96] R.B. Melrose, Differential analysis on manifolds with corners. http://www-math.mit.edu/~rbm/book.html (1996)

[Mol88] P. Molino, *Riemannian Foliations*, Progress in Mathematics, vol. 73 (Birkhäuser Boston, Boston, MA, 1988), Translated from the French by G. Cairns, With appendices by Cairns, Y. Carrière, É. Ghys, E. Salem and V. Sergiescu. https://doi.org/10.1007/978-1-4684-8670-4. MR 932463

[MS88] C.C. Moore, C. Schochet, *Global Analysis on Foliated Spaces*, Mathematical Sciences Research Institute Publications, vol. 9 (Springer, New York, 1988), With appendices by S. Hurder, Moore, Schochet and Robert J. Zimmer. MR 0918974

[MU08] R.B. Melrose, G. Uhlmann, *An Introduction to Microlocal Analysis* (Department of Mathematics, Massachusetts Institute of Technology, Cambridge, MA, 2008), https://books.google.es/books?id=Os2jswEACAAJ, http://www-math.mit.edu/~rbm/book.html

[Müm06] B. Mümken, On tangential cohomology of Riemannian foliations. Amer. J. Math. **128**(6), 1391–1408 (2006). MR 2275025

[NB11] L. Narici, E. Beckenstein, *Topological Vector Spaces*, Pure and Applied Mathematics (Boca Raton), second edn., vol. 296 (CRC Press, Boca Raton, FL, 2011). MR 2723563

[Nov81] S.P. Novikov, Multivalued functions and functionals. An analog of the Morse theory. Soviet. Math. Dokl. **24**, 222–226 (1981). MR 630459

[Nov82] S.P. Novikov, The Hamiltonian formalism and a multivalued analogue of Morse theory. Russian Math. Surv. **37**, 1–56 (1982). MR 676612

[Nov02] S.P. Novikov, On the exotic De-Rham cohomology. Perturbation theory as a spectral sequence, arXiv:math-ph/0201019 (2002)

[O'N66] B. O'Neill, The fundamental equations of a submersion. Michigan Math. J. **13**, 459–469 (1966). MR 0200865

[Paj87] A.V. Pajitnov, An analytic proof of the real part of Novikov's inequalities. Soviet Math. Dokl. **35**, 456–457 (1987). MR 891557

[Pet98] P. Petersen, *Riemannian Geometry*, Graduate Texts in Mathematics, vol. 171 (Springer, New York, 1998). MR 1480173

[Poo81] W.A. Poor, *Differential Geometric Structures* (McGraw-Hill Book Co., New York, 1981). MR 647949

[Roe87] J. Roe, Finite propagation speed and Connes' foliation algebra. Math. Proc. Camb. Philos. Soc. **102**(3), 459–466 (1987). MR 906620

[Roe88] J. Roe, An index theorem on open manifolds. I. J. Differential Geom. **27**(1), 87–113 (1988). MR 918459

[Roe98] J. Roe, *Elliptic Operators, Topology and Asymptotic Methods*, second edn., Pitman Research Notes in Mathematics, vol. 395 (Longman, Harlow, 1998). MR 1670907

[San08] L. Sanguiao, L^2-invariants of Riemannian foliations. Ann. Global Anal. Geom. **33**(3), 271–292 (2008). MR 2390835

[Sch71] H.H. Schaefer, *Topological Vector Spaces*, Graduate Texts in Mathematics, vol. 3 (Springer, New York, 1971)

[Sch96] T. Schick, Analysis on ∂-manifolds of bounded geometry, Hodge-De Rham isomorphism and L^2-index theorem, Ph.D. thesis, Johannes Gutenberg Universität Mainz, Mainz, 1996

[Sch01] T. Schick, Manifolds with boundary and of bounded geometry. Math. Nachr. **223**(1), 103–120 (2001). MR 1817852

[See64] R.T. Seeley, Extension of C^∞ functions defined in a half space. Proc. Amer. Math. Soc. **15**, 625–626 (1964). MR 0165392

[Shu92] M.A. Shubin, Spectral theory of elliptic operators on noncompact manifolds. Astérisque **207**, 35–108 (1992), Méthodes semi-classiques, Vol. 1 (Nantes, 1991). MR 1205177

[Sim90] S.R. Simanca, *Pseudo-differential Operators*, Pitman Research Notes in Mathematics Series, vol. 236 (Longman Scientific & Technical, Harlow; copublished in the United States with Wiley, New York, 1990). MR 1075017

[Tay81] M.E. Taylor, *Pseudodifferential Operators*, Princeton Mathematical Series, vol. 34 (Princeton University Press, Princeton, NJ, 1981). MR 618463

[Val89] M. Valdivia, A characterization of totally reflexive Fréchet spaces. Math. Z. **200**(3), 327–346 (1989). MR 978594

[Wen03] J. Wengenroth, *Derived Functors in Functional Analysis*, Lecture Notes in Mathematics, vol. 1810 (Springer, Berlin, 2003). MR 1977923

[Wit82] E. Witten, Supersymmetry and Morse theory. J. Differ. Geom. **17** (1982), 661–692. MR 683171

Index

J. A. Álvarez López et al., *A Trace Formula for Foliated Flows*, Lecture Notes in Mathematics 2387, https://doi.org/10.1007/978-3-032-15413-2

LECTURE NOTES IN MATHEMATICS 

Editorial Policy

1. Lecture Notes aim to report new developments in all areas of mathematics and their applications – quickly, informally and at a high level. Mathematical texts analysing new developments in modelling and numerical simulation are welcome.

   Manuscripts should be reasonably self-contained and rounded off. Thus they may, and often will, present not only results of the author but also related work by other people. They may be based on specialised lecture courses. Furthermore, the manuscripts should provide sufficient motivation, examples and applications. This clearly distinguishes Lecture Notes from journal articles or technical reports which normally are very concise. Articles intended for a journal but too long to be accepted by most journals, usually do not have this "lecture notes" character. For similar reasons it is unusual for doctoral theses to be accepted for the Lecture Notes series, though habilitation theses may be appropriate.

2. Besides monographs, multi-author manuscripts resulting from SUMMER SCHOOLS or similar INTENSIVE COURSES are welcome, provided their objective was held to present an active mathematical topic to an audience at the beginning or intermediate graduate level (a list of participants should be provided).

   The resulting manuscript should not be just a collection of course notes, but should require advance planning and coordination among the main lecturers. The subject matter should dictate the structure of the book. This structure should be motivated and explained in a scientific introduction, and the notation, references, index and formulation of results should be, if possible, unified by the editors. Each contribution should have an abstract and an introduction referring to the other contributions. In other words, more preparatory work must go into a multi-authored volume than simply assembling a disparate collection of papers, communicated at the event.

3. Manuscripts should be submitted either online at www.editorialmanager.com/lnm to Springer's mathematics editorial in Heidelberg, or electronically to one of the series editors. Authors should be aware that incomplete or insufficiently close-to-final manuscripts almost always result in longer refereeing times and nevertheless unclear referees' recommendations, making further refereeing of a final draft necessary. The strict minimum amount of material that will be considered should include a detailed outline describing the planned contents of each chapter, a bibliography and several sample chapters. Parallel submission of a manuscript to another publisher while under consideration for LNM is not acceptable and can lead to rejection.

4. In general, **monographs** will be sent out to at least 2 external referees for evaluation.

   A final decision to publish can be made only on the basis of the complete manuscript, however a refereeing process leading to a preliminary decision can be based on a pre-final or incomplete manuscript.

   Volume Editors of **multi-author works** are expected to arrange for the refereeing, to the usual scientific standards, of the individual contributions. If the resulting reports can be forwarded to the LNM Editorial Board, this is very helpful. If no reports are forwarded or if other questions remain unclear in respect of homogeneity etc, the series editors may wish to consult external referees for an overall evaluation of the volume.

5. Manuscripts should in general be submitted in English. Final manuscripts should contain at least 100 pages of mathematical text and should always include

   - a table of contents;
   - an informative introduction, with adequate motivation and perhaps some historical remarks: it should be accessible to a reader not intimately familiar with the topic treated;
   - a subject index: as a rule this is genuinely helpful for the reader.
   - For evaluation purposes, manuscripts should be submitted as pdf files.

6. Careful preparation of the manuscripts will help keep production time short besides ensuring satisfactory appearance of the finished book in print and online. After acceptance of the manuscript authors will be asked to prepare the final LaTeX source files (see LaTeX templates online: https://www.springer.com/gb/authors-editors/book-authors-editors/manuscriptpreparation/5636) plus the corresponding pdf- or zipped ps-file. The LaTeX source files are essential for producing the full-text online version of the book, see http://link.springer.com/bookseries/304 for the existing online volumes of LNM). The technical production of a Lecture Notes volume takes approximately 12 weeks. Additional instructions, if necessary, are available on request from lnm@springer.com.

7. Authors receive a total of 30 free copies of their volume and free access to their book on SpringerLink, but no royalties. They are entitled to a discount of 33.3 % on the price of Springer books purchased for their personal use, if ordering directly from Springer.

8. Commitment to publish is made by a *Publishing Agreement*; contributing authors of multiauthor books are requested to sign a *Consent to Publish form*. Springer-Verlag registers the copyright for each volume. Authors are free to reuse material contained in their LNM volumes in later publications: a brief written (or e-mail) request for formal permission is sufficient.

Addresses:

Professor Jean-Michel Morel, CMLA, École Normale Supérieure de Cachan, France
E-mail: moreljeanmichel@gmail.com

Professor Bernard Teissier, Equipe Géométrie et Dynamique,
Institut de Mathématiques de Jussieu – Paris Rive Gauche, Paris, France
E-mail: bernard.teissier@imj-prg.fr

Springer: Ute McCrory, Mathematics, Heidelberg, Germany,
E-mail: lnm@springer.com

The manufacturer's authorised representative in the EU is Springer Nature Customer Service Centre GmbH, Europaplatz 3, 69115 Heidelberg, Germany. If you have any concerns regarding our products, please contact ProductSafety@springernature.com

Printed and bound by CPI Group (UK) Ltd, Croydon, CR0 4YY

13/07/2026

02165297-0002